Control and Optimization of Power Converters

Control and Optimization of Power Converters

Guest Editors

Fei Deng
Xiangke Li
Ziheng Xiao
Zhigang Yao
Lei Zhang

MDPI

Basel • Beijing • Wuhan • Barcelona • Belgrade • Novi Sad • Cluj • Manchester

Guest Editors

Fei Deng
School of Automation
Northwestern Polytechnical University
Xi'an
China

Xiangke Li
Department of Electrical Engineering
Northwestern Polytechnical University
Xi'an
China

Ziheng Xiao
Energy Research Institute
Nanyang Technological University
Singapore
Singapore

Zhigang Yao
School of Electrical Engineering
Chongqing University
Chongqing
China

Lei Zhang
Institute of Electrical Engineering
Chinese Academy of Sciences
Beijing
China

Editorial Office
MDPI AG
Grosspeteranlage 5
4052 Basel, Switzerland

This is a reprint of the Special Issue, published open access by the journal *Energies* (ISSN 1996-1073), freely accessible at: https://www.mdpi.com/journal/energies/special_issues/X86OD1D66G.

For citation purposes, cite each article independently as indicated on the article page online and as indicated below:

Lastname, A.A.; Lastname, B.B. Article Title. *Journal Name* **Year**, *Volume Number*, Page Range.

ISBN 978-3-7258-7388-3 (Hbk)
ISBN 978-3-7258-7389-0 (PDF)
https://doi.org/10.3390/books978-3-7258-7389-0

Contents

About the Editors

Fei Deng

Fei Deng is a full professor with the School of Automation, Northwestern Polytechnical University, Xi'an, China. His main research interests include power management and energy optimization in electric aircraft and renewable energy microgrids. He has expertise in virtual-impedance-based strategies, including iterative impedance regulation and two-dimensional impedances in ac microgrids, a small ac signal injection technique in dc microgrids, heteroid-impedance-based inertia emulation, and aggregated-modeling-based hierarchical power management in aircraft electric propulsion systems. Based on the above strategies, he has published over 50 academic articles.

Xiangke Li

Xiangke Li is a Professor at Northwestern Polytechnical University. He received his Ph.D. degree from Northwestern Polytechnical University and was a joint Ph.D. student at Nanyang Technological University, Singapore, as well as a Postdoctoral Fellow at The Hong Kong Polytechnic University. His research interests mainly include modeling, stability analysis, advanced control and intelligent optimization of power-electronic-based land, shipboard, and airborne power systems. He has been the Principal Investigator of the Youth Science Fund (Class C) of the National Natural Science Foundation of China and the General Program of the Natural Science Foundation of Guangdong Province. He has also participated in various research projects supported by the Hong Kong Innovation and Technology Commission, The Hong Kong Polytechnic University, and the Key Program of the National Natural Science Foundation of China, and has co-authored more than 50 academic papers.

Ziheng Xiao

Ziheng Xiao (Member, IEEE) received his B.S. degree in electrical engineering and automation and his Ph.D. degree in electrical engineering from the College of Electrical and Information Engineering, Hunan University, Changsha, China, in 2017 and 2022, respectively. Since 2022, he has been a Research Fellow with the Energy Research Institute at Nanyang Technological University, Singapore. Since 2023, he has also served as a part-time lecturer in the School of Electrical and Electronic Engineering at the same university. His main research interests include the medium-voltage dc system, dual-active bridge converters, resonant converters, energy routers in renewable energy systems, and the application of artificial intelligence in power electronics. Dr. Xiao was a recipient of the Best Paper Award First Prize in ECCE Asia in 2024. He also served as the Session Chair for the 49th Annual Conference of the IEEE Industrial Electronics Society.

Zhigang Yao

Zhigang Yao (Senior Member, IEEE) received his Ph.D. degree in electrical engineering from Chongqing University, Chongqing, China. From 2020 to 2022, he was an Assistant Professor with Southwest Jiaotong University, Chengdu, China. From 2022 to 2025, he was a Research Fellow with the Energy Research Institute, Nanyang Technological University, Singapore. He is currently an Associate Professor with the School of Electrical Engineering, Chongqing University, Chongqing, China. He has authored or co-authored over 60 papers and holds 10 Chinese patents. He first proposed the trapezoidal current mode (TZCM) for multilevel converters to achieve ZVS while reducing peak current. His current research interests include high-power multiphase/multilevel DC–DC converters, three-level inverters, resonant converters, and ZVS soft-switching techniques. Dr. Yao received the Best Paper Award First Prize at CIYCEE in 2021 and at WiPDA Asia in 2025.

Lei Zhang

Lei Zhang is an Associate Researcher at the Institute of Electrical Engineering, Chinese Academy of Sciences (IEE-CAS). He received his Ph.D. in Electrical Engineering from Arizona State University, USA, and subsequently served as a Research Fellow at Nanyang Technological University, Singapore. His research focuses on modular multilevel converters (MMCs), energy storage systems, and hybrid AC–DC microgrids. Dr. Zhang has made significant contributions to the development of battery energy storage systems (BESSs) and hybrid energy storage systems based on MMC architectures. He has led or participated in multiple major research initiatives, including the BR-B Talent Fund, Key National Science & Technology Specific Project, and the IEE-CAS Talent Fund. To date, he has authored or co-authored 16 papers indexed by SCI and was awarded the Best Paper Award at the 8th IEEE Energy Conversion Congress & Exposition (ECCE) Asia. Additionally, he serves as a reviewer for the Chilean National Science Funding Agency.

energies

MDPI

Article

Adaptive Harmonic Impedance Control and Flexible Compensation Method for AI Data Centers

Jinsong Li, Bo Yang, Hao Li, Zhigang Yao, Qiwei Xu and Shuai Lu *

School of Electrical Engineering, Chongqing University, Chongqing 400044, China; 202311021106t@stu.cqu.edu.cn (J.L.); 20163625@cqu.edu.cn (B.Y.); 20104365@cqu.edu.cn (H.L.); zgyao@cqu.edu.cn (Z.Y.); xuqw@cqu.edu.cn (Q.X.)

* Correspondence: shuai.lu@cqu.edu.cn

Abstract

The stochastic fluctuations of AI computational loads inject harmonic currents into the DC bus, amplifying bus voltage ripples and weakening the power quality. Existing strategies typically rely on high-gain control strategies to minimize harmonic output impedance, aiming at full absorption of harmonic currents. However, such designs rarely consider engineering constraints such as capacity and current boundaries, which impose inherent limits on harmonic absorption. To address these issues, this paper proposes an adaptive harmonic impedance control and flexible compensation method for AI data centers. By integrating DC bus voltage feedforward with output current feedback, a virtual harmonic impedance control channel is constructed to enable real-time impedance shaping. Then, an adaptive gain regulation mechanism is developed to adjust harmonic impedance according to the available capacity and current margin. Compared with traditional strategies relying on fixed high gains or resonant links, the proposed method allows for the continuous regulation of harmonic impedance over a wide range. This enables the dynamic matching of harmonic absorption capability with the available capacity, effectively suppressing the risks of overcurrent, saturation, and stability degradation. Simulation and 8 kW experimental results verify the correctness and effectiveness of the proposed analysis and control strategy.

Keywords: AI data center; adaptive harmonic impedance; flexible compensation; impedance reshaping

1. Introduction

In recent years, the rapid advancement of artificial intelligence (AI) has driven a continuous increase in the power density and capacity of AI data centers (AIDC) [1–3]. Large-scale GPU/TPU computing clusters exhibit pronounced stochastic and fast-varying load characteristics, which inject wideband harmonic currents into the DC bus [4–7]. This harmonic injection amplifies DC bus voltage ripple, degrades power quality, and poses increasing challenges to the stability and reliability of the power delivery system [8,9].

To mitigate DC bus ripple and absorb harmonic currents, existing research has primarily developed along three technical routes: passive filtering, additional compensation devices, and impedance reshaping.

Firstly, passive approaches, e.g., large electrolytic capacitors or LC filters, are straightforward but suffer from large volume, high cost, and limited effectiveness across wide frequency ranges, which makes them less attractive for high-density AIDC applications [10].

As for additional compensation devices, one class of research achieves active ripple throughput by constructing additional energy channels, including active capacitor technologies [11–14] and active power decoupling (APD) [15–20]. These methods can significantly reduce DC bus capacitance requirements, but they suffer from complex hardware structures, limited modulation depth, and compensation capabilities that easily reach limits under large-amplitude load disturbances. Another class of research relies on DC active power filters (DC-APF) to reduce bus ripples through harmonic current detection and injection compensation. However, conventional DC-APF is susceptible to measurement noise, filtering-induced phase lag, and DC bus coupling, causing ripple suppression performance and stability to vary with operating conditions. Meanwhile, under severe disturbances or low-frequency pulsations, the required peak compensating current and energy buffering increase, further leading to higher device stress and losses. Ref. [21] introduces a first-order inertial link in the voltage loop to achieve synergy between voltage control and harmonic extraction, but the inertial link may sacrifice dynamic response speed, limiting suppression capability under severe operating conditions. Ref. [22] employs sliding mode active disturbance rejection control to enhance DC-APF disturbance rejection performance and dynamic response, but the complexity of sliding mode chattering and parameter tuning still constrains its engineering application.

Thirdly, with the development of impedance modeling and frequency-domain control theory, harmonic suppression methods based on impedance reshaping have received widespread attention in recent years. These methods realize the diversion and redistribution of harmonic currents by regulating the output impedance in specific frequency bands. Specifically, Ref. [23] incorporates a bandpass filter into the inductor current feedback to increase impedance in the target frequency band, guiding harmonic currents to the bus capacitor. Ref. [24] reduces bus voltage ripples by lowering the output impedance of the bidirectional DC/DC converter (BDC), but the associated optimization algorithms present difficulties in engineering implementation. Ref. [25] proposes an adaptive sliding mode control-based output impedance shaping method that enhances the source-side converter's output impedance at twice the line frequency. However, this approach relies on sparse communication networks and does not account for communication delays. Additionally, the chattering inherent in sliding mode control and the complexity of parameter tuning increase the difficulty of engineering implementation. Ref. [26] proposes a frequency-selective droop strategy to provide high gain at specific frequencies for precise impedance reshaping, but the parameter design is complex. For comparison, this paper summarizes the characteristic differences between typical DC bus harmonic suppression methods in Table 1.

Table 1. Comparison of different DC bus voltage harmonic suppression methods.

Method Category	Additional Power Devices	Adjustable Output Impedance	Risk of Modulation Saturation	Operating Constraints Considered
Passive filtering	×	×	×	×
Additional compensation devices	√	√	√	×
Impedance Shaping	×	√	√	×
Proposed method	×	√	×	√

https://doi.org/10.3390/en19030862

Overall, existing active harmonic suppression methods mostly employ high-gain controllers, resonant links, or intelligent optimization strategies to minimize the equivalent harmonic impedance of the compensation device, thereby forming a "low impedance point" on the bus side to absorb harmonic currents. These impedance-shaping methods typically use fixed gains, which result in a fixed equivalent harmonic impedance and, consequently, an unchanged harmonic absorption capability. When the data center capacity continues to expand and load disturbances become more severe, compensation strategies with fixed or excessively high gains easily trigger problems like overcurrent, over-modulation, or control saturation, further causing a reduction in stability margin or even system instability. Therefore, for high-dynamic, high-power-density AIDC scenarios, there is an urgent need for a novel control method that explicitly considers engineering constraints and allows for the flexible regulation of compensation capability.

To bridge the above-mentioned gap, this paper proposes an adaptive harmonic impedance control method for flexible compensation in AIDC. Unlike DC-APF solutions that require additional power devices, the proposed approach reuses the energy-storage BDC already deployed in the data center as the actuator. By integrating DC bus voltage feedforward with output current feedback, a virtual harmonic impedance control channel is constructed to enable real-time impedance shaping. In addition, an adaptive gain regulation mechanism is developed. Different from fixed-gain impedance-shaping methods, the proposed approach adjusts the harmonic impedance according to the converter's available capacity and current margin, so that harmonic absorption capability can be flexibly allocated while avoiding overcurrent, modulation saturation, and stability degradation. The main contributions are summarized as follows:

(1) Control framework: An adaptive harmonic impedance control architecture is proposed for AI data center DC buses, which realizes flexible ripple compensation through impedance shaping without requiring additional power devices.
(2) Constraint-aware regulation: A capacity- and current-margin-aware adaptive gain regulation mechanism is established to continuously tune the equivalent harmonic impedance, enabling dynamic matching between harmonic absorption capability and available converter margins, and improving safety and stability under large disturbances.
(3) Verification: The simulation and 8 kW experiment results demonstrate that the proposed method can regulate the equivalent harmonic output impedance of the BDC over a wide range, achieving dynamic matching between compensation capability and converter available capacity, and effectively suppressing DC bus voltage ripples while ensuring system stability.

The remainder of this paper is organized as follows: Section 2 details the principle and implementation structure of adaptive harmonic impedance control; Section 3 presents the unified design framework for the adaptive gain coefficient k; Section 4 verifies the effectiveness of the proposed strategy through simulation and experimental results; Section 5 concludes the paper and looks forward to future work.

2. Proposed Adaptive Harmonic Impedance Control

2.1. Typical Architecture and Control Strategies of AIDC

Figure 1 illustrates a typical DC microgrid power supply topology for AIDC. The system consists of distributed power sources, energy storage units, and AIDC loads. In Figure 1, the 54 V (and 12 V) rails are load-side buses, typically generated by dedicated load-side DC-DC modules in the AIDC power chain. It should be noted that the topology adopted in this paper is not intended for the AI power-supply stage that steps down an 800 V DC bus to 54 V; instead, it is designed for the DC/DC converter in the energy storage

https://doi.org/10.3390/en19030862

unit within an AIDC architecture. In an AIDC architecture, some DC/DC or AC/DC converters are responsible for regulating and maintaining the DC bus voltage, and these units can be regarded as equivalent DC voltage sources, also called "power source side" [27]. In addition, components such as AI loads, AC loads, DC loads, and certain bidirectional DC/DC converters in energy storage systems typically achieve power regulation by controlling their output currents; these units can therefore be treated as equivalent DC current sources [27]. To reduce DC bus voltage ripples, this paper integrates the ripple suppression function as an additional control objective into the BDC, equipped with active filtering capabilities, achieving "flexible compensation without adding power devices".

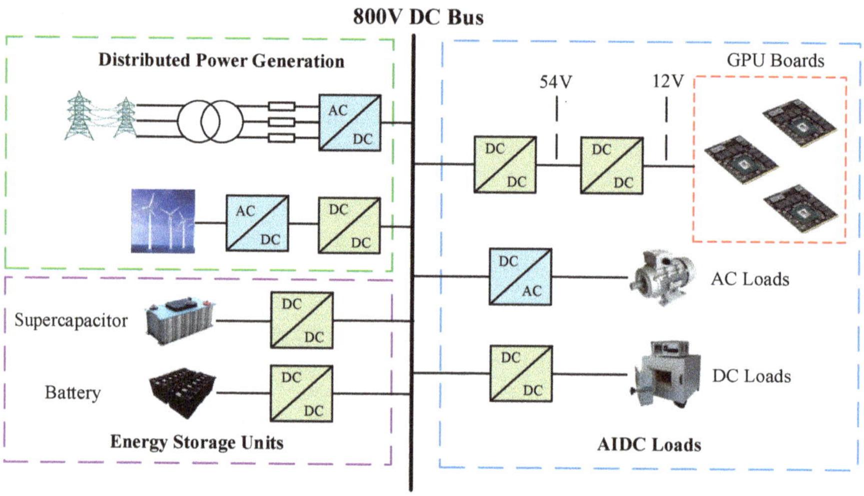

Figure 1. Typical topology of the power supply system in an AIDC.

To enhance power density while balancing low ripple, low loss, and high reliability, BDCs in AIDC typically employ an interleaved parallel structure as the energy hub between the DC bus and the storage side. The basic principle is to electrically parallel multiple identical power phases and apply a fixed phase difference in modulation to drive them interleaved. This increases the equivalent switching frequency on the bus side to N times, significantly reducing inductor current ripple and the ripple current stress on the bus capacitor [28,29]. In addition, advanced interleaved topologies, such as the interleaved cascaded non-inverting buck–boost converter, have also been reported to further improve power density and performance [30]. However, the focus of this paper is the harmonic suppression strategy and its impedance-shaping mechanism. Therefore, to simplify the modeling and highlight the control principle, the converter stage is represented by the simplest three-phase interleaved Buck-based consideration in this work, without loss of generality for the proposed control concept. As shown in Figure 2, the three-phase interleaved parallel BDC consists of three power phases. Each phase includes synchronous half-bridge switching devices S_i and an inductor L_i. The input side connects to the storage voltage u_s, and the output side parallels the DC bus voltage u_{dc}, where C_1 is the input-side capacitor, C_2 is the bus-side capacitor, and i_{Li} is the inductor current of each phase.

https://doi.org/10.3390/en19030862

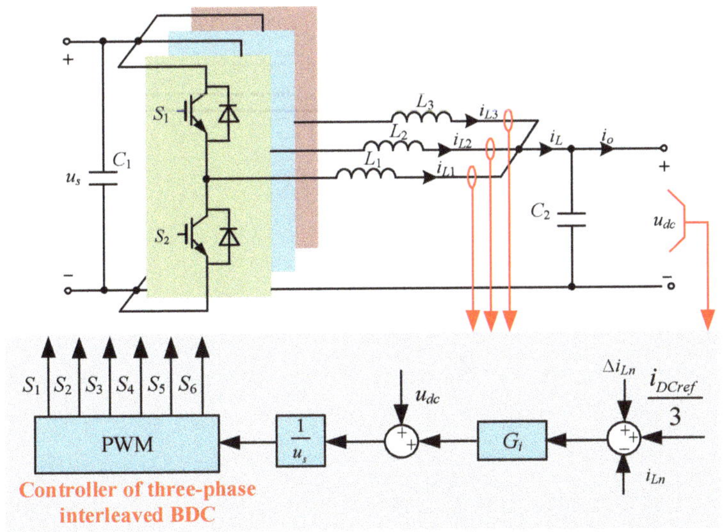

Figure 2. Circuit topology and control strategy of the three-phase interleaved BDC.

In terms of control strategy, this paper adopts the current source control mode shown in Figure 2, using an inductor current closed loop for fast current tracking. A current sharing loop is configured for each phase to ensure $i_{Li} \approx i_{DCref}/N$, where G_i is the current loop transfer function, i_{DCref} is the current reference value, and Δi_{Ln} denotes the current-balancing compensation term. As shown in Figure 3, the three-phase interleaved DC/DC converter adopted in this paper employs three PWM modules to drive and control the three phase legs, respectively. For each phase, the corresponding PWM module triggers the inductor current sampling. The sampling trigger is placed at the peak/valley of the PWM carrier waveform, so that the mid-point value of the phase inductor current can be captured, and this sampled value is used as the feedback signal for control. Interleaved modulation is achieved by setting the phase difference in the carrier or switching function of each phase to $2\pi/N$. Under ideal current-sharing conditions, the N-phase inductor current ripples cancel each other out on the bus side, theoretically reducing the output current ripple amplitude as N increases; meanwhile, the increased equivalent ripple frequency helps reduce bus capacitance requirements. It is noted that, in practical interleaved systems, nonidealities such as inductance tolerances, device parameter dispersion, and sensor gain/offset errors may lead to imperfect current sharing; dedicated current-balancing/equalization strategies have been reported to enhance the robustness of branch current distribution [31].

To characterize the formation mechanism of DC bus ripples from the perspective of "impedance reshaping", this paper establishes an equivalent circuit model in the frequency domain, as illustrated in Figure 4. The equivalent circuit modeling in Figure 4 follows the commonly adopted impedance-based/Thevenin–Norton representation for DC microgrids and converter-interfaced systems [32]. Specifically, the impedances used in this paper are derived from a formal small-signal linearization around a given steady-state operating point, yielding the corresponding port output/input impedances in the frequency domain. In practice, these impedances can also be obtained (or validated) via standard impedance measurement/identification procedures, e.g., injecting small-amplitude sinusoidal (or broadband) perturbations and measuring the resulting port voltage/current responses.

https://doi.org/10.3390/en19030862

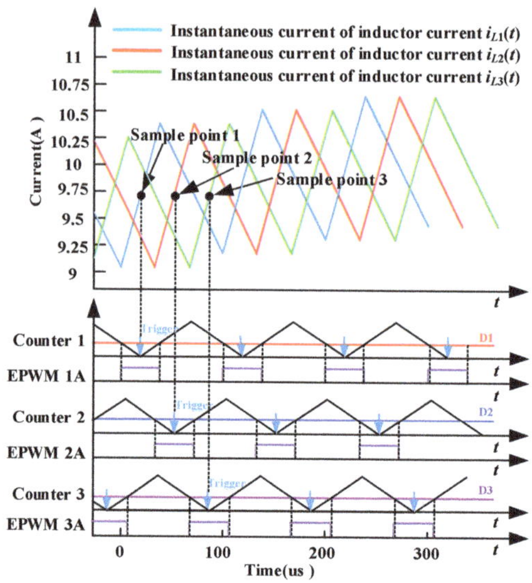

Figure 3. Timing diagram illustrating the instantaneous current waveform and the sampled current sequence.

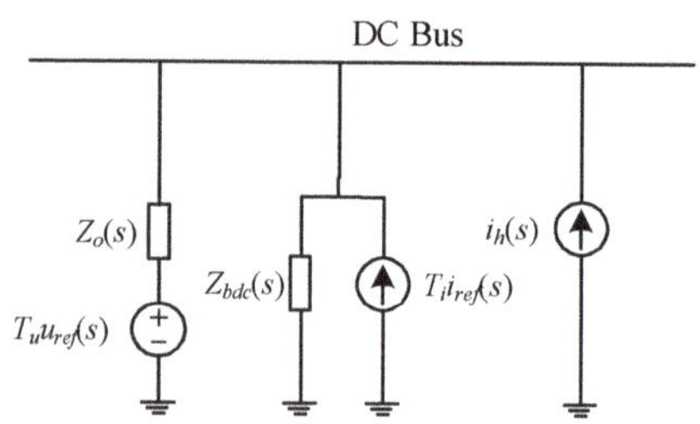

Figure 4. System equivalent circuit.

The power source side is modeled as a series connection of a controlled voltage source $T_u u_{ref}(s)$ and an output impedance Z_o(s), the AI load is equivalent to a current source $i_h(s)$ being injected into the Point of Common Coupling (PCC) in the harmonic frequency band, and the BDC is equivalent to a parallel connection of a controlled current source $T_i i_{ref}(s)$ and an output impedance $Z_{bdc}(s)$. When the system lacks a ripple compensation device, the ripple current is shared by the load connected to the DC bus and the converter. Analyzing the schematic of Figure 4 in the harmonic frequency domain yields the equivalent circuit shown in Figure 5.

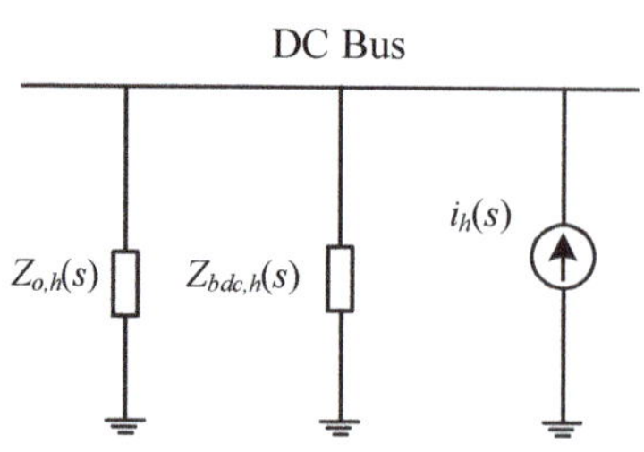

Figure 5. System equivalent circuit in the harmonic frequency domain.

https://doi.org/10.3390/en19030862

This model provides a unified description of "how harmonic currents are distributed among $Z_{o,h}(s)$ and $Z_{bdc,h}(s)$", laying the foundation for subsequent stability and compensation capability analysis. Under this equivalent model, the harmonic voltage ripple at the bus can be written as

$$V_h(s) = Z_{eq}(s) \cdot i_h(s) \tag{1}$$

where

$$Z_{eq}(s) = \frac{1}{\left(\frac{1}{Z_{o,h}(s)} + \frac{1}{Z_{bdc,h}(s)}\right)} \tag{2}$$

Therefore, one of the fundamental ways to reduce bus ripple is to make $Z_{bdc,h}(s)$ present a "controllable low impedance" in the target frequency band, thereby prioritizing the diversion of $i_h(s)$ to the BDC. Conversely, when the system's available margin is insufficient or participation in compensation is undesirable, $Z_{bdc,h}(s)$ can be raised to a "high impedance" to reduce harmonic current sharing.

2.2. The Proposed Adaptive Harmonic Impedance Control Method

Existing control approaches based on the DC bus harmonic voltage or the converter harmonic output current typically introduce a tunable factor to generate the harmonic current or harmonic voltage reference, which is then tracked by the current or voltage controller. By adjusting this factor, the harmonic output impedance of the BDC can be continuously regulated. However, to achieve highly flexible tuning, the factor often needs to vary over a wide range, which substantially amplifies the feedforward or feedback gains; excessive gains may lead to overcurrent, saturation, and a reduced stability margin. To address these issues, this paper proposes an adaptive harmonic impedance control and flexible compensation method for AIDC. The proposed strategy constructs a virtual-impedance path within the control loop using the DC bus harmonic voltage and the converter output harmonic current, thereby enabling online dynamic reconstruction of the reference voltage. With this method, the harmonic output impedance of the BDC can be flexibly set over a wide range without excessively amplifying feedforward or feedback gains, thus avoiding overmodulation and instability. The modified harmonic voltage command can be expressed as

$$u^*_{dc\_h} = k(u_{dc\_h} - Z_v i_{o\_h}) \tag{3}$$

where Z_v represents the virtual impedance and k is the gain coefficient.

To achieve precise suppression of bus voltage ripples, a multi-resonant controller is employed in the control channel to track the reference voltage:

$$G_h = \sum_h K_r \frac{2\omega_c s}{s^2 + 2\omega_c s + \omega_r^2} \tag{4}$$

where K_r is the resonant gain, ω_r is the resonant frequency, and the cutoff frequency ω_c determines the control bandwidth. In this work, the target frequency set is explicitly selected as $f_i \in \{20,50,100,150\}$ Hz, and the resonant branches are designed with $\omega_{r,i} = 2\pi f_i$ to provide targeted suppression at these representative disturbance components. The resonant gains $K_{r,i}$ and the bandwidth parameter ω_c are selected according to widely adopted tuning guidelines [33]. This controller can respond rapidly to harmonic components at specific frequencies, effectively mitigating periodic voltage ripple on the DC bus.

Figure 6 depicts the control scheme of the BDC based on adaptive harmonic impedance. The specific implementation of the adaptive harmonic impedance control module is shown in Figure 7.

https://doi.org/10.3390/en19030862

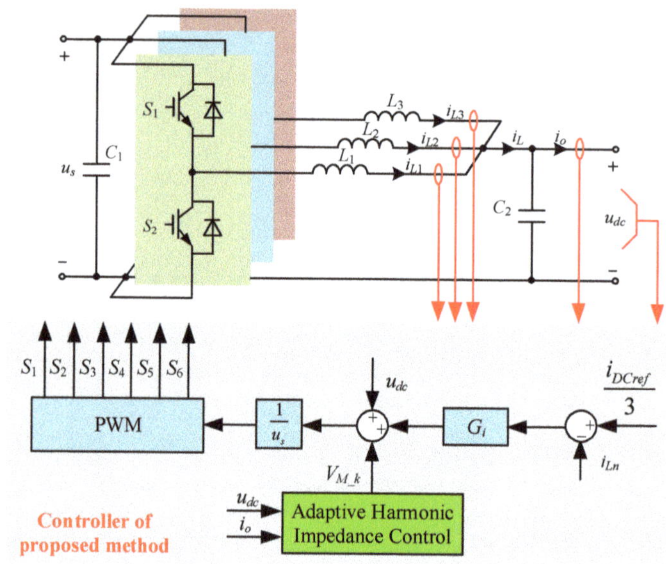

Figure 6. Control scheme of the BDC based on adaptive harmonic impedance control.

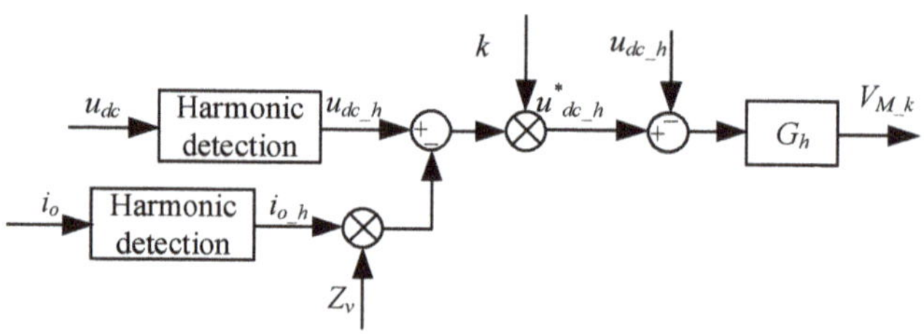

Figure 7. Adaptive harmonic impedance control module.

2.3. Equivalent Harmonic Output Impedance Analysis

The harmonic voltage reference value obtained in Equation (3) will be tracked by a voltage controller based on a resonant loop. It should be noted that an ideal resonant element in the continuous-time domain provides theoretically high gain at ω_r; however, in digital implementation, this gain becomes finite due to ω_c, sampling/computation delays, and discretization effects. Therefore, ω_c is introduced to limit the resonant peak and bandwidth, ensuring implementation feasibility and robustness. Under zero-static-error tracking conditions at this resonant frequency, the harmonic voltage reference becomes equal to the actual harmonic voltage, which leads to the following relation:

$$u_{dc\_h} = u^*_{dc\_h} = k(u_{dc\_h} - Z_v i_{o\_h}) \tag{5}$$

Consequently, the equivalent harmonic output impedance of the BDC can be expressed as

$$Z_{o\_h} = \frac{u_{dc\_h}}{-i_{o\_h}} = \frac{k}{1-k} Z_v \tag{6}$$

From Equation (6), the BDC's equivalent harmonic impedance is mainly determined by the gain coefficient k and the virtual impedance Z_v, where the design of Z_v can follow

https://doi.org/10.3390/en19030862

Ref. [34]. It should be noted that the virtual impedance Z_v adopted in this paper is a purely resistive type. The corresponding design guideline is as follows:

$$Z_v = \frac{(\alpha U_{dcref})^2}{\eta P_{bn}} \tag{7}$$

where α is the maximum ripple coefficient on the DC bus, P_{bn} is the rated power of the BDC, and η is the harmonic compensation coefficient.

When k approaches 0, the BDC equivalent harmonic output impedance approaches zero, allowing the BDC to actively absorb harmonic currents in the system and thus effectively suppress DC bus voltage ripple; physically, this corresponds to a "strong-compensation" operating mode, but in a real converter, the achievable absorption is ultimately limited by current ratings, the modulation margin, and thermal constraints. When k approaches 1, the equivalent harmonic output impedance tends to infinity, and the BDC response to DC bus voltage disturbances becomes negligible; this can be interpreted as a "weak-compensation" mode in which the converter intentionally reduces harmonic participation to avoid saturation or overcurrent when the available margin is insufficient. Therefore, the BDC equivalent harmonic output impedance can be continuously tuned over a wide range. Furthermore, as indicated by Equation (6), when k varies within the interval (0, 1), the gains of the voltage feedforward and current feedback paths remain within a reasonable range without excessive amplification, which improves system stability and helps avoid oscillations and instability caused by control-loop overmodulation or gain coupling.

When the BDC is interfaced to the DC bus of the AIDC as a current-source-type unit and the proposed control strategy is applied, its equivalent control block diagram is as shown in Figure 8.

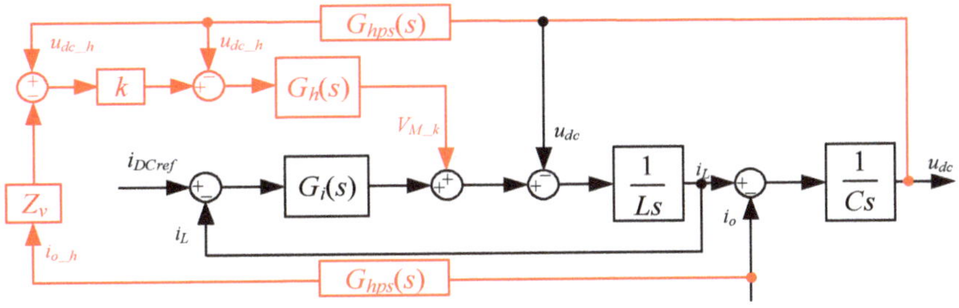

Figure 8. Equivalent control block diagram of the proposed compensation strategy.

From the control block diagram, the expression of the output impedance of the BDC can be derived as

$$Z_{bdc}(s) = \frac{u_{dc}}{-i_o} = \frac{Ls + G_i(s) + G_h(s)G_{hps}(s)kZ_v}{1 + Cs(Ls + G_i(s)) - G_h(s)G_{hps}(s)(k-1)} \tag{8}$$

where $G_{hps}(s)$ is the high-pass filter used for harmonic extraction, and its transfer function is given by

$$G_{hps} = \frac{s}{s + \omega_{hpf}} \tag{9}$$

where ω_{hpf} is the cutoff frequency and is typically selected as 15–30 rad/s.

To investigate the influence of the gain coefficient k on the BDC output impedance, taking the 20 Hz harmonic component as an example, the Bode plots of the output impedance under different values of k are drawn, as shown in Figure 9. Meanwhile, to illustrate the

https://doi.org/10.3390/en19030862

influence of the virtual impedance Z_v on the impedance-shaping effect, the Bode plots of the BDC output impedance under different Z_v values are presented in Figure 10.

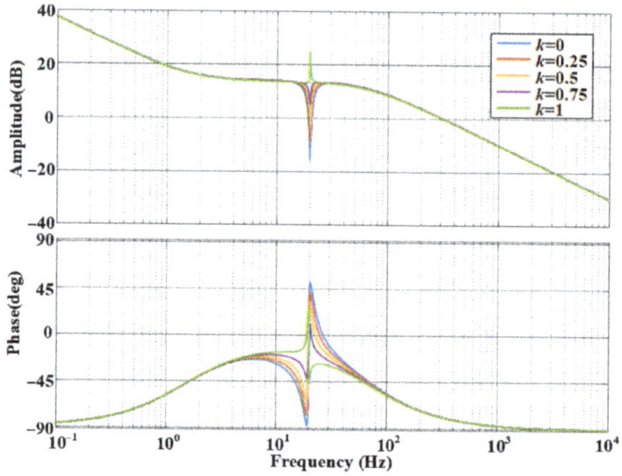

Figure 9. Bode plots of the output impedance of the current-source-type BDC as gain k varies from 0 to 1.

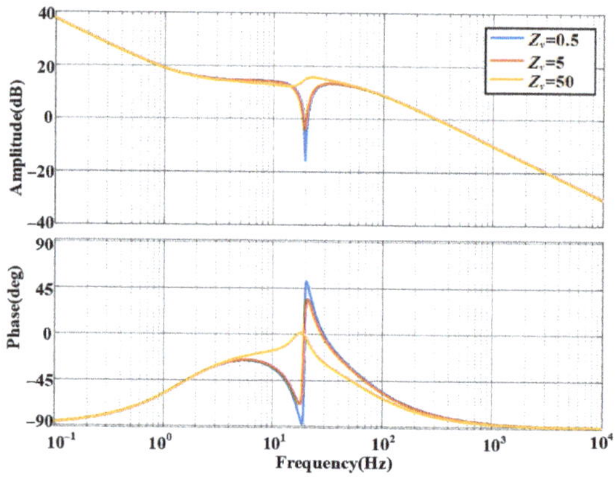

Figure 10. Bode plots of the output impedance of the current-source-type BDC for different Z_v values.

From the Bode plots in Figure 9, it can be seen that as the gain k gradually increases from 0 to 1, the output impedance of the BDC at the frequency point increases accordingly. This indicates that the gain k can regulate the BDC output impedance, thereby changing the magnitude of the harmonic current absorbed by the BDC. As can be observed from Figure 10, as the virtual impedance Z_v increases, the BDC's equivalent harmonic impedance at the selected frequency correspondingly increases. Consequently, the ripple suppression capability is weakened, and the compensation current decreases. Therefore, in practical design, Z_v should not be chosen excessively large. Simultaneously, the phase in Figures 9 and 10 remains consistently between $-90°$ and $90°$, demonstrating that the impedance shaping process does not introduce abnormal phase characteristics.

According to the characteristic equation of the system output impedance, the loop gain of the system can be obtained as

$$D(s) = \frac{kZ_v G_h(s) G_{hps}(s)}{Ls + G_i(s)} \tag{10}$$

The stability of the system is determined by the loop gain $D(s)$. From Equation (10), it can be observed that the poles of the loop gain are independent of k and Z_v. First, the pole distribution of $D(s)$ is examined to evaluate the open-loop stability of the BDC; when all poles of the transfer function $D(s)$ lie in the left half of the complex plane, the open-loop system is stable. Figure 11 shows the pole distribution of $D(s)$; it can be observed that there are no poles with positive real parts, indicating that the open-loop system is stable under the selected parameters. From Figure 11 and the denominator term of Equation (10), it can be seen that the factors affecting open-loop stability include only the bandwidth of the resonant controller, the parameters of the current-loop controller, and the circuit parameters of the BDC. Assuming the BDC component parameters (e.g., capacitance and inductance) are properly designed, the open-loop stability condition is satisfied only when the resonant-controller bandwidth is positive and the current-loop controller parameters are appropriately selected; this premise always holds for the method proposed in this paper.

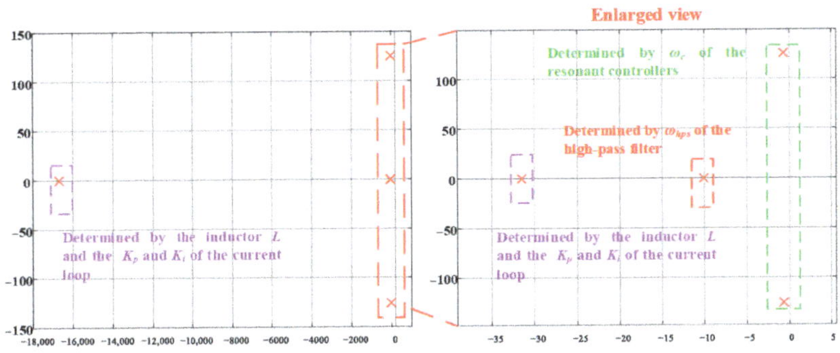

Figure 11. Pole map of $D(s)$.

Therefore, within the variation range of $k \in [0, 1]$, the system remains open-loop stable, and no unstable poles are introduced, nor is stability compromised by tuning k. Meanwhile, the bandwidth of the adaptive update channel for k is designed to be far lower than that of the inner current loop, meaning that the variation in k is a slow-variable adjustment and will not cause a "rapid gain jump" on the fast dynamic timescale dominated by the current loop, thereby avoiding transient instability or a significant reduction in stability margin due to fast gain variations.

2.4. Stability Analysis of the Proposed Strategy in a Parallel-Converter System

To analyze the stability of the proposed strategy in a parallel-converter scenario, a parallel system consisting of a voltage-source BDC and a current-source BDC is constructed, and the proposed strategy is applied to this system. The impedance model of the parallel system is shown in Figure 12.

The control block diagram of the voltage-source BDC is shown in Figure 13.

Its equivalent output impedance is given by

$$Z_o(s) = \frac{u_{dc}}{-i_o} = \frac{Ls + G_i(s)}{1 + Cs(Ls + G_i(s)) + G_i(s)G_v(s)} \tag{11}$$

The Nyquist criterion based on the impedance ratio $T_m(s)$ is an effective method for assessing interaction stability between two subsystems in a cascaded network. If the Nyquist plot of $T_m(s)$ encircles the point $(-1, 0)$, the overall system becomes unstable. The expression of $T_m(s)$ is given by

$$T_m(s) = \frac{Z_o(s)}{Z_{bdc}(s)} \tag{12}$$

https://doi.org/10.3390/en19030862

Figure 14 shows the Nyquist plots of $T_m(s)$ under the proposed strategy for $k = 0, 0.25, 0.75$, and 1. As observed in Figure 14, when k varies from 0 to 1, none of the Nyquist curves of the parallel system encircle the point $(-1, 0)$. Therefore, the choice of k within this range does not compromise the system's stability.

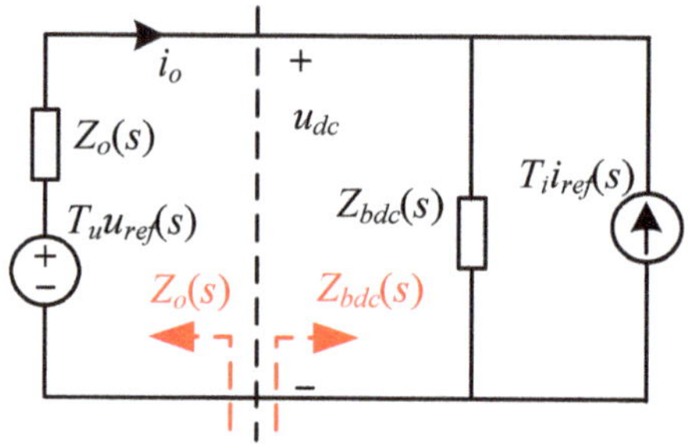

Figure 12. Impedance model of the parallel system comprising a voltage-source BDC and a current-source BDC.

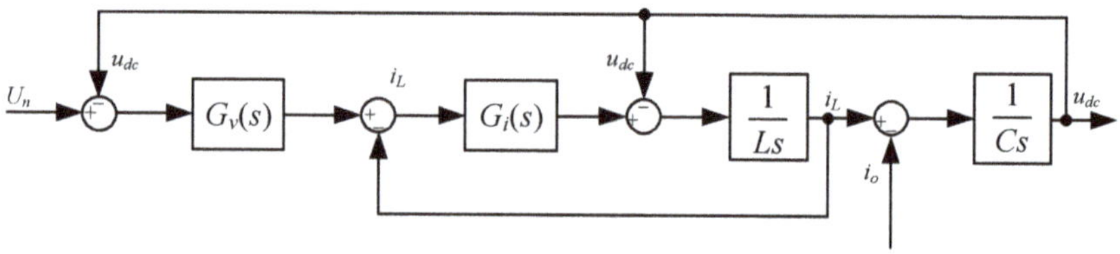

Figure 13. Control block diagram of the voltage-source BDC.

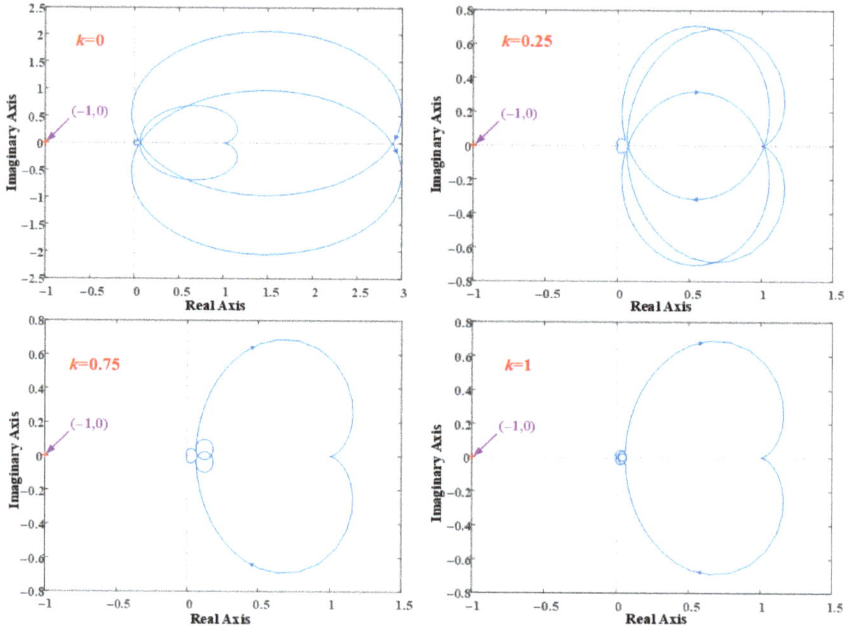

Figure 14. Nyquist Plots of $T_m(s)$ for $k = 0, 0.25, 0.75$, and 1.

https://doi.org/10.3390/en19030862

3. Develop an Adaptive Gain Adjustment Mechanism

In practical engineering applications, converters are typically subject to constraints such as power limits, current limits, modulation margin, and thermal losses. To explicitly associate the gain coefficient k with these constraints, this paper proposes a unified design framework for an adaptive gain coefficient and introduces a capacity and current boundary-based adaptive gain regulation mechanism, thereby enabling real-time shaping and on-demand tuning of the equivalent harmonic impedance of the BDC. The proposed framework exhibits good scalability, allowing new influencing factors to be incorporated flexibly without changing the overall architecture. It mainly consists of four modules: normalization and available margin calculation, piecewise-differentiable gating mapping, multi-indicator geometric aggregator, and linear mapping. The detailed flowchart is shown in Figure 15.

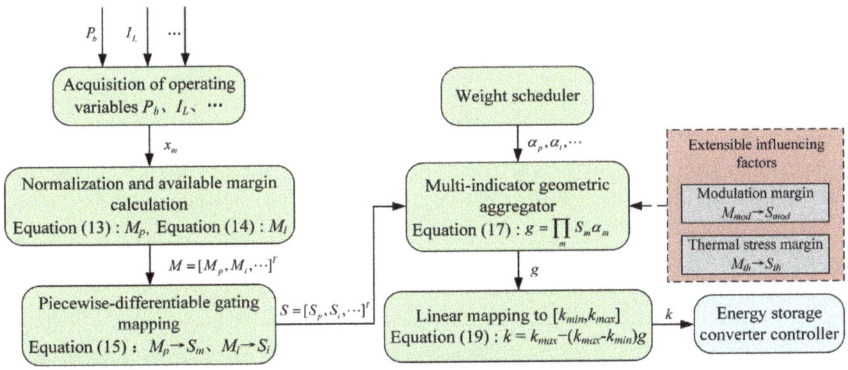

Figure 15. Design flowchart of the adaptive gain coefficient k.

3.1. Normalization and Available Margin Calculation

First, dimensionless "available margin" indices $M_m \in [0,1]$ are computed for various constraint quantities to characterize the safety degree of the current operating point relative to the rated limits. We take power and current available margins as examples.

The power availability margin is defined as

$$M_p = \frac{P_n - |P_b|}{P_n}, 0 \le M_p \le 1 \tag{13}$$

where P_n is the rated power of the device and P_b is the current active power (or power command). Specifically, P_b is obtained by multiplying the measured instantaneous DC bus voltage and output current to compute the instantaneous power, and then applying a low-pass filter to generate a smoothed estimate, so as to mitigate the influence of switching ripple and measurement noise on the adaptive decision. When the output power approaches the rated value, M_p approaches 0, indicating that the power capacity is gradually being exhausted; when the output is far below the rated power, M_p approaches 1, characterizing sufficient regulation space on the power side.

Similarly, the current availability margin is defined as

$$M_i = \frac{I_{L\max} - |I_L|}{I_{L\max}}, 0 \le M_i \le 1 \tag{14}$$

where I_{Lmax} is the maximum allowable current and I_L is the inductor current. Likewise, M_i decreases from 1 to 0 as the current constraint is gradually approached. Specifically, I_L is

https://doi.org/10.3390/en19030862

obtained by applying a low-pass filter to the measured inductor current signal, so as to reflect the available current margin and the proximity to the current constraint.

The filtering stages used to compute P_b and I_L, together with the update channel of k, form a slow adaptive loop whose bandwidth is designed to be much lower than that of the inner current loop. As a result, the adaptation does not introduce rapid gain jumps on the fast time scale dominated by the current loop, thereby avoiding transient instability or a significant reduction in stability margins caused by fast parameter variations.

3.2. Piecewise-Differentiable Gating Mapping

To effectively suppress chattering or abrupt changes near the threshold caused by measurement noise or small perturbations, thereby preventing high-frequency switching or oscillations of the gain k in the vicinity of the operating boundary, a smooth cubic gating function $S(x; x_0, x_1)$ is introduced to nonlinearly map the available margin:

$$S(x; x_0, x_1) = \begin{cases} 0, x \le x_0 \\ \left(\frac{x-x_0}{x_1-x_0}\right)^2 \left(3 - 2\frac{x-x_0}{x_1-x_0}\right), x_0 < x < x_1 \\ 1, x \ge x_1 \end{cases} \tag{15}$$

where x_0 and x_1 are the thresholds at which the regulation starts to take effect and is fully released, respectively. Within the interval (x_0, x_1), the function is a cubic smooth curve, and both the function value and its first derivative are continuous at (x_0, x_1). This guarantees a smooth transition of the tuning coefficient k and its rate of change, thus preventing control chattering induced by small fluctuations in the margins.

For each available margin M_m (such as M_p, M_i, etc.), a corresponding adjustable margin factor S_m is obtained through the gating function

$$S_m = S(M_m; m_{m0}, m_{m1}) \in [0, 1] \tag{16}$$

The thresholds (m_{m0}, m_{m1}) can be flexibly set according to the importance of each physical quantity, thereby determining the start point and width of the transition region for its regulation effect.

3.3. Multi-Indicator Geometric Aggregator

After obtaining the adjustable margin factors S_m for each constraint, they are combined using a geometric-weighted form:

$$g = \prod_m {S_m}^{\alpha_m} \tag{17}$$

where $\alpha_m > 0$ is the weight exponent of each influencing factor, typically chosen such that $\sum_m \alpha_m = 1$, ensuring that the aggregated factor g remains within [0, 1]. For example, when considering only power and current available margins, one has

$$g = S(M_p; m_{p0}, m_{p1})^{\alpha_p} \cdot S(M_i; m_{i0}, m_{i1})^{\alpha_i} \tag{18}$$

The proposed geometric weighting method exhibits excellent consistency, sensitivity, and scalability. When any critical available margin is severely insufficient, causing the corresponding S_m to approach 0, the weighted product g will decrease rapidly. This makes the final regulation coefficient more sensitive to the "weakest link", enabling the timely reinforcement of protection or limiting actions. If other constraint factors, such as modulation margin or thermal loss margin, need to be considered subsequently, it is only necessary to define new M_j and S_j, following the same steps, and introduce the

https://doi.org/10.3390/en19030862

corresponding $S_j^{\alpha_j}$ term into the product, without altering the original control framework. Meanwhile, by adjusting the weight coefficients α_m, the relative importance of different constraints to system safety can be flexibly characterized based on engineering experience or optimization results.

3.4. Linear Mapping

The comprehensive influence factor g is linearly mapped to the regulation coefficient interval $[k_{min}, k_{max}]$:

$$k = k_{\max} - (k_{\max} - k_{\min})g \tag{19}$$

Since $g \in [0, 1]$, it directly follows that

$$k \in [k_{\min}, k_{\max}] \tag{20}$$

When the system operates with sufficient available margins, each M_m is large, resulting in $S_m \approx 1$ and thus g approaches 1. In this case,

$$k \approx k_{\min} \tag{21}$$

At this point, the equivalent harmonic output impedance of the BDC is very small, providing strong equivalent harmonic absorption capability for the DC bus. Conversely, when certain constraints approach their limits and the available margin decreases, the relevant S_m decreases significantly, causing g to approach 0. It can be obtained that

$$k \approx k_{\max} \tag{22}$$

At this point, the regulation coefficient automatically increases, resulting in a larger harmonic output impedance of the BDC and reduced harmonic absorption capability. This forces the system operation to retreat towards the safe region, achieving adaptive protection against operational risks under multi-constraint conditions.

In summary, this method constructs normalized available margins, employs smooth gate functions for interval mapping, and utilizes geometric weighting to fuse multiple influence factors. Ultimately, the comprehensive quantity g is mapped to the adaptive gain coefficient k, flexibly altering the equivalent harmonic output impedance of the BDC.

3.5. Example of Extended Constraints

In practical PWM converters, the harmonic compensation path is implemented as an auxiliary signal superimposed on the modulation waveform (or the duty-cycle command). When the compensation demand becomes large, the modulation command may move closer to the saturation boundary, which can induce overmodulation and nonlinear distortion. Consequently, the achievable impedance-shaping capability is inherently bounded in practice. To explicitly capture this engineering limitation and to enable an online constraint-aware capping of the compensation strength, the modulation margin is introduced as an illustrative constraint and incorporated into the unified framework.

For PWM modulation (or duty-cycle control), the unsaturated modulation command is denoted as $m^*(t)$, and it is constrained within the admissible range $[m_{min}, m_{max}]$. Based on this, a normalized index describing the available modulation margin, $M_{mod}(t)$, is defined as

$$M_{\text{mod}}(t) = sat_{[0,1]}\left(\frac{m_{\max} - |m^*(t)|}{m_{\max} - m_{safe}}\right) \tag{23}$$

where m_{safe} represents a reserved safety margin threshold selected to avoid pronounced nonlinear distortion when operating near saturation. With this definition, $M_{mod}(t)$ decreases as

https://doi.org/10.3390/en19030862

$|m^*(t)|$ approaches m_{max}, indicating an insufficient modulation margin; conversely, $M_{mod}(t)$ remains close to unity when $|m^*(t)|$ is well below m_{max}, indicating a sufficient margin.

To prevent abrupt variations and potential chattering when the operating point is close to the margin boundary, $M_{mod}(t)$ is further transformed into a modulation margin factor, S_{mod}, through a smooth gating function:

$$S_{\text{mod}} = S(M_{\text{mod}}) \tag{24}$$

The gating function $S(M_{mod})$ can adopt the cubic smooth form used in the unified framework, ensuring consistency with the gating design applied to the power and current margin constraints.

Subsequently, the modulation margin factor is incorporated into the multi-indicator geometric aggregation to form the composite influence factor g:

$$g = \prod_m S_m{}^{\alpha_m} \Rightarrow g = S_p^{\alpha_p} S_i^{\alpha_i} S_{\text{mod}}^{\alpha_{\text{mod}}} \tag{25}$$

where $\alpha_{mod} > 0$ denotes the weighting exponent associated with the modulation margin constraint and reflects its relative importance. The established linear mapping from g to k over the range $[k_{min}, k_{max}]$ is preserved. Under a reduced modulation margin, the value of $M_{mod}(t)$ becomes smaller, which leads to a smaller S_{mod} and consequently a smaller aggregated factor g. In this case, the coefficient k is adaptively increased, which raises the equivalent harmonic impedance and weakens the participation of harmonic compensation. In this way, the proposed formulation provides an explicit, constraint-aware cap on the compensation strength and helps avoid overmodulation and saturation.

The online computation of the modulation margin index typically requires no additional sensors and primarily relies on controller-internal variables and routinely measured signals. Specifically, the controller can directly provide the modulation command $m^*(t)$, the corresponding limits m_{max} and m_{min}, and the safety threshold m_{safe} determined by engineering settings. If a formulation based on the modulation index and DC bus voltage is adopted, the DC bus voltage u_{dc} (which is normally sampled) and the relevant normalization base values or per-unit scaling factors of the modulator may also be used. These signals support a complete online closed-loop procedure that sequentially performs margin evaluation, smooth gating, multi-indicator aggregation, and coefficient update, thereby integrating modulation margin constraints into the unified framework.

4. Simulation and Experimental Verification

To verify the effectiveness and practical feasibility of the proposed adaptive harmonic impedance control and flexible compensation strategy, a simulation model of a DC microgrid with two parallel-connected BDCs is first established in Matlab/Simulink based on the parameters listed in Table 2, as shown in Figure 16. Here, U_{dcref} denotes the reference voltage of the DC bus; U_b is the reference voltage on the high-voltage side of the BDC; C_b and L_b are the BDC filter capacitor and filter inductor, respectively; P_{bn} is the rated power of the BDC; m_{p0} and m_{p1} represent the lower and upper limits of the available power margin, respectively; R_{ac} is the resistance of the single-phase AC load; P_{acl} is the rated power of the single-phase AC load; C_{dc} is the DC-link filter capacitor; f_{sw} and f_s are the switching frequency and sampling frequency, respectively; U_{acref} is the reference voltage of the single-phase AC output; f_{ac} is the rated frequency of the single-phase AC load; and C_{ac} and Lac are the AC-side filter capacitor and filter inductor, respectively (used in the AC-side filtering stage). In this setup, BDC 1 adopts voltage source control to establish and regulate the DC bus voltage, whereas BDC 2 employs current source control and incorporates the proposed strategy to provide harmonic compensation capability. It should

https://doi.org/10.3390/en19030862

be noted that the proposed harmonic mitigation method is applied only to BDC 2, while the control strategy of BDC 1 remains unchanged.

Table 2. System parameters.

Symbol	Value	Symbol	Value
U_{dcref}	350 V	U_b	500 V
C_b	500 μF	L_b	300 μH
P_{bn}	8 kW	m_{p0}	0.1
m_{p1}	0.3	P_{acl}	8 kW
R_{ac}	6.05 Ω	C_{dc}	1000 μF
f_{sw}	10 kHz	f_s	20 kHz
U_{acref}	220 V	f_{ac}	50 Hz
C_{ac}	25 μF	L_{ac}	300 μH

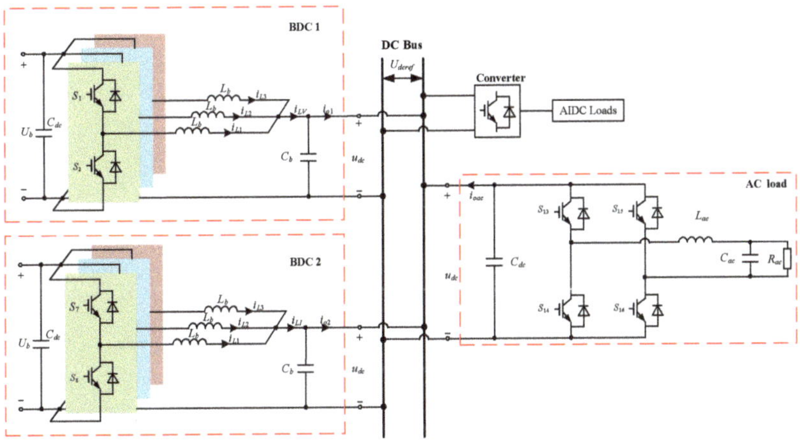

Figure 16. System simulation structure.

In the simulations, harmonic currents at 20 Hz, 50 Hz, 100 Hz, and 150 Hz are injected into the DC bus to emulate random fluctuations of the AIDC load, where the selection of these frequency points is justified as follows:

(i) Public engineering analyses indicate that large-scale AI training loads, due to the synchronized "computation–communication" iterations of GPU clusters, exhibit second-scale low-frequency power swings whose spectral energy can extend over approximately the 0.1–20 Hz band [35,36]; thus, 20 Hz is chosen as a representative point within this low-frequency disturbance range to cover relatively faster components and to evaluate the adaptive harmonic suppression capability of the proposed method under different compensation strengths.

(ii) In practical AIDC power-supply chains, power electronic interfaces such as UPSs, PFC stages, and inverters may couple the AC-side fundamental frequency-related disturbances and their integer-multiple-related components into the DC side [37,38]. When the DC bus supplies a single-phase AC load through an inverter, a typical double-line-frequency ripple appears on the DC side; moreover, among low-order harmonics, the third harmonic corresponds to 150 Hz in a 50 Hz system and is commonly observed in scenarios involving nonlinear loads such as computers/UPSs [39].

This set of frequency points is selected for representative validation. The proposed control framework is not limited to these specific frequencies; in practical applications, it can be

https://doi.org/10.3390/en19030862

extended and its parameters can be tuned according to the measured harmonic/disturbance spectra on site. Accordingly, the above-outlined injected frequency components are used to evaluate the harmonic suppression performance and adaptive regulation capability under different compensation strengths and operating conditions.

4.1. Simulation Results Analysis

Without the proposed control strategy, the DC bus voltage ripple amplitude is 27.4 V, and the voltage THD is 3.06%. The ripple current mainly flows into BDC 1 with an amplitude of 71.5 A, whereas the ripple current of BDC 2 is only 7.7 A (see Figure 17). This is because, in a paralleled converter system, the sharing of harmonic currents is mainly determined by the closed-loop output impedances of the branches, and ripple currents tend to flow into the branch with the smaller output impedance. In the low-frequency range of interest, the closed-loop output impedance of the current-source-type converter BDC 2 is much higher than that of the voltage-source-type converter BDC 1. Therefore, the harmonic current primarily flows into BDC 1, which has the smaller closed-loop output impedance.

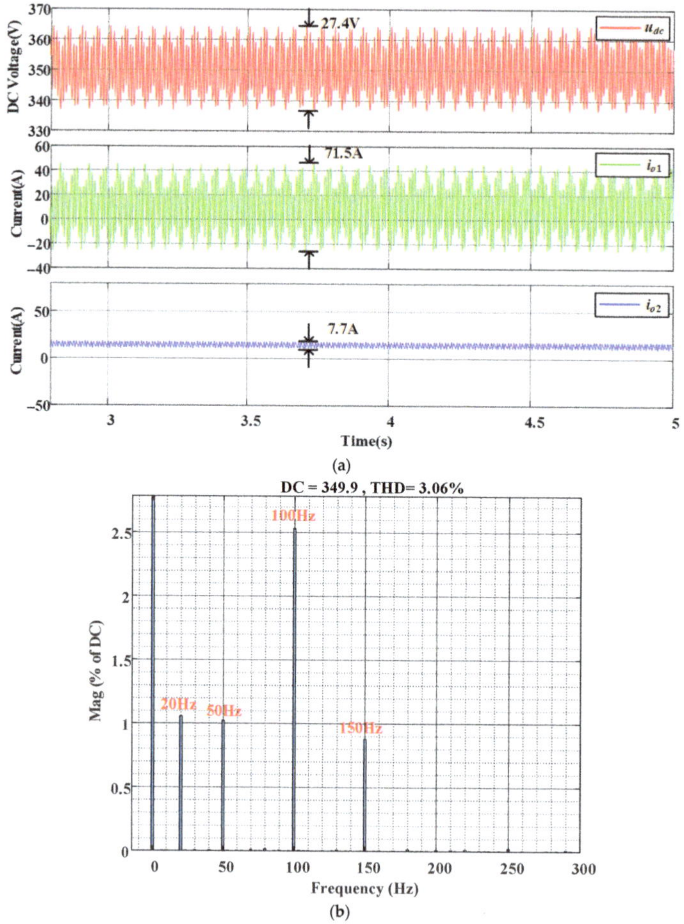

Figure 17. Simulation results and FFT analysis without the proposed compensation strategy. (**a**) DC bus voltage and BDC current waveforms without the proposed compensation strategy. (**b**) FFT analysis of DC bus voltage without the proposed compensation strategy.

https://doi.org/10.3390/en19030862

4.1.1. Harmonic Suppression Performance

When $k = 0$, the equivalent harmonic output impedance of the BDC 2 approaches zero and is much smaller than the output impedance of BDC 1, so BDC 2 assumes the main harmonic current compensation. At this time, the bus ripple amplitude drops from 27.4 V to 8.4 V, voltage THD drops from 3.06% to 0.89%, and BDC 1 ripple current drops from 71.5 A to 19.3 A, while the ripple current of BDC 2 increases from 7.7 A to 56 A (see Figure 18), demonstrating an extremely strong ripple suppression effect.

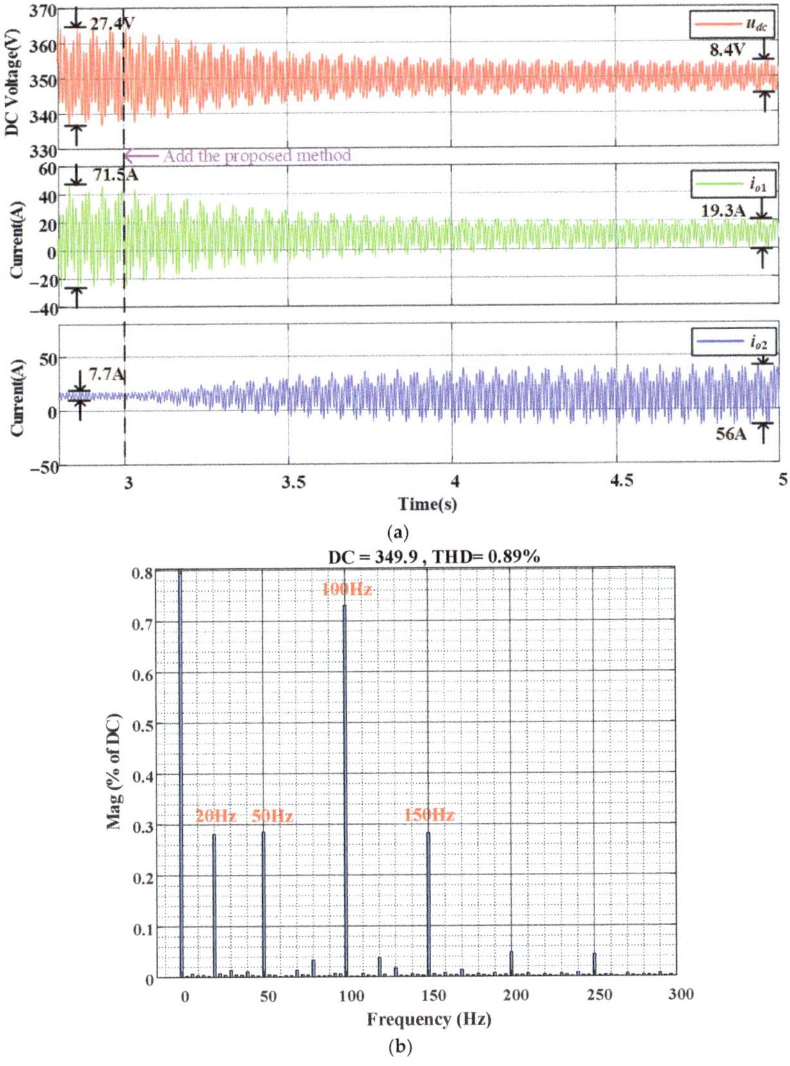

Figure 18. Simulation results and FFT analysis with the proposed compensation strategy ($k = 0$). (**a**) DC bus voltage and BDC current waveforms with the proposed compensation strategy ($k = 0$). (**b**) FFT analysis of DC voltage with the proposed compensation strategy ($k = 0$).

To demonstrate the steady-state suppression performance, Figure 19 presents the simulated waveforms obtained using the frequency-selective control strategy in [26]. As shown, the DC bus voltage ripple is reduced from 27.4 V to 10.5 V, and the voltage THD decreases from 3.06% to 1.19%. Since the frequency-selective control reduces the output impedance of BDC 2, most of the ripple current is guided to the BDC 2 branch. As a result, the ripple current of BDC 1 is reduced from 71.5 A to 25.1 A, while the ripple current absorbed by BDC 2 increases from 7.7 A to 48.7 A.

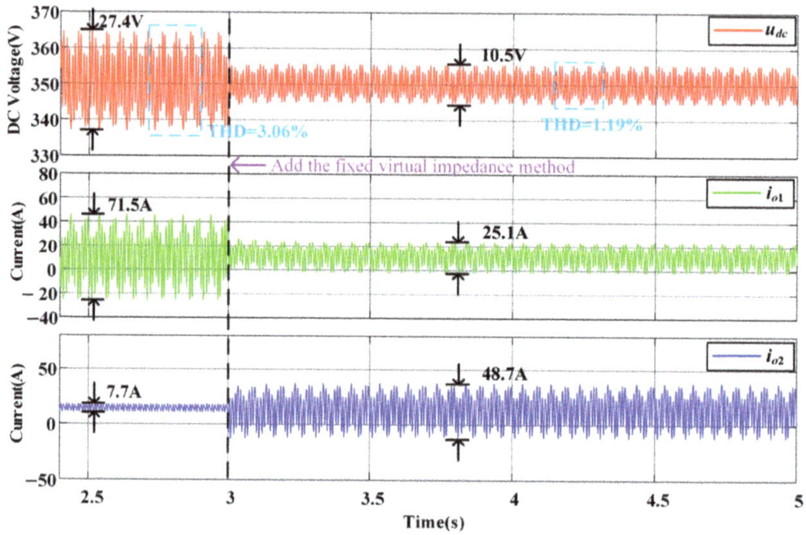

Figure 19. Simulation results for comparison using the previous control method reported in [26].

To further compare the implementation performance of the proposed strategy with the method in [26], Figure 20 presents a quantitative comparison of key metrics for different methods with respect to the conventional current-source control baseline. The metrics include the bus-voltage ripple amplitude, voltage THD, and the peak-to-peak current of BDC 1. As shown in Figure 20, the proposed strategy achieves better improvement in all the above metrics than the method in [26].

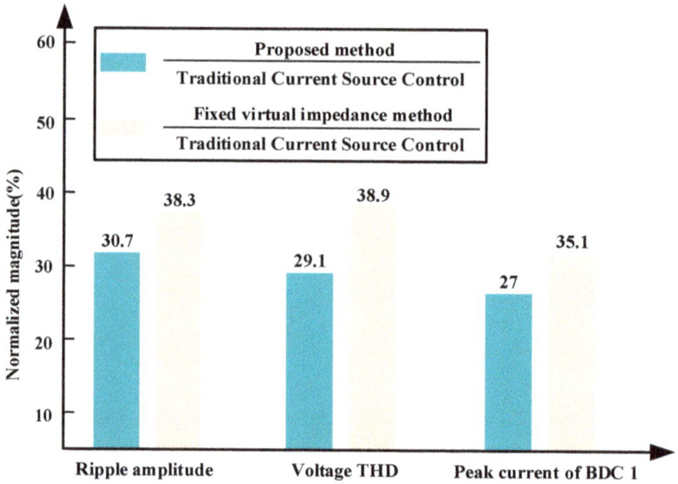

Figure 20. Comparison results of various performance metrics between the proposed method and the method in [26].

https://doi.org/10.3390/en19030862

When $k = 0.5$, the BDC 2 equivalent harmonic impedance increases moderately. The bus ripple amplitude rises to 12.9 V, THD is 1.45%, and the BDC 1 ripple current amplitude rises to 29.4 A (see Figure 21). Results indicate that as gain k increases, harmonic absorption capability weakens, and bus power quality declines accordingly.

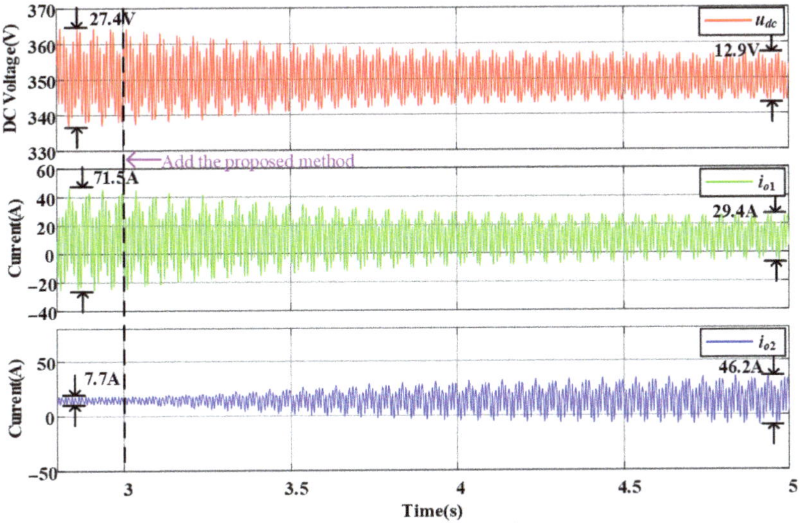

(**a**)

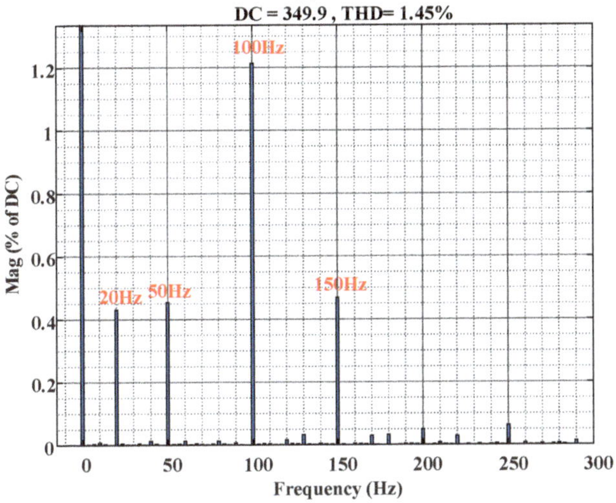

(**b**)

Figure 21. Simulation results and FFT analysis with the proposed compensation strategy ($k = 0.5$). (**a**) DC bus voltage and BDC current waveforms with the proposed compensation strategy ($k = 0.5$). (**b**) FFT analysis of DC voltage with the proposed compensation strategy ($k = 0.5$).

When $k = 0.8$, the output impedance of BDC 2 becomes close to that of BDC 1, but it remains smaller than the output impedance of BDC 1. The bus ripple amplitude is 19.4 V, THD is 2.22%, and the BDC 1 ripple current amplitude is 47.8 A (see Figure 22). Although compensation capability is markedly reduced, the overall performance remains superior to the baseline condition without the compensation strategy.

https://doi.org/10.3390/en19030862

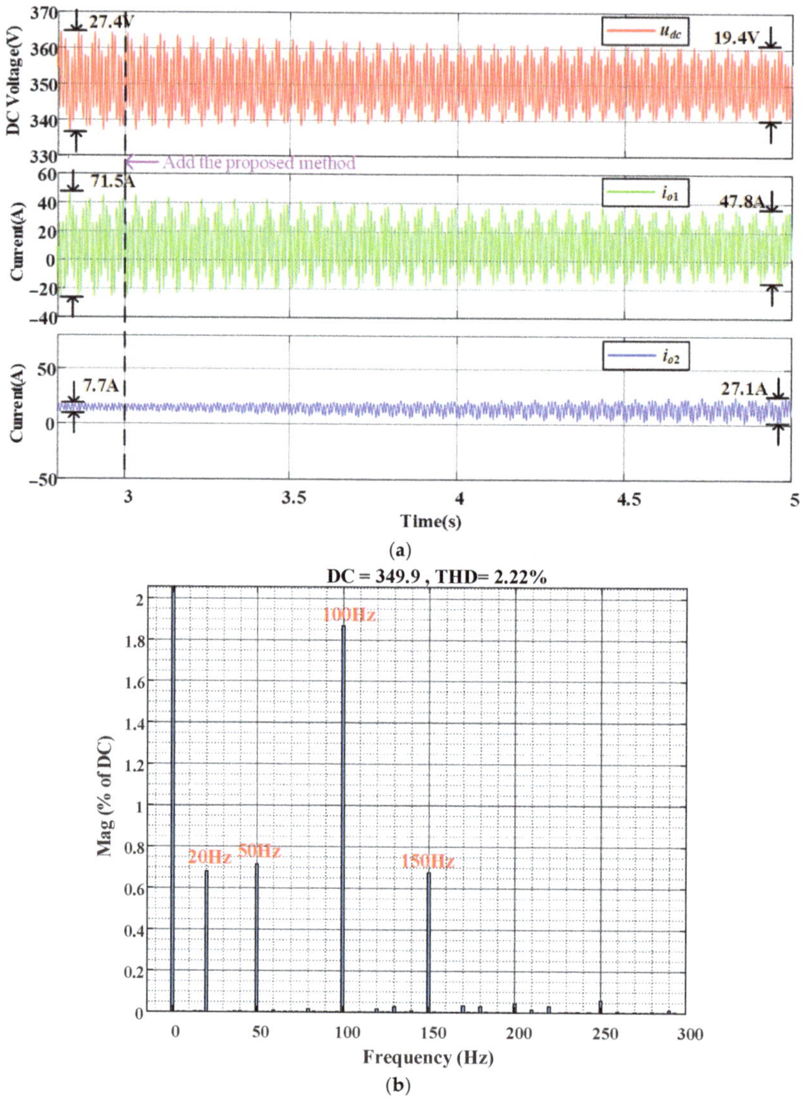

Figure 22. Simulation results and FFT analysis with the proposed compensation strategy ($k = 0.8$). (**a**) DC bus voltage and BDC current waveforms with the proposed compensation strategy ($k = 0.8$). (**b**) FFT analysis of DC voltage with the proposed compensation strategy ($k = 0.8$).

4.1.2. Dynamic Adaptive Adjustment Performance

To further verify the online regulation capability of the proposed adaptive mechanism under varying available capacity, a step change in the output power of BDC 2 is set.

Power Surge ($t = 5$ s): The system initially operates in strong compensation mode ($k = 0$), with bus voltage THD maintained at 0.89%. At $t = 5$ s, a sudden increase in BDC 2 output power causes the power available margin to decrease. The adaptive mechanism drives k to rise smoothly from 0 and stabilize at 0.38 (Figure 23b). Correspondingly, the converter's equivalent harmonic impedance increases, the compensation current is actively limited, and the bus ripple amplitude increases from 8.4 V to 10.5 V, with THD rising to 1.17% (Figure 23a). This process demonstrates that when the converter's power

https://doi.org/10.3390/en19030862

available margin is insufficient, the controller prioritizes ensuring the converter does not overcurrent or overload at the cost of sacrificing some power quality, achieving dynamic matching between compensation capability and available capacity.

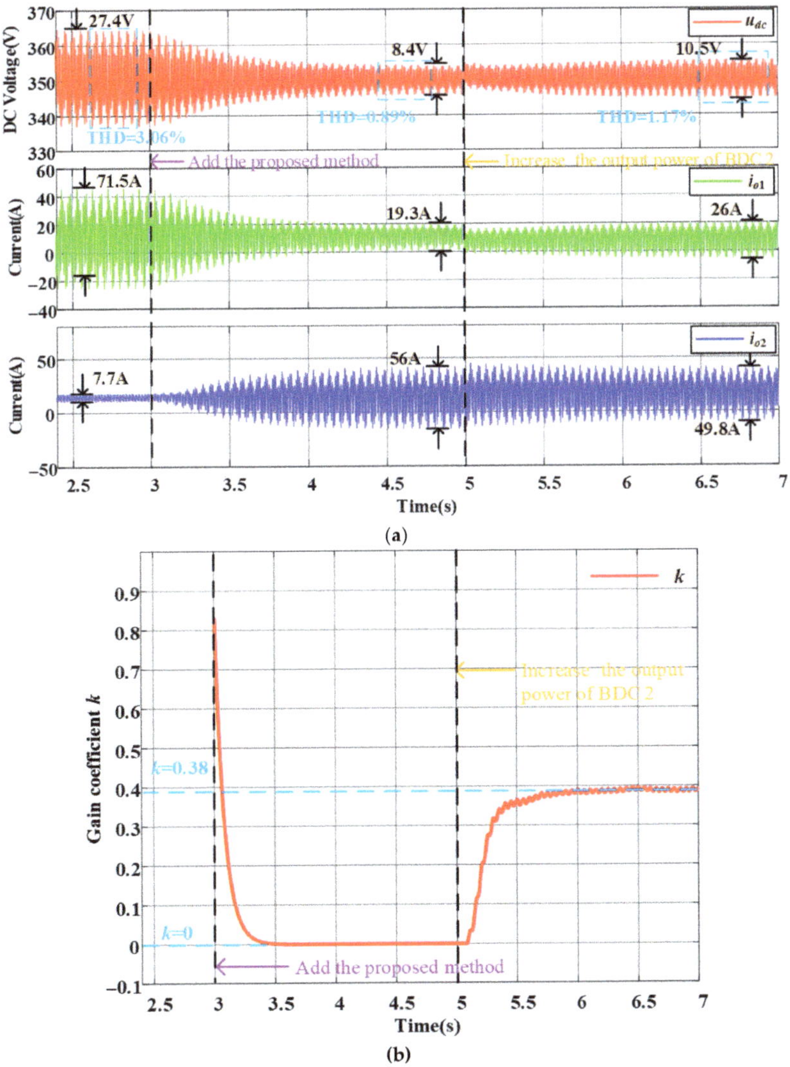

Figure 23. Simulation results and variation curve of gain coefficient k when BDC 2 output power increases at $t = 5$ s. (**a**) DC bus voltage and BDC current waveforms when BDC 2 output power increases at $t = 5$ s. (**b**) Variation curve of gain coefficient k when BDC 2 output power increases at $t = 5$ s.

Power Drop ($t = 5$ s): The system runs in semi-compensation mode with $k = 0.54$, and bus voltage THD is 1.59%. At $t = 5$ s, a sudden decrease in BDC 2 output power causes the power available margin to increase. The adaptive mechanism drives k to decrease smoothly from 0.54 and finally fall back to 0 (Figure 24b). After compensation capability is enhanced, the bus ripple amplitude drops from 14.2 V to 8.4 V (Figure 24a). This result indicates that

https://doi.org/10.3390/en19030862

when device capacity is sufficient, the proposed method automatically enhances harmonic absorption capability, unifying power quality improvement with margin utilization.

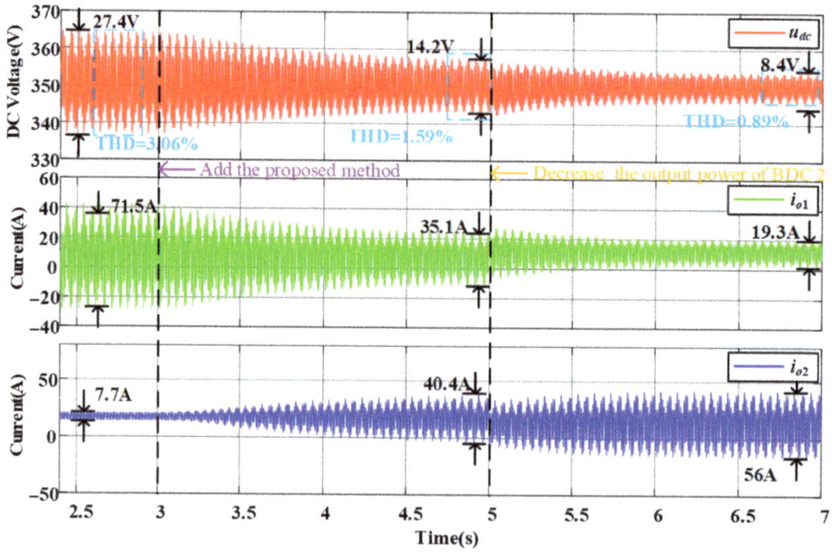

(a)

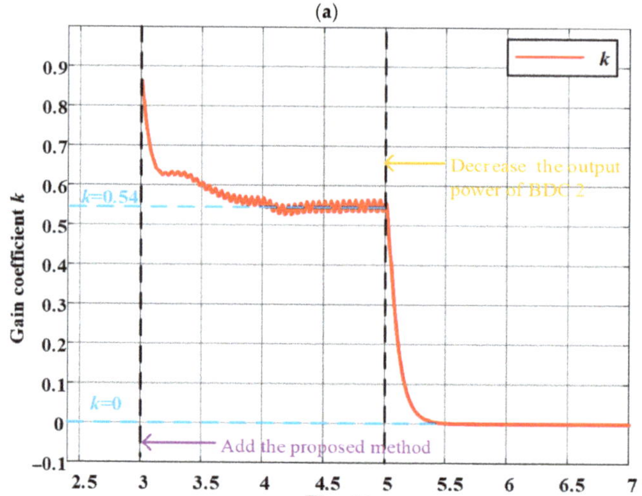

(b)

Figure 24. Simulation results and variation curve of gain coefficient k when BDC 2 output power decreases at $t = 5$ s. (**a**) DC bus voltage and BDC current waveforms when BDC 2 output power decreases at $t = 5$ s. (**b**) Variation curve of gain coefficient k when BDC 2 output power decreases at $t = 5$ s.

To highlight the advantage of the proposed method over the impedance-shaping approach in [26] (which uses a fixed gain to reduce the output impedance of BDC 2), the output power of BDC 2 is increased at $t = 6$ s in the simulation. As shown in Figure 25a, because a fixed gain is used in [26], the output impedance remains constant, and the peak value of the absorbed harmonic current is also fixed. When the active current is superimposed with the harmonic current, the inductor current can easily reach the current

https://doi.org/10.3390/en19030862

limit and trigger overcurrent protection. In contrast, as shown in Figure 25b, the proposed method automatically computes the current margin and adaptively increases k when the current is about to hit the limit (see Figure 25c), thereby increasing the output impedance of BDC 2 and reducing the absorbed harmonic current.

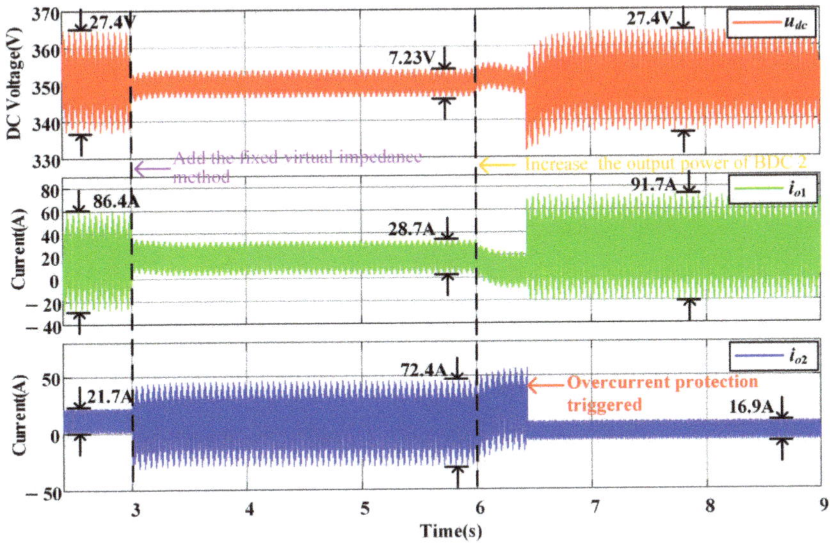

(a)

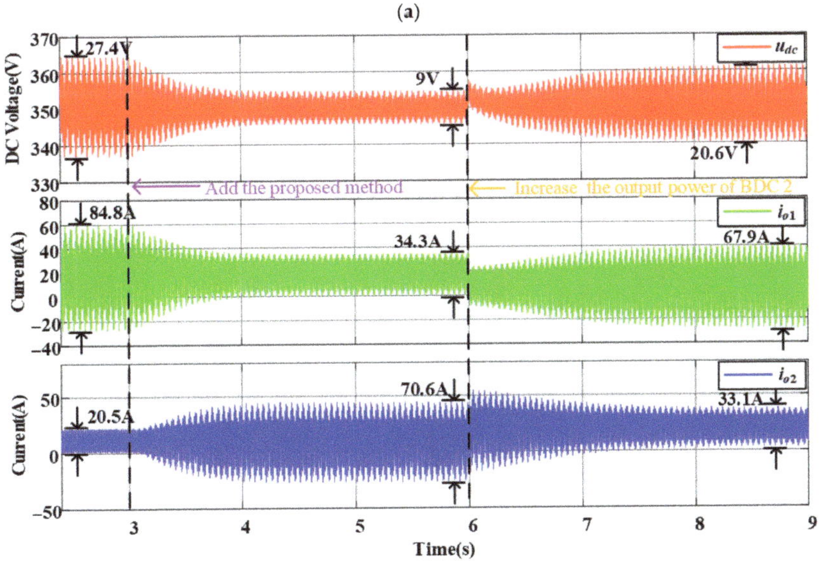

(b)

Figure 25. *Cont.*

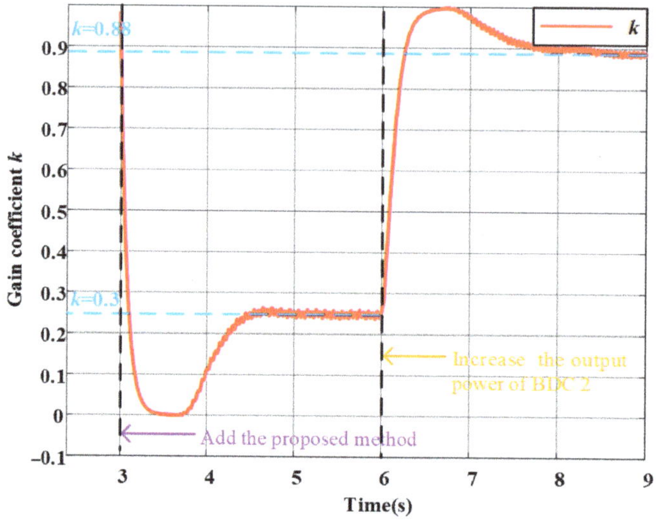

(c)

Figure 25. Simulation results comparing the method in [26] and the proposed method, and the variation curve of the gain coefficient k when the output power of BDC 2 decreases at t = 6 s. (**a**) DC bus voltage and BDC current waveforms using the method in [26] when the output power of BDC 2 decreases at t = 6 s. (**b**) DC bus voltage and BDC current waveforms using the proposed method when the output power of BDC 2 decreases at t = 6 s. (**c**) Variation curve of gain coefficient k when BDC 2 output power increases at t = 6 s.

4.2. Experimental Results Analysis

To further validate the engineering applicability of the proposed method, experiments are conducted on a hardware platform, with the experimental parameters kept consistent with the simulation settings. Since the random fluctuations of an AIDC load are difficult to reproduce in a laboratory environment, a single-phase load is connected to the DC bus through an inverter to intentionally introduce a second-harmonic voltage ripple on the bus, thereby verifying the effectiveness of the proposed strategy. Figure 26 illustrates the configuration of the experimental platform, where the core control algorithm is implemented and executed in real time on a TI TMS320F28376S digital signal processor (DSP) manufactured by Texas Instruments (Dallas, TX, USA).

From Figure 27, without the proposed control strategy, the DC bus voltage ripple amplitude is 28 V, and the voltage THD is 4.14%; the ripple current mainly flows into BDC 1, with the BDC 1 ripple current amplitude at 26 A and BDC 2 at 13.7 A.

4.2.1. Harmonic Suppression Performance

From Figure 28, when the proposed strategy is introduced with k = 0, the DC bus voltage ripple amplitude is suppressed to 4 V, the BDC 1 ripple current amplitude is suppressed to 5.6 A, the BDC 2 ripple current amplitude increases to 44 A, and the voltage THD decreases to 0.76%. Experimental results verify that BDC 2's equivalent harmonic impedance approaches zero at this time, presenting ideal voltage source characteristics and demonstrating extremely strong ripple suppression effects.

https://doi.org/10.3390/en19030862

Figure 26. Configuration of the experimental platform.

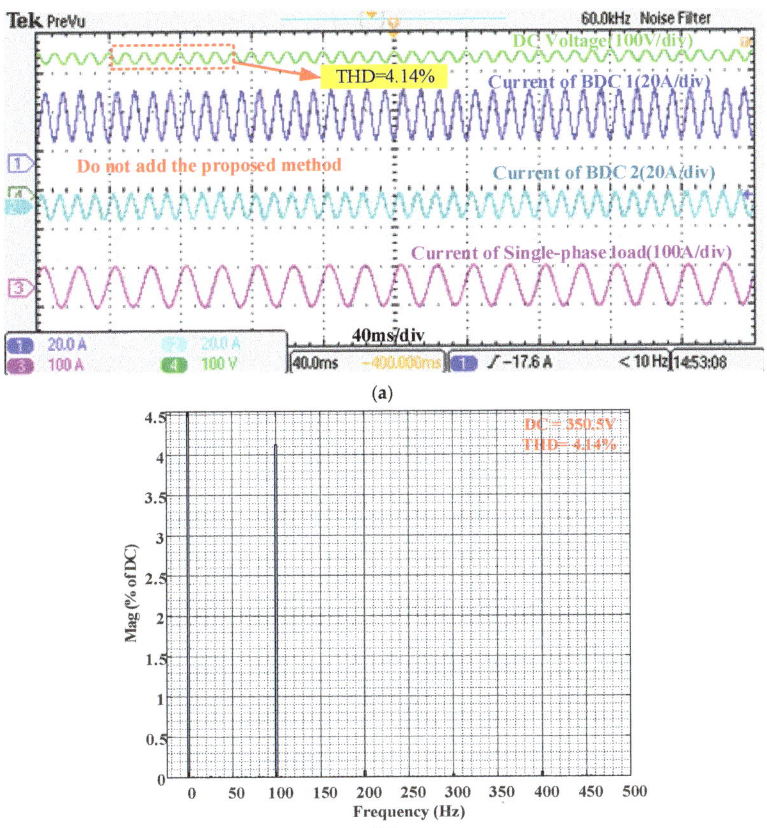

Figure 27. Experimental waveforms and FFT analysis without the proposed strategy. (**a**) Waveforms of DC bus voltage, BDC output current, and single-phase AC load current without the proposed strategy. (**b**) FFT analysis of DC bus voltage without the proposed strategy.

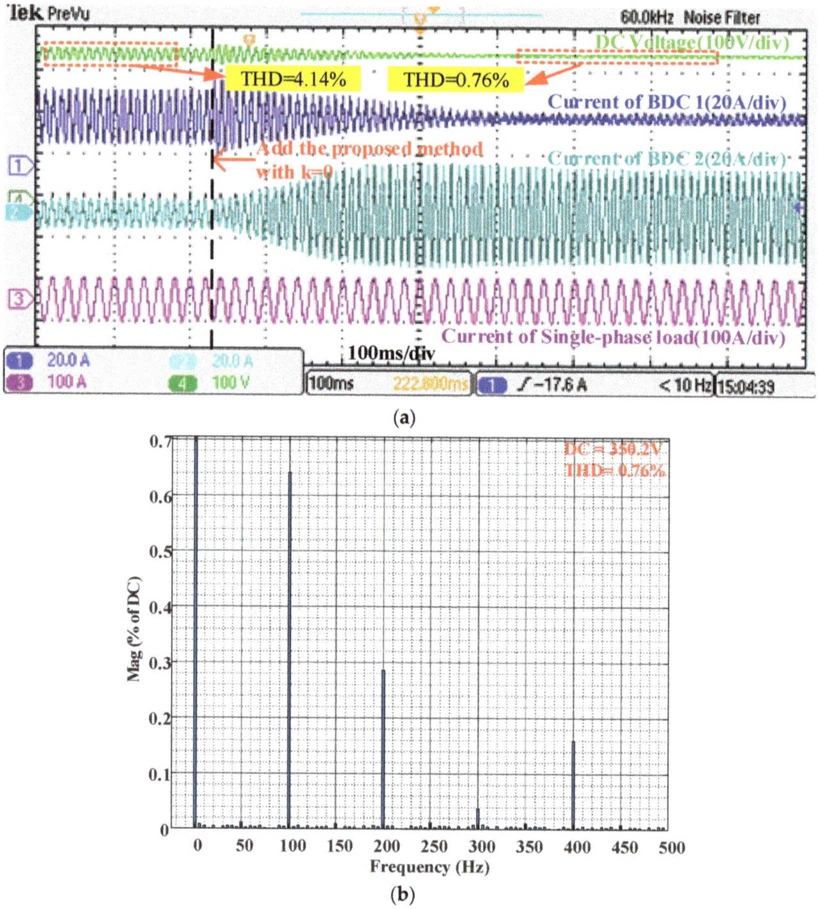

Figure 28. Experimental waveforms and FFT analysis with the proposed strategy ($k = 0$). (**a**) Waveforms of DC bus voltage, BDC output current, and single-phase AC load current with the proposed strategy ($k = 0$). (**b**) FFT analysis of DC bus voltage with the proposed strategy ($k = 0$).

From Figure 29, when k increases to 0.5, the converter's equivalent harmonic impedance increases moderately. DC bus voltage ripple amplitude rises to 8 V, voltage THD increases to 1.26%, BDC 1 ripple current amplitude rises to 8.7 A, and BDC 2 ripple current amplitude decreases to 41.6 A. Experimental results show that as k increases, harmonic absorption capability weakens, and bus power quality declines.

From Figure 30, when $k = 0.9$, the converter approaches a high-impedance state, and its suppression effect on bus harmonics weakens. Bus ripple amplitude increases to 20 V, voltage THD increases to 3.1%, BDC 1 ripple current amplitude rises to 19.7 A, and BDC 2 ripple current amplitude decreases to 20.4 A. Experimental results indicate that although compensation capability is significantly reduced, the overall performance is still better than the baseline condition without compensation.

https://doi.org/10.3390/en19030862

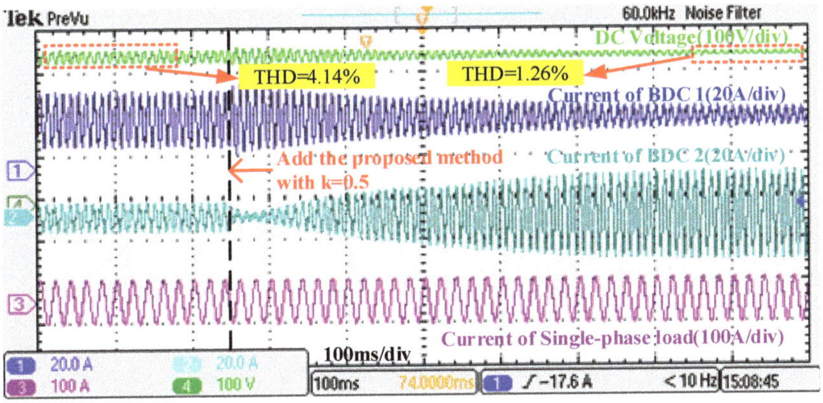

(a)

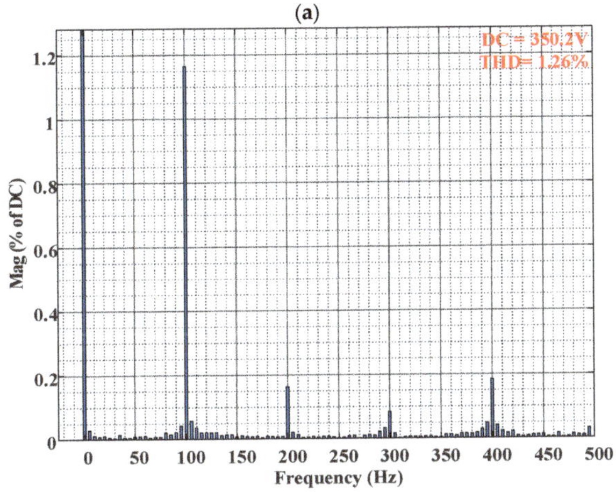

(b)

Figure 29. Experimental waveforms and FFT analysis with the proposed strategy (k = 0.5). (**a**) Waveforms of DC bus voltage, BDC output current, and single-phase AC load current with the proposed strategy (k = 0.5). (**b**) FFT analysis of DC bus voltage with the proposed strategy (k = 0.5).

4.2.2. Dynamic Adaptive Regulation Performance

Figures 31 and 32 show experimental results under a sudden increase and decrease in BDC 2 output power, respectively, verifying the online regulation capability under varying available capacity.

From Figure 31, when the BDC 2 output current is 5 A with the proposed strategy, BDC 2 has maximum remaining capacity and the lowest equivalent harmonic impedance. When the bus ripple amplitude is suppressed to 4 V, THD decreases to 0.76%, the ripple current into BDC 1 decreases to 5.6 A, and the ripple current absorbed by BDC 2 increases to 44 A. When the BDC 2 output current is suddenly increased to 15 A, BDC 2 adjusts its equivalent harmonic output impedance according to remaining capacity. At this point, the ripple voltage amplitude increases to 12 V, THD increases to 1.91%, the ripple current into BDC 1 increases to 8 A, and the ripple current absorbed by BDC 2 decreases to 35.4 A.

From Figure 32, when the BDC 2 output current is 15 A with the proposed strategy, BDC 2 adjusts its equivalent harmonic output impedance according to the remaining capacity. When the bus ripple amplitude is suppressed to 8.1 V, THD decreases to 1.36%, the ripple current into BDC 1 decreases to 7.2 A, and the ripple current absorbed by BDC 2 increases to 39.8 A. When the BDC 2 output current is suddenly decreased to 5 A, BDC 2

https://doi.org/10.3390/en19030862

has the maximum remaining capacity and the lowest equivalent harmonic impedance. At this point, the ripple voltage amplitude is further reduced to 4 V, THD decreases to 0.76%, the ripple current into BDC 1 decreases to 5.6 A, and the ripple current absorbed by BDC 2 increases to 44 A.

Experimental results demonstrate that when the converter's available capacity is restricted, the proposed strategy prioritizes ensuring that the converter does not overcurrent or overload at the cost of sacrificing some power quality; when the converter's available capacity is sufficient, the proposed method automatically enhances harmonic absorption capability, thereby achieving dynamic matching between compensation capability and available capacity.

Due to the limitations of the experimental setup, the experiments in this paper mainly validate the DC bus power-pulsation condition caused by a single-phase load. To complement operating conditions that are difficult to realize on the current test bench, the simulation study further considers a harmonic source composed of multiple superimposed frequency components for a more comprehensive evaluation.

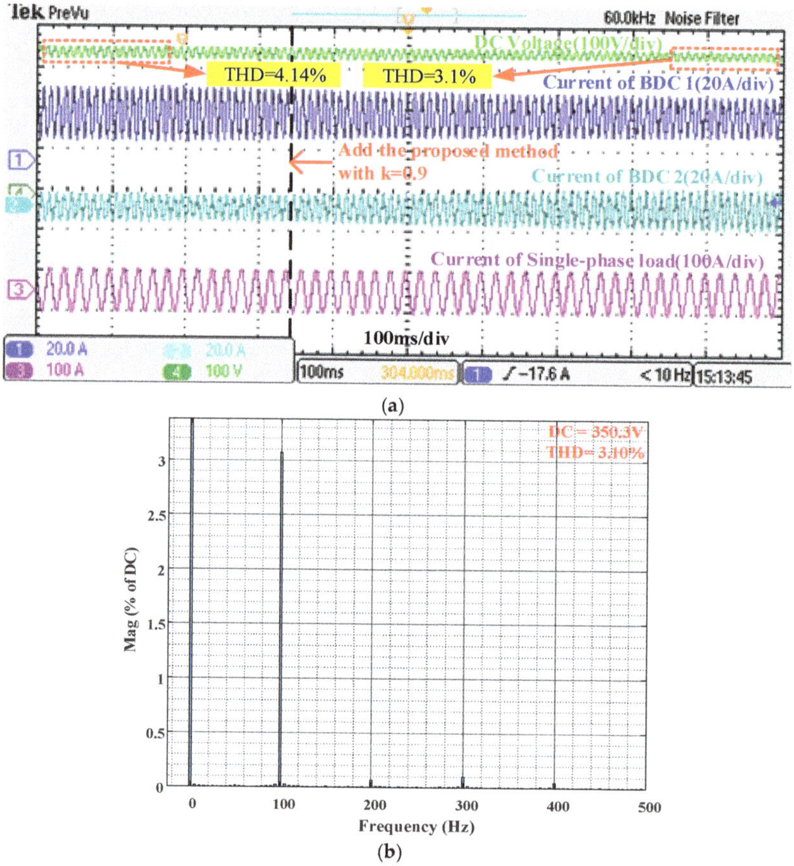

Figure 30. Experimental waveforms and FFT analysis with the proposed strategy (k = 0.9). (**a**) Waveforms of DC bus voltage, BDC output current, and single-phase AC load current with the proposed strategy (k = 0.9). (**b**) FFT analysis of DC bus voltage with the proposed strategy (k = 0.9).

https://doi.org/10.3390/en19030862

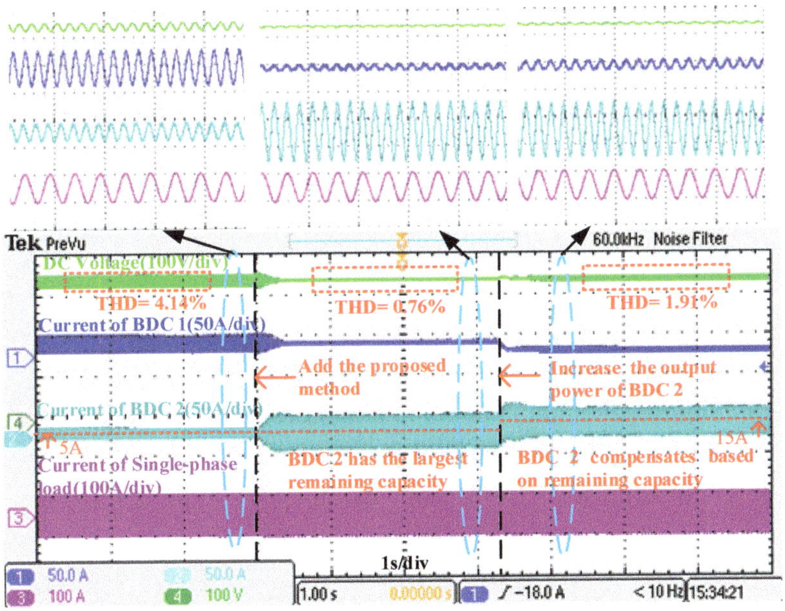

Figure 31. Experimental waveforms and FFT analysis results when BDC 2 output power suddenly increases.

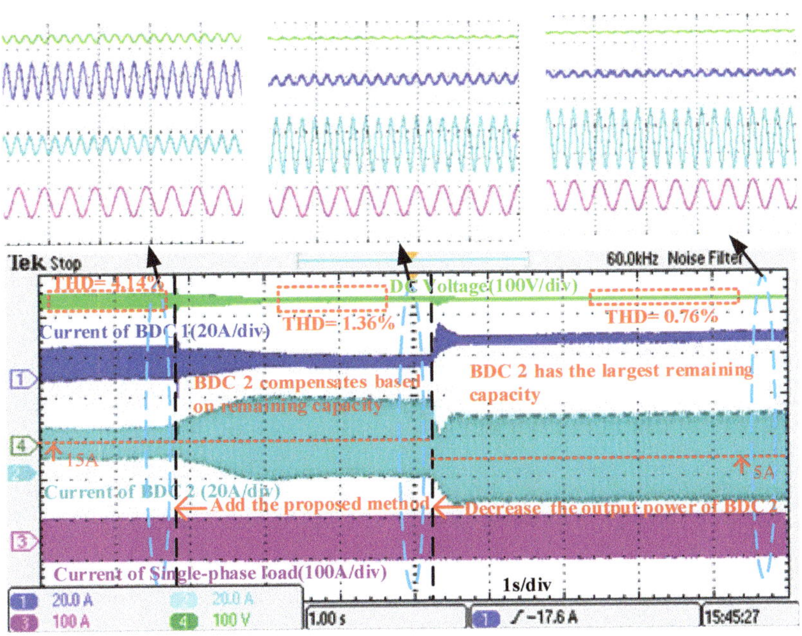

Figure 32. Experimental waveforms and FFT analysis results when BDC 2 output power suddenly decreases.

https://doi.org/10.3390/en19030862

5. Conclusions

Addressing the issues of DC bus voltage ripple and power quality degradation caused by the stochastic fluctuation of AI computing loads, this paper proposes an adaptive harmonic impedance control and flexible compensation method. The main conclusions are as follows:

(1) A virtual impedance control architecture combining bus voltage feedforward and output current feedback is constructed. Without relying on extremely high gains or narrow-band resonant links, controllable harmonic output impedance in the target frequency band is achieved through virtual impedance reshaping.

(2) An adaptive adjustment mechanism considering converter capacity and current boundaries is established. By monitoring the power margin and current margin in real-time and dynamically adjusting the gain coefficient k, on-demand shaping of the BDC's equivalent harmonic impedance is realized.

(3) Simulation and experimental results verify that the proposed method can smoothly transition between "strong compensation/low impedance" and "weak compensation/high impedance" modes according to the converter's operating state. This method not only effectively suppresses voltage ripples under steady-state conditions but also automatically limits compensation intensity under extreme conditions, achieving a good balance between improving power quality and ensuring system stability.

At present, the proposed method targets a control improvement for a single converter, and its applicability and performance in multi-converter systems remain to be further validated.

Author Contributions: Conceptualization, J.L. and B.Y.; methodology, H.L.; validation, J.L. and B.Y.; formal analysis, Z.Y.; investigation, J.L. and B.Y.; writing—original draft preparation, J.L.; writing—review and editing, S.L.; supervision, S.L.; project administration, Q.X. All authors have read and agreed to the published version of the manuscript.

Funding: This research received no external funding.

Data Availability Statement: Data are contained within the article.

Conflicts of Interest: The authors declare no conflicts of interest.

References

1. Gutiérrez-Escalona, J.; Roncero-Clemente, C.; Husev, O.; Matiushkin, O.; Blaabjerg, F. Artificial Intelligence in the Hierarchical Control of AC, DC, and Hybrid AC/DC Microgrids: A Review. *IEEE Access* **2024**, *12*, 157227–157246. [CrossRef]
2. Yao, Z.; He, X.; Liu, M.; Liu, J.; Xiao, Z.; Tang, Y. Fixed Switching Frequency Control Using Trapezoidal Current Mode to Achieve ZVS in Three-Level DC–DC Converters. *IEEE Trans. Ind. Electron.* **2025**, *72*, 4175–4185. [CrossRef]
3. Liao, J.; Guo, C.; Huang, Y.; Qiu, G.; Zhou, N.; Wang, Q. Active Damping Control of Cascaded DC Converter in DC Microgrids Based on Optimized Parallel Virtual Resistance. *IEEE J. Emerg. Sel. Top. Ind. Electron.* **2023**, *4*, 560–570. [CrossRef]
4. Liu, G.; Caldognetto, T.; Mattavelli, P.; Magnone, P. Suppression of Second-Order Harmonic Current for Droop-Controlled Distributed Energy Resource Converters in DC Microgrids. *IEEE Trans. Ind. Electron.* **2020**, *67*, 358–368. [CrossRef]
5. Li, X.; Guo, L.; Zhang, S.; Wang, C.; Li, Y.W.; Chen, A.; Feng, Y. Observer-Based DC Voltage Droop and Current Feed-Forward Control of a DC Microgrid. *IEEE Trans. Smart Grid* **2018**, *9*, 5207–5216. [CrossRef]
6. Tu, Z.; Fan, B.; Zhang, W.; Peng, J.; Liu, W. Optimal State-Constrained Control of DC Shipboard Power Systems for Online Pulsed Power Load Accommodation. *IEEE Trans. Smart Grid* **2022**, *13*, 96–105. [CrossRef]
7. Vitorino, M.A.; Alves, L.F.S.; Wang, R.; de Rossiter Corrêa, M.B. Low-Frequency Power Decoupling in Single-Phase Applications: A Comprehensive Overview. *IEEE Trans. Power Electron.* **2017**, *32*, 2892–2912. [CrossRef]
8. Huang, K.-P.; Wang, Y.; Wai, R.-J. Design of Power Decoupling Strategy for Single-Phase Grid-Connected Inverter Under Nonideal Power Grid. *IEEE Trans. Power Electron.* **2019**, *34*, 2938–2955. [CrossRef]
9. Kan, S.; Ruan, X.; Li, X. Topology and Control for Second Harmonic Current Reduction in Two-Stage Single-Phase Inverter Without Electrolytic Capacitors. *IEEE Trans. Power Electron.* **2024**, *39*, 9051–9058. [CrossRef]

https://doi.org/10.3390/en19030862

10. Shimizu, T.; Jin, Y.; Kimura, G. DC ripple current reduction on a single-phase PWM voltage source rectifier. In *Conference Record of the 1999 IEEE Industry Applications Conference, Proceedings of the Thirty-Forth IAS Annual Meeting (Cat. No.99CH36370), Phoenix, AZ, USA, 3–7 October 1999*; IEEE: Piscataway, NJ, USA, 2002; Volume 2, pp. 810–817.
11. Lin, J.; Weiss, G. An Indirect Approach to Control an Active Capacitor. *IEEE J. Emerg. Sel. Top. Power Electron.* **2020**, *8*, 2898–2906. [CrossRef]
12. Bhus, V.; Lin, J.; Weiss, G. The modular active capacitor for high power ripple attenuation. *CPSS Trans. Power Electron. Appl.* **2021**, *6*, 251–262. [CrossRef]
13. Chen, Y.-C.; Chen, L.-R.; Lai, C.-M.; Lin, Y.-C.; Kuo, T.-J. Development of a DC-Side Direct Current Controlled Active Ripple Filter for Eliminating the Double-Line-Frequency Current Ripple in a Single-Phase DC/AC Conversion System. *Energies* **2020**, *13*, 4772. [CrossRef]
14. Zhang, H.; Ma, S.; Zhu, B.; Li, Z.; Du, Z.; Zhang, M.; Pan, Y. A modular multilevel DC active power filter scheme based on optimal parameter design for the CFETR N-NBI accelerator grid power supply. *Fusion Eng. Des.* **2023**, *193*, 113804. [CrossRef]
15. Su, M.; Pan, P.; Long, X.; Sun, Y.; Yang, J. An Active Power-Decoupling Method for Single-Phase AC–DC Converters. *IEEE Trans. Ind. Inform.* **2014**, *10*, 461–468. [CrossRef]
16. Qiu, M.; Wang, P.; Bi, H.; Wang, Z. Active Power Decoupling Design of a Single-Phase AC–DC Converter. *Electronics* **2019**, *8*, 841. [CrossRef]
17. Rodríguez-Benítez, O.; Ponce-Silva, M.; Aqui-Tapia, J.A.; Rodríguez-Benítez, Ó.M.; Lozoya-Ponce, R.E.; Adamas-Pérez, H. Active Power-Decoupling Methods for Photovoltaic-Connected Applications: An Overview. *Processes* **2023**, *11*, 1808. [CrossRef]
18. Liu, Y.; Zhang, W.; Sun, Y.; Su, M.; Xu, G.; Dan, H. Review and Comparison of Control Strategies in Active Power Decoupling. *IEEE Trans. Power Electron.* **2021**, *36*, 14436–14455. [CrossRef]
19. Chen, M.; Liu, S.; Cao, Q.; Wang, H. Sensor-Reduced Active Power Decoupling Method for Single-Phase Rectifiers. *Energies* **2025**, *18*, 3711. [CrossRef]
20. Li, P.; Liu, C.; Tong, X.; Li, X.; Zheng, G. Improved direct ripple power predictive control of single-phase rectifier based on ripple separation. *Control Eng. Pract.* **2025**, *155*, 106173. [CrossRef]
21. Li, G.; Wu, H.; Chen, J.; Liu, X. A Harmonic Current Self-Compensation Control Strategy of DC Active Power Filter. *IEEE J. Emerg. Sel. Top. Power Electron.* **2025**, *13*, 5252–5261. [CrossRef]
22. Li, G.; Wang, H.; Liu, X.; Chen, J. A sliding mode active disturbance rejection control of DC active power filter for enhancing DC microgrid voltage robustness. *Electr. Power Syst. Res.* **2025**, *241*, 111388. [CrossRef]
23. Zhang, L.; Ren, X.; Ruan, X. A Bandpass Filter Incorporated Into the Inductor Current Feedback Path for Improving Dynamic Performance of the Front-End DC–DC Converter in Two-Stage Inverter. *IEEE Trans. Ind. Electron.* **2014**, *61*, 2316–2325. [CrossRef]
24. Kan, S.; Ruan, X.; Dang, H.; Zhang, L.; Huang, X. Second Harmonic Current Reduction in Front-End DC–DC Converter for Two-Stage Single-Phase Photovoltaic Grid-Connected Inverter. *IEEE Trans. Power Electron.* **2019**, *34*, 6399–6410. [CrossRef]
25. Chaturvedi, S.; Fulwani, D.; Guerrero, J.M. Adaptive-SMC Based Output Impedance Shaping in DC Microgrids Affected by Inverter Loads. *IEEE Trans. Sustain. Energy* **2020**, *11*, 2940–2949. [CrossRef]
26. Sevostyanov, N.A.; Gorbunov, R.L. Control Strategy to Mitigate Voltage Ripples in Droop-Controlled DC Microgrids. *IEEE Trans. Power Electron.* **2023**, *38*, 15377–15389. [CrossRef]
27. Rashidirad, N.; Hamzeh, M.; Sheshyekani, K.; Afjei, E. A Simplified Equivalent Model for the Analysis of Low-Frequency Stability of Multi-Bus DC Microgrids. *IEEE Trans. Smart Grid* **2018**, *9*, 6170–6182. [CrossRef]
28. Yao, Z.; Lu, S. A Simple Approach to Enhance the Effectiveness of Passive Currents Balancing in an Interleaved Multiphase Bidirectional DC–DC Converter. *IEEE Trans. Power Electron.* **2019**, *34*, 7242–7255. [CrossRef]
29. Yao, Z.; He, X.; Xiao, Z.; Deng, F.; Dai, C.; Lu, S.; Tang, Y. Current Ripple Prediction and ZVS-Based Variable Switching Frequency Control for Interleaved Multiphase Three-Level DC–DC Converter. *IEEE Trans. Power Electron.* **2025**, *40*, 5109–5119. [CrossRef]
30. Alajmi, B.N.; Abdelsalam, I.; Marei, M.I.; Ahmed, N.A. Two Stage single-phase EV On-Board Charger Based on Interleaved Cascaded Non-Inverting Buck-Boost Converter. In Proceedings of the 2023 IEEE Conference on Power Electronics and Renewable Energy (CPERE), Luxor, Egypt, 19–21 February 2023; pp. 1–6.
31. Alajmi, B.N.; Abdelsalam, I.; Marei, M.I.; Ahmed, N.A. A Three-Phase Interleaved AC-DC Converter Based off-Board Battery Charger Station With Balanced Supply and Submodules Currents. *IEEE Trans. Ind. Appl.* **2025**, *62*, 2937–2947. [CrossRef]
32. Zhang, Q.; Li, J.; Zhu, R.; Deng, F.; Sun, X.; An, S.; Liserre, M. Output Impedance Modeling and High-Frequency Impedance Shaping Method for Distributed Bidirectional DC–DC Converters in DC Microgrids. *IEEE Trans. Power Electron.* **2020**, *35*, 7001–7014. [CrossRef]
33. Tian, H.; Li, Y. Virtual Resistor Based Second-Order Ripple Sharing Control for Distributed Bidirectional DC–DC Converters in Hybrid AC–DC Microgrid. *IEEE Trans. Power Electron.* **2021**, *36*, 2258–2269. [CrossRef]
34. Wu, Y.; Wang, W.; Tang, F.; Wu, X. Adaptive virtual admittance control of bidirectional DC/DC converters for DC bus voltage ripple suppression in DC microgrids. *Int. J. Electr. Power Energy Syst.* **2024**, *159*, 110045. [CrossRef]
35. Ko, M.-S.; Zhu, H. Wide-Area Power System Oscillations from Large-Scale AI Workloads. *arXiv* **2025**, arXiv:2508.16457.

https://doi.org/10.3390/en19030862

36. Mishra, C.; Vanfretti, L.; Delaree, J., Jr.; Purcell, T.J.; Jones, K.D. Understanding the inception of 14.7 Hz oscillations emerging from a data center. *Sustain. Energy Grids Netw.* **2025**, *43*, 101735. [CrossRef]
37. Rodriguez, E.; Leyva, R.; Farivar, G.G.; Townsend, C.D.; Pou, J. Negative-Sequence Current Capability in Low-Capacitance Cascaded H-Bridge Static Compensators with Optimal Third-Harmonic Circulating Current Injection. *IEEE Trans. Power Electron.* **2023**, *38*, 4544–4558.
38. Zhu, X.; Wang, S.; Chen, X.; Cheng, A.; Li, H.; Zhang, Z.; Ren, X. Input current ripple reduction of DC/DC converters. *Energy Rep.* **2023**, *9*, 444–452. [CrossRef]
39. Nuñez-Ramírez, V.; Guerrero-Rodríguez, N.F.; Batista-Jorge, R.O.; Mercado-Ravelo, R.; Ramírez-Rivera, F.; Ferreira, J.A.; Ramos-Ciprian, R.D. Harmonic distortion caused by non-linear household loads: Measurement and modelling. *Results Eng.* **2025**, *25*, 104483. [CrossRef]

https://doi.org/10.3390/en19030862

energies

MDPI

Article

An Adaptive Extended Kalman Filter with Passive Control for DC-DC Converter Supplying Constant Power and Constant Voltage Loads

Peng Wang [1], Zhenlong Ma [1], Junfeng Tian [1], Zhe Li [1], Yani Li [1], Panbao Wang [2,*] and Yang Zhou [2]

[1] Daxinganling Power Supply Company of State Grid Heilongjiang Electric Power Co., Ltd., Daxinganling 165099, China; wangpeng19830919@163.com (P.W.); 13555088420@163.com (Z.M.); 13846572369@139.com (J.T.); 15245759428@163.com (Z.L.); liyani24@163.com (Y.L.)

[2] School of Electrical Engineering and Automation, Harbin Institute of Technology, Harbin 150000, China; nxzy303@163.com

* Correspondence: wangpanbao@hit.edu.cn

Abstract

This article introduces an integrated control scheme combining an Adaptive Extended Kalman Filter (AEKF) with a Passivity-Based Control (PBC) approach to stabilize a DC-DC boost converter feeding both constant voltage and constant power loads (CPLs) in DC microgrids. Unlike conventional observers, the AEKF adapts its covariance matrices in real time to accurately estimate both system states and the unknown load dynamics introduced by CPLs, thereby eliminating the need for additional sensors and enhancing estimation convergence. Coupled with the PBC, the estimated disturbances are compensated via a feedforward path, significantly improving the system's resilience to input voltage fluctuations and load variations. Through a Lyapunov-based stability analysis, the combined strategy is proven to ensure large-signal stability while maintaining a rapid transient recovery profile, even under significant parametric uncertainties. The simulation of this algorithm was implemented using PLECS, thoroughly validating the effectiveness and robustness of the proposed method.

Keywords: DC-DC boost converter; Extended Kalman Filter; DC microgrids; constant power loads; passivity-based control

1. Introduction

In recent years, amid the technological evolution of power systems, DC microgrids have emerged as a promising integrated architecture incorporating renewable energy sources, DC loads, and energy storage systems. Compared to their AC counterparts, DC microgrids offer distinct advantages, including a simpler structure, relatively lower control complexity, and the absence of reactive power and phase-related issues [1–3]. Their distributed control schemes further contribute to enhanced response speeds. Within the DC microgrid framework, the DC-DC boost converter is a critical component, necessitating high bidirectional energy conversion efficiency [4–6]. Consequently, developing novel control strategies for Boost converters that account for the influence of constant power loads (CPLs), mitigate disturbances, and enable rapid voltage regulation is of paramount importance.

Power electronic loads in DC microgrids are primarily categorized into resistive loads and constant power loads. Resistive loads, due to their passive nature, introduce positive damping into the system [7–9]. In contrast, CPLs, which maintain constant power draw, exhibit negative impedance characteristics at their input terminals. This negative impedance interaction with other system converters can reduce overall system damping, compromise stability, and introduce oscillation risks [10–13]. The increasing penetration of renewable energy sources further exacerbates these stability challenges posed by CPLs. Power system stability analysis is typically divided into large-signal [14–18] and small-signal [19–21] stability, based on operating scenarios and disturbance magnitudes, with current research predominantly focused on the latter. Primary mitigation strategies include passive damping [20,21], which adds passive components to improve damping directly but increases cost and weight, counter to lightweight trends; and active damping [22,23], which modifies the control structure to reshape system impedance. However, active damping is often limited to linearized small-signal models around an equilibrium point, restricting its applicability.

Given the inherent nonlinearity of power electronic converters and the negative impedance of CPLs, employing nonlinear control strategies is essential to enhance the global stability of DC microgrid systems. To improve converter stability, nonlinear control methods such as Passivity-Based Control (PBC) [24], Back-stepping Control (BSC) [25], Model Predictive Control (MPC) [26], Sliding Mode Control (SMC) [27], and PWM-based nonlinear control [28] have been progressively applied. However, challenges remain in improving output voltage tracking accuracy and system efficiency, leading to extensive research on combined control strategies. The Back-stepping Control method proposed in [29] is an effective nonlinear approach for enhancing tracking precision and system efficiency, but it requires accurate estimation of system uncertainties and disturbances to minimize tracking error. Observer technology, particularly the Nonlinear Disturbance Observer (NDO) [24], offers a promising solution for estimating disturbances without additional sensors, providing effective disturbance rejection and uncertainty compensation in nonlinear systems. For instance, ref. [30] employed an NDO combined with Back-stepping Control to enhance the large-signal stability of a conventional Boost converter system. While these strategies effectively ensure large-signal stability, their high dependency on accurate system modeling makes them susceptible to inevitable steady-state deviations under random source and load disturbances. Integrating compensation loops at relevant stages can reduce or eliminate these deviations, but often at the expense of dynamic response speed.

To address the aforementioned challenges, this paper proposes a Passivity-Based Control (PBC) method for DC-DC converters based on an Adaptive Extended Kalman Filter (AEKF). This approach not only significantly enhances the large-signal stability and robustness of the system but also incorporates precise current limiting control and rapid voltage regulation capabilities, thereby effectively ensuring the stability and reliability of DC microgrids under various complex operating conditions. The proposed control methodology primarily integrates PBC with Kalman filtering techniques. Firstly, an AEKF is designed for the real-time and accurate estimation of time-varying system parameters, notably load power and input voltage. Subsequently, an adaptive PBC technique, utilizing the state variables estimated by the filter, is developed to control the DC-DC boost converter with a CPL. Furthermore, the design of the proposed nonlinear control loop circumvents the difficulties associated with obtaining nominal parameters and performing complex multi-degree-of-freedom parameter tuning inherent in traditional methods, resulting in a relatively simple and practical control structure.

2. System Modeling and Problem Formulation

The object of this study is a typical DC microgrid architecture and its equivalent circuit, depicted in Figure 1. This system features a source bus that is intricately connected to a diversified array of power generation and storage technologies through appropriate power

https://doi.org/10.3390/en19030682

electronic interfaces. These include intermittent renewable sources like photovoltaic arrays and wind turbines, dispatchable generators, battery energy storage systems (BESS), and a point of common coupling to an external AC grid via an AC-DC converter. This multi-faceted integration strategy empowers the source bus to dynamically and flexibly draw power from the most available or economical sources, thereby significantly enhancing the overall reliability, resilience, and operational adaptability of the microgrid. The system's energy management system (EMS) employs sophisticated optimization algorithms to coordinate power flows, ensuring optimal utilization of renewable resources while maintaining the state of charge of the BESS for grid support services.

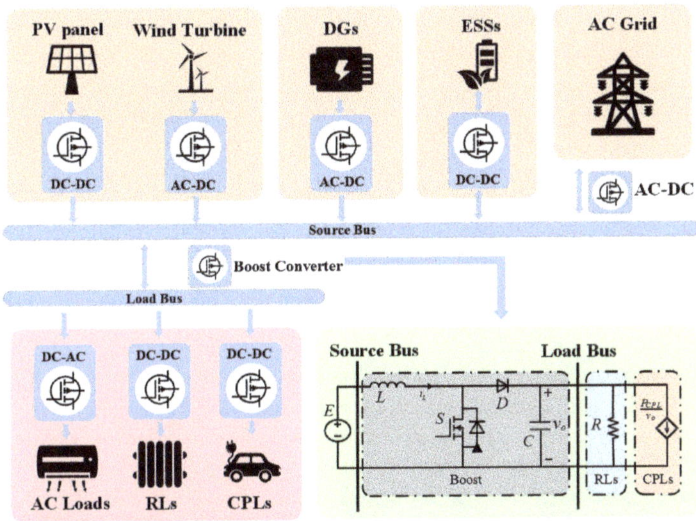

Figure 1. Typical DC microgrid architecture and its equivalent boost circuit.

A pivotal component in this architecture is the DC-DC boost converter, which interfaces the source bus with the load bus. Its primary function is twofold: first, to ensure precise and stable voltage regulation on the load bus, which is fundamental for high-quality power delivery; and second, to efficiently step up the variable, often lower voltage from the source side to a higher, regulated voltage level required by the loads. The converter's design must account for wide input voltage variations characteristic of renewable sources, necessitating advanced control techniques to maintain stability across the entire operating range. The performance of this converter is therefore critical, demanding high conversion efficiency and robust control capabilities to maintain a continuous and stable power supply under varying generation and consumption scenarios. Its dynamic response must be sufficiently fast to mitigate disturbances propagating through the system, particularly those originating from load-side perturbations.

The load bus supplies power to a mix of load types, primarily categorized into resistive loads and CPLs. Resistive loads present a linear relationship between current and voltage, contributing naturally to system damping and stable operation. In stark contrast, CPLs, such as tightly regulated power electronic converters found in electric vehicle charging stations, server farms, and variable-speed drive systems, maintain a constant power draw. This behavior manifests as a negative impedance characteristic at their input terminals, meaning an increase in input voltage results in a decrease in input current. This negative impedance introduces non-linearity and can significantly degrade system damping through positive feedback mechanisms. The interaction between this negative impedance and the

https://doi.org/10.3390/en19030682

source-side converters can lead to voltage oscillations, instability, and even system collapse, presenting a major challenge for DC microgrid stability. This instability phenomenon is further exacerbated in cascaded converter systems, where the impedance interaction between source and load converters can create forbidden operating regions. Consequently, the design and implementation of sophisticated, robust control strategies for the interface DC-DC converter are not just beneficial but imperative to mitigate these destabilizing effects and ensure the stable, reliable, and efficient operation of the entire DC microgrid system. Advanced control methodologies must incorporate both large-signal stability guarantees and adaptive capabilities to handle the time-varying nature of modern load profiles.

A mathematical model of the DC-DC boost converter needs to be established to serve as the foundation for subsequent controller design and stability analysis [31]. Without loss of generality, and assuming continuous conduction mode, the average dynamic model of the DC-DC boost converter can be represented by the following state-space equations:

$$\begin{cases} L\frac{di_L}{dt} = E - (1-\mu)v_o \\ C\frac{dv_o}{dt} = (1-\mu)i_L - \frac{v_o}{R} - \frac{P_{CPL}}{v_o} \end{cases} \tag{1}$$

where L, C, E, v_o, i_L, μ, R, and P_{CPL} represent the inductance, output capacitance, input voltage, output voltage, inductor current, switching duty cycle, resistive load, and constant power load of the DC-DC boost converter, respectively. Furthermore, the input voltage and the power of the constant power load are state variables that require estimation.

This model captures the essential nonlinear dynamics of the system. The term $\frac{P_{CPL}}{v_o}$ explicitly embodies the negative impedance characteristic of the CPL, as the current it draws decreases with increasing voltage. It is crucial to note that within the context of the proposed control strategy, the power of the constant power load P_{CPL} are treated as unknown time-varying parameters or state variables that require real-time estimation for precise control synthesis.

3. Adaptive Extended Kalman Filter (AEKF) and Stability Analysis

3.1. Adaptive Extended Kalman Filter (AEKF)

To accurately estimate the state variables using an Extended Kalman Filter, the dynamic model (1) is rewritten and can be expressed in the kinetic form of state-space representation as [32]:

$$\begin{cases} \dot{x} = Sx + d\rho + BE \\ y = hx \end{cases} \tag{2}$$

In Equation (2), the state variable matrix is $x = \begin{bmatrix} i_L & v_0 \end{bmatrix}^T$.

E is the input voltage, S is the system matrix, B is the input matrix, d is the nonlinear term matrix, h is the output matrix, y is the output quantity, and ρ is the influence factor based on the capacitor voltage. Then:

$$S = \begin{bmatrix} 0 & -\frac{1-\mu}{L} \\ \frac{1-\mu}{C} & -\frac{1}{RC} \end{bmatrix}, B = \begin{bmatrix} 0 & \frac{1}{L} \\ 0 & 0 \end{bmatrix}, d = \begin{bmatrix} 0 \\ -\frac{P_{CPL}}{C} \end{bmatrix}, h = \begin{bmatrix} 1 & 1 \end{bmatrix} \tag{3}$$

$$\rho = \frac{1}{v_0} \tag{4}$$

Based on the proposed Extended Kalman Filter method, the power of the constant power load needs to be considered as one of the state variables and added to the state-space expression, resulting in a new state variable matrix:

https://doi.org/10.3390/en19030682

$$x_{aekf} = \begin{bmatrix} x \\ P \end{bmatrix} \tag{5}$$

Due to the power characteristics of the constant power load, it follows that P equals 0. At this point, the state variable matrix becomes:

$$x_{aekf} = \begin{bmatrix} x \\ 0 \end{bmatrix} = f\left(x_{aekf}, u(t)\right) \tag{6}$$

where $u(t)$ is the time-varying system input, and f is a function of $u(t)$ and x_{aekf}, representing the Euler equation of the dynamic model after incorporating the output power into the state variable matrix. Furthermore, the output of the EKF algorithm is updated based on changes in the state variables:

$$y = [h|0] \begin{bmatrix} x \\ P \end{bmatrix} \tag{7}$$

After adding noise, the new form of the equation set is obtained, ω and ν can be considered as Gaussian white noise with zero mean. Their covariance matrices are denoted as Q and R, respectively, and their specific values need to be specified by the user during application. The state noise Q can be determined through adaptive training, while the observation noise is generally related to sensor accuracy. The following equations are obtained:

$$\begin{cases} \dot{x}_{aekf} = f\left(x_{aekf}, u(t)\right) + \omega \\ y = hx_{aekf} + \nu \end{cases} \tag{8}$$

Discretization is performed using the Euler method:

$$\begin{cases} x_{aekf,k+1} = x_{aekf,k} + T_s f\left(x_{aekf,k}, u(t)_k\right) + \omega_k \\ y_k = [h|0] x_{aekf,k} + \nu_k \end{cases} \tag{9}$$

where $f\left(x_{aukf,k}, u(t)_k\right)$ represents the system dynamic model at time k, and T_s is the sampling time. The nonlinear function $f\left(x_{aukf,k}, u(t)_k\right)$ is incompatible with the EKF. Instead, its Jacobian, given in (10), is employed within the EKF algorithm.

$$F_k = \frac{\partial f}{\partial x_{\text{aekf}}} \Big|_{(\hat{x}_{\text{aekf},k}, u_k, k)} \tag{10}$$

Here, $\hat{x}_{\text{ackf},k}$ denotes the parameter estimated by the EKF at the k-th step. The term $f_{\hat{x}_{\text{ackf},k}, u_k, k}$ represents the dynamic model utilizing these estimated values. As a result, for the initial values of $\hat{x}_0$, p_0, R_0, and Q_0, the EKF algorithm consists of two stages: the time update and the measurement update, as given below.

$$\begin{cases} \hat{x}^-_{\text{aekf},k} = T_s f(\hat{x}_{\text{aekf},k-1}, u_{k-1}) + \hat{x}_{aekf,k-1} \\ P^-_{\text{aekf},k} = F_{k-1} p_{\text{aekf},k-1} F^T_{k-1} + Q_{k-1} \\ Q_k = \lambda Q_{k-1} + (1-\lambda)\left(K_{\text{aekf},k} d_{\text{aekf},k} d^T_{\text{aekf},k} K^T_{\text{aekf},k}\right) \end{cases} \tag{11}$$

$$\begin{cases} K_{\text{aekf},k} = p^-_{\text{aekf},k} H^T_k \left(H_k p^-_{\text{aekf},k} H^T_k + R_K\right)^{-1} \\ \hat{x}_{\text{aekf},k} = K_{k,aekf}\left(y_k - H\hat{x}^-_{\text{aekf},k}\right) + \hat{x}^-_{\text{aekf},k} \\ p_{\text{aekf},k} = (I - K_{k,\text{aekf}} H_k) p^-_{aekf,k} \\ R_k = \lambda R_{k-1} + (1-\lambda)\left(\varepsilon_{\text{aekf},k} \varepsilon^T_{aekf,k} + H_k p^-_{\text{aekf},k} H^T_k\right) \end{cases} \tag{12}$$

The calculations for $\varepsilon_{\text{aekf},k}$ and $d_{\text{aekf},k}$ are as follows:

https://doi.org/10.3390/en19030682

$$\begin{cases} d_{\text{ekf},k} = [y_k - H\hat{x}^-_{\text{ekf},k}] \\ \hat{x}^+_{\text{ekf},k} = \hat{x}^-_{\text{ekf},k} + K_{\text{ekf},k} d_{\text{ekf},k} \\ \varepsilon_{\text{ekf},k} = [y_k - H\hat{x}^+_{\text{ekf},k}] \end{cases} \tag{13}$$

In the proposed algorithm, the prediction step prior to measurement utilization yields the predicted state $\hat{x}^-_{ackf,k}$ and its covariance $P^-_{ackf,k}$, computed via the dynamic model $f\left(\hat{x}_{ackf,k-1}, u_{k-1}\right)$. Following the measurement update, these are refined to the estimated state $\hat{x}_{ackf,k}$ and covariance $P_{ackf,k}$. The update magnitude is governed by the EKF gain $K_{ackf,k}$. A forgetting factor $\lambda(0 < \lambda \leq 1)$ is introduced to weight previous estimates, where a larger λ increases this influence. This factor directly modulates R_k, inducing adaptive time delays to compensate for changes. The linearized system is subsequently derived as follows:

$$F_k = I + T_s \frac{\partial f(x_{\text{aekf},k}, u_k)}{\partial x_{\text{aekf},k}} \tag{14}$$

The Jacobian matrix F_k is computed as $\frac{\partial f\left(x_{ackf,k}, u_k\right)}{\partial x_{ackf,k}}$ for all state variables, where $f\left(x_{ackf,k}, u_k\right)$ is the dynamic model and I is the identity matrix.

Ultimately, the AEKF algorithm processes the measured current i_L and voltage v_C to estimate the Constant Power Load (CPL) power $\hat{P}$ from the state vector $\hat{x}_{aekf,k}$. This estimated power is then fed directly into the PBC controller. A key advantage of this method is its ability to accurately estimate CPL power, thereby eliminating the need for dedicated power sensors. This reduction in hardware not only lowers system cost but also inherently increases robustness against measurement noise. Furthermore, the proposed AEKF eliminates the need for trial-and-error in tuning the covariance matrices.

3.2. Stability Analyzes

The stability of the Adaptive Extended Kalman Filter (AEKF) estimation error is analyzed as follows [23]. Beginning with the state prediction and update equations omitting Euler expansion:

$$\begin{aligned} \hat{x}^-_{aekf,k} &= f(\hat{x}_{aekf,k-1}, u_{k-1}) \\ \hat{x}_{aekf,k} &= \hat{x}^-_{aekf,k} + K_k H(x_{aekf,k} - \hat{x}^-_{aekf,k}) \end{aligned} \tag{15}$$

The dynamic error is defined as $\varepsilon_k = x_{aekf,k} - \hat{x}^-_{aekf,k}$. Applying Taylor's expansion and grouping high-order terms into φ_k, the error dynamics are linearized as:

$$\varepsilon_k = (A_{k-1} - K_k H A_{k-1})\varepsilon_{k-1} + \varphi_k \tag{16}$$

To assess exponential behavior, consider a Lyapunov function $v(\varepsilon_k)$ bounded by $C_1 \parallel \varepsilon_k \parallel^2 \leq v(\varepsilon_k) \leq C_2 \parallel \varepsilon_k \parallel^2$. If its difference $\Delta v(\varepsilon_k) \leq -C_3 \parallel \varepsilon_k \parallel^2$ for positive constants C_1, C_2, C_3, and the inequality $\tilde{F}_k^{-T} P_k^{-1} \tilde{F}_k^{-} \leq (1-\gamma) P_{k-1}^{-1}$ holds, then the system is exponentially stable, satisfying $\parallel \varepsilon_k \parallel \leq \beta \parallel \varepsilon_0 \parallel \gamma^k$ for $\beta > 0, 0 < \gamma < 1$.

4. Passivity-Based Control (PBC) and Stability Analysis

4.1. Adaptive PBC

Adaptive Passivity-Based Control Voltage Loop: To facilitate the design of the passivity-based controller, the state-space model (1) is expressed in the following matrix form:

$$\boldsymbol{H}\dot{\boldsymbol{X}} + [\boldsymbol{G} + \boldsymbol{R}(x)]\boldsymbol{X} = \boldsymbol{\Gamma} \tag{17}$$

https://doi.org/10.3390/en19030682

In the equation, $\boldsymbol{H}$ is the transient quantity in the passivity-based controller, $\boldsymbol{X}$ is the state variable matrix in the passivity-based controller, $\boldsymbol{\Gamma}$ is the input voltage matrix in the passivity-based controller, $\boldsymbol{G}$ is the voltage transformation coefficient matrix in the passivity-based controller and $\boldsymbol{R}(x)$ is the system impedance.

$$\begin{gathered} \boldsymbol{X}=\begin{pmatrix} x_1 \\ x_2 \end{pmatrix}=\begin{pmatrix} i_L \\ v_o \end{pmatrix}, \boldsymbol{H}=\begin{pmatrix} L & 0 \\ 0 & C \end{pmatrix}, \boldsymbol{\Gamma}=\begin{pmatrix} E \\ 0 \end{pmatrix} \\ \boldsymbol{G}=\begin{pmatrix} 0 & (1-\mu) \\ -(1-\mu) & 0 \end{pmatrix}, \boldsymbol{R}(x)=\begin{pmatrix} 0 & 0 \\ 0 & (\frac{1}{R}+\frac{P_{CPL}}{x_2^2}) \end{pmatrix} \end{gathered} \tag{18}$$

Obviously, $\boldsymbol{G}=-\boldsymbol{G}^{\mathrm{T}}$, $\boldsymbol{R}=\boldsymbol{R}^{\mathrm{T}}$. Equation (17) satisfies the characteristics of the Euler–Lagrange model, providing a model basis for the design of the passivity-based controller. The design of the passivity-based control voltage loop is divided into two stages:

(1) Energy Shaping Stage: Using $\boldsymbol{X}=\tilde{\boldsymbol{X}}+\boldsymbol{X}_d$ for transformation, (17) can be reshaped into

$$\boldsymbol{H}\dot{\tilde{\boldsymbol{X}}}+[\boldsymbol{G}+\boldsymbol{R}(x_m)]\tilde{\boldsymbol{X}}=\boldsymbol{\Gamma}-\left(\boldsymbol{H}\dot{\boldsymbol{X}}_d+[\boldsymbol{G}+\boldsymbol{R}(x_d)]\boldsymbol{X}_d\right) \tag{19}$$

In the above equation, $\boldsymbol{X}_d$ and $\tilde{\boldsymbol{X}}$ represent the reference value and the disturbance value of $\boldsymbol{X}$, respectively, where $\boldsymbol{R}(x_m)$ is the actual system impedance matrix, and $\boldsymbol{R}(x_d)$ is the reference value of the system impedance at the equilibrium point.

(2) Damping Injection Stage: This stage modifies the system dissipation function by injecting the virtual damping matrix $\boldsymbol{R}_V\boldsymbol{X}$ into both sides of (3), which can be expressed as

$$\boldsymbol{H}\dot{\tilde{\boldsymbol{X}}}+[\boldsymbol{G}+\boldsymbol{R}_p]\tilde{\boldsymbol{X}}=\boldsymbol{\Gamma}-\left(\boldsymbol{H}\dot{\boldsymbol{X}}_d+[\boldsymbol{G}+\boldsymbol{R}(x_d)]\boldsymbol{X}_d-\boldsymbol{R}_V\tilde{\boldsymbol{X}}\right) \tag{20}$$

$$\boldsymbol{R}_V=\begin{pmatrix} \boldsymbol{R}_{1V} & 0 \\ 0 & \frac{1}{R_V} \end{pmatrix}, \boldsymbol{R}(x_d)=\begin{pmatrix} \boldsymbol{R}_{1V} & 0 \\ 0 & \frac{1}{R}+\frac{P_{CPL}}{x_{2d}^2} \end{pmatrix}, \boldsymbol{R}_p=\boldsymbol{R}(x_m)+\boldsymbol{R}_V=\begin{pmatrix} 0 & 0 \\ 0 & (\frac{1}{R_V}+\frac{1}{R}-\frac{P_{CPL}}{x_2\cdot x_{2d}}) \end{pmatrix} \tag{21}$$

To ensure that the system with the injected virtual damping matrix satisfies Lyapunov stability, thus possessing a globally stable equilibrium point for the system ($\mathbf{X}=0$). The system's energy function is selected as follows:

$$W(\tilde{\boldsymbol{X}})=0.5\tilde{\boldsymbol{X}}^T\boldsymbol{H}\tilde{\boldsymbol{X}} \tag{22}$$

Then, the time derivative of the energy function (22) can be expressed as

$$\begin{aligned} \dot{W}(\tilde{\boldsymbol{X}}) &= \tilde{\boldsymbol{X}}^T\boldsymbol{H}\dot{\tilde{\boldsymbol{X}}} \\ &= \tilde{\boldsymbol{X}}^T\left(\boldsymbol{\Gamma}-\left(\boldsymbol{H}\dot{\boldsymbol{X}}_d+[\boldsymbol{G}+\boldsymbol{R}(x_d)]\boldsymbol{X}_d-\boldsymbol{R}_V\tilde{\boldsymbol{X}}\right)-[\boldsymbol{G}+\boldsymbol{R}_p]\tilde{\boldsymbol{X}}\right) \\ &= \tilde{\boldsymbol{X}}^T\left(\boldsymbol{\Gamma}-\left(\boldsymbol{H}\dot{\boldsymbol{X}}_d+[\boldsymbol{G}+\boldsymbol{R}(x_d)]\boldsymbol{X}_d-\boldsymbol{R}_V\tilde{\boldsymbol{X}}\right)\right)-\tilde{\boldsymbol{X}}^T\boldsymbol{R}_p\tilde{\boldsymbol{X}} \end{aligned} \tag{23}$$

To ensure global stable convergence of the system, the injected virtual damping coefficient $\boldsymbol{R}_\mathbf{v}$ should make all elements of the matrix $\boldsymbol{R}_\mathbf{p}$ positive. Under this premise, the control law for the passivity-based control voltage loop can be designed as

$$\boldsymbol{\Gamma}-\left(\boldsymbol{H}\dot{\boldsymbol{X}}_d+[\boldsymbol{G}+\boldsymbol{R}(x_d)]\boldsymbol{X}_d-\boldsymbol{R}_V\tilde{\boldsymbol{X}}\right)=0 \tag{24}$$

Although the control law for the passivity-based control voltage loop obtained from Equation (24) can operate with multiple rated parameters, the randomness and time-varying nature of source and load side parameters can lead to steady-state deviations from

https://doi.org/10.3390/en19030682

the control objective. Therefore, substituting the estimated values of output power from AEKF into (24) can be expressed as

$$i_{L,ref} = \frac{\hat{P}_{CPL}}{E} - \frac{v_{ref}}{R_V E}(v_o - v_{ref}) + \frac{v^2{}_{ref}}{RE} \tag{25}$$

Based on the inherent properties of the converter and Equation (24), the control result for the switching duty cycle μ can be obtained.

$$\mu = 1 - \left[\frac{E + R_{1v}\left(i_L - i_{L,ref}\right)}{v_{ref}} \right] \tag{26}$$

Furthermore, by applying a current limiting block after the current reference, the peak current during startup and disturbances can be suppressed, thereby reducing the current stress on the equipment. Combining the passivity-based control voltage loop and the Extended Kalman Filter forms a nonlinear controller with a simple structure, high robustness, and low computational burden. Integrating the above controller design, the final proposed adaptive Extended Kalman Filter passivity-based control method is shown in Figure 2. It should be noted that while the stability condition requires $\boldsymbol{R_p}$ to be positive definite, the practical selection of $\boldsymbol{R_{1v}}$ is also constrained by the PWM switching frequency and modulator dynamics, as discussed in [33].

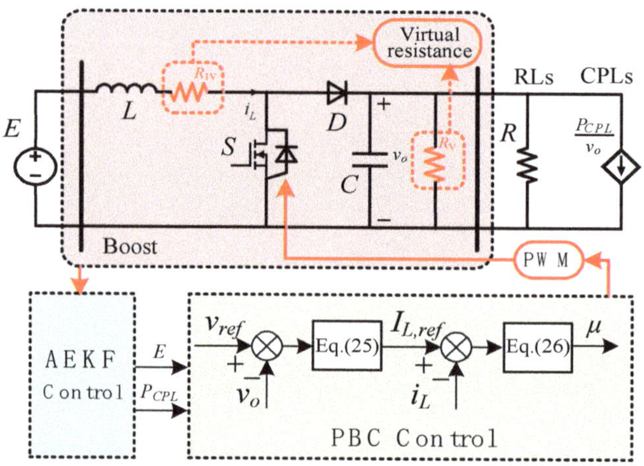

Figure 2. Boost converter with proposed adaptive passivity Control with AEKF.

4.2. Stability Analyzes

The stability analysis of the proposed passivity-based control (PBC) algorithm is conducted within the formal framework of Lyapunov stability theory. Following the completion of energy reshaping and damping injection phases, the closed-loop system dynamics are characterized by the following differential equation:

$$\boldsymbol{H}\dot{\tilde{\boldsymbol{X}}} + [\boldsymbol{G} + \boldsymbol{R}_i(x)]\tilde{\boldsymbol{X}} = 0 \tag{27}$$

The system energy function is selected as the Lyapunov function candidate:

$$V = \frac{1}{2}\tilde{\boldsymbol{X}}^T \boldsymbol{H}\tilde{\boldsymbol{X}} > 0 \ (\forall \tilde{\boldsymbol{X}} \neq 0) \tag{28}$$

https://doi.org/10.3390/en19030682

Since the inertia matrix $\boldsymbol{H}$ is positive definite and symmetric, $\boldsymbol{V}$ is consequently positive definite. To examine system stability, we compute the time derivative of the Lyapunov function along the trajectories of the system:

$$\dot{V} = \frac{d}{dt}\left(\frac{1}{2}\tilde{X}^T H\tilde{X}\right) = \tilde{X}^T H\dot{\tilde{X}} \tag{29}$$

Substituting the system equation yields:

$$\dot{\tilde{X}} = -H^{-1}[G + R_i(x)]\tilde{X} \tag{30}$$

Expanding this expression:

$$\dot{V} = -\tilde{X}^T G\tilde{X} - \tilde{X}^T R_i(x)\tilde{X} \tag{31}$$

Given that matrix $\boldsymbol{G}$ is skew-symmetric ($\boldsymbol{G} = -\boldsymbol{G}^T$), it follows that:

$$\tilde{X}^T G\tilde{X} = 0 \tag{32}$$

$$\dot{V} = -\tilde{X}^T R_i(x)\tilde{X} < 0 \tag{33}$$

According to Lyapunov stability theory, the system is globally asymptotically stable at the equilibrium point $\tilde{X} = 0$. Regardless of initial conditions, the system state error $\tilde{X}$ converges to zero over time.

The AEKF is designed to estimate the unknown, time-varying CPL power in real time by treating it as an augmented state variable (Equations (4) and (5)). This accurate estimate $\widehat{\mathrm{P}}$ is then fed forward into the PBC law (Equations (25) and (26)). The PBC controller, through its energy shaping and damping injection stages, inherently compensates for the negative impedance effect of the CPL. The injected virtual damping restores system passivity, ensuring large-signal stability despite the nonlinear load characteristic.

This analysis demonstrates that the passivity-based control algorithm ensures system stability under all operating conditions, thereby providing theoretical assurance for practical engineering applications.

5. Simulation Results

To comprehensively assess both the steady-state and transient performance of the proposed Adaptive Extended Kalman Filter-based Passivity-Based Control strategy for DC microgrid DC-DC converters, a detailed simulation model was implemented within the PLECS simulation environment. The system configuration comprises a single Boost converter supplying power to parallel-connected constant power load and resistive load, where the CPL is practically realized through a current source combined with resistive components.

The converter control system was rigorously evaluated under multiple training scenarios designed to simulate realistic operational challenges. These included step changes in load conditions, abrupt variations in reference voltage commands, and sudden disturbances in the Boost converter's input voltage. Through these comprehensive tests, the dynamic performance and robustness of the proposed control strategy were systematically evaluated under diverse operating conditions, with control parameters being progressively refined through this process. The simulation parameters employed for this validation study are detailed in Table 1.

https://doi.org/10.3390/en19030682

Table 1. System configuration parameters.

Symbol	Representation	Value
Boost converter		
C	Capacitance	940 μF
L	Inductance	2 mH
E	Input Voltage	100 V
R	Load Resistance	100 Ω
v_{ref}	Reference Output Voltage	200 V
P	Initial CPL Power	500 W
f_{sw}	Switching Frequency	20 kHz
Proposed Strategy		
T_S	Discretizing Time	5×10^{-5}
R_{Vi}	Virtual Resistor	1.5
$Q_{i,m,1}, Q_{i,m,2}, Q_{i,m,3}$	Initial Covariance of ProcessNoise	$10^{-3}, 10^{-4}, 0.6$
$P_{i,m,1}, P_{i,m,2}, P_{i,m,3}$	Initial Covariance of States	0.05, 0.05, 0.05
$\hat{x}_0$	Initial Value of States	0.1, 0.1, 0.1

5.1. Case 1: Constant Power and Constant Impedance Load Changes

The proposed control algorithm demonstrates exceptional dynamic performance and robustness when subjected to sudden changes in constant power load (CPL). As illustrated in Figure 3, when the system experiences a step change in CPL power from 500 W to 900 W at t = 0.2 s, both the output voltage and inductor current exhibit minor, brief transients. Critically, the system recovers to its steady-state values rapidly. A subsequent step reduction in CPL power from 900 W back to 500 W occurs at t = 0.6 s, followed by a step decrease in the resistive load R to 50 Ω at t = 1.0 s. In both cases, the output voltage and inductor current revert to their steady-state values promptly, with a remarkably short settling time of 8 ms and a minimal voltage overshoot constrained within 2 V. The output voltage profile maintains a smooth trajectory during these transients, devoid of significant oscillations, which verifies the algorithm's efficacy in suppressing power oscillations induced by load disturbances. The swift response of the inductor current further underscores the controller's superior capability for rapid power balance regulation.

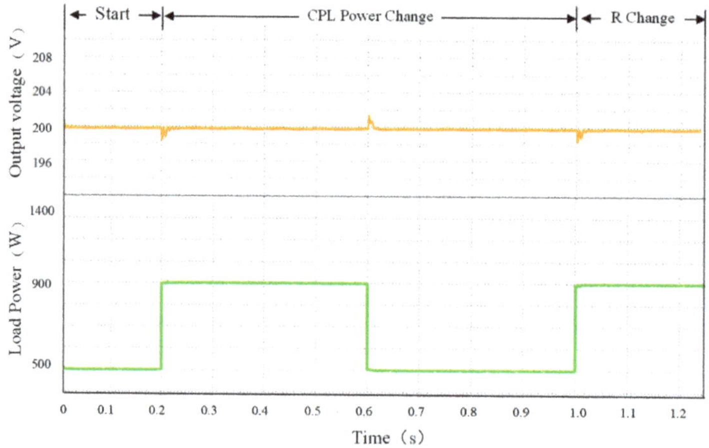

Figure 3. Constant power and constant impedance load changes.

https://doi.org/10.3390/en19030682

5.2. Case 2: Constant Power and Input Voltage Changes

The proposed control algorithm also exhibits outstanding dynamic regulation and robustness in response to input voltage disturbances. As shown in Figure 4, the system was tested under a sequence of step changes. Initially, a CPL power step from 500 W to 900 W at t = 0.2 s causes brief deviations in the output voltage and inductor current, which are quickly stabilized within a settling time of 10 ms and a voltage overshoot below 1 V. When the CPL power steps back down from 900 W to 500 W at t = 0.6 s, the system demonstrates a faster recovery, with a settling time of 8 ms and an overshoot controlled within 1 V. Furthermore, when subjected to an input voltage step change from 100 V to 80 V at t = 1.0 s, the system again manages the transient effectively, stabilizing within 9 ms. Throughout these tests, the voltage waveform transitions smoothly, and all fluctuations in load voltage and power remain within 10%. These results collectively highlight the controller's strong capability for rapid compensation of input-side disturbances.

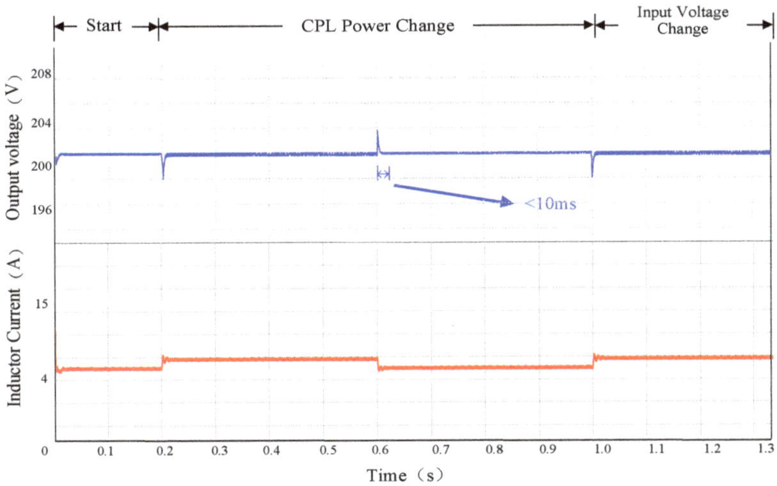

Figure 4. Constant power and input voltage changes.

5.3. Case 3: Comparative Analysis with Conventional Methods

The proposed control method demonstrates rapid bus voltage stabilization, achieving a mere 1 V overshoot and an 8 ms settling time. In contrast, the PBC + NDO control method yields a larger overshoot of 1.7 V and a longer settling time of 20 ms. As shown in Figure 5, these comparative results highlight the exceptional robustness of the proposed approach to load disturbances.

https://doi.org/10.3390/en19030682

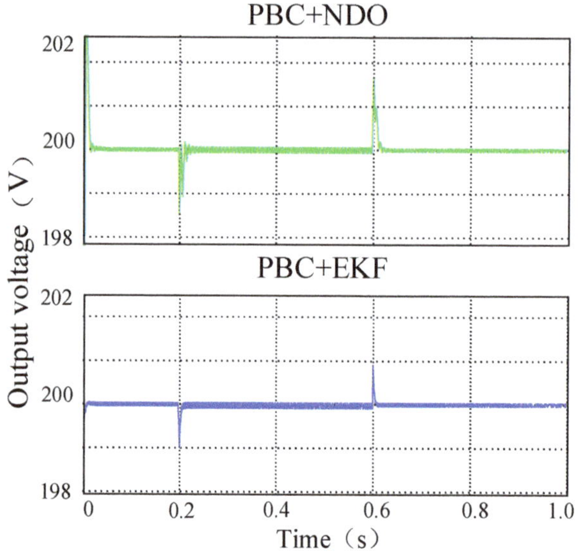

Figure 5. Comparative simulation results.

6. Experimental Results

To validate the superior performance of the proposed converter, an experimental setup was constructed for testing. The setup consists of an oscilloscope (Yokogawa DL850), a DC power supply (ITECH IT-M3433,150 V/12 A), an electronic load (Chroma 63804, 350 V/45 A/4500 W), a sampling circuit, a DC-DC boost converter, and an RT-Box1. The Chroma electronic load can be configured in either resistive mode or constant power load (CPL) mode to accommodate different load types. The control algorithm is developed in PLECS on a host computer, which generates code for the RT-Box. This code enables the RT-Box to acquire voltage and current signals via its analog-to-digital converters (ADCs) and execute the control algorithm, outputting a 20 kHz PWM signal to the DC-DC boost converter. Variations in source-side and load-side operating conditions are implemented using the programmable DC power supply and the electronic load. Furthermore, to faithfully reflect the control performance, the parameters of the experimental platform are kept consistent with those used in the simulation studies. where v_o denotes the output voltage, i_L represents the inductor current, and i_o is the output current.

6.1. Case 1: Constant Power Load Changes

The system was initialized with an input voltage of 100 V and a constant power load of 200 W. The experimental waveforms are shown in Figure 6. During the test, the constant power load was stepped up to 400 W and, after a period of time, stepped back down to 200 W. Accordingly, the inductor current increased from 2 A to 4 A and then returned to 2 A, while the output current rose from 1 A to 2 A and then recovered to 1 A. The settling time was measured as 23 ms, and the overshoot was limited to 2.1 V. The experimental results demonstrate that the proposed control method can rapidly stabilize the bus voltage under abrupt load changes.

https://doi.org/10.3390/en19030682

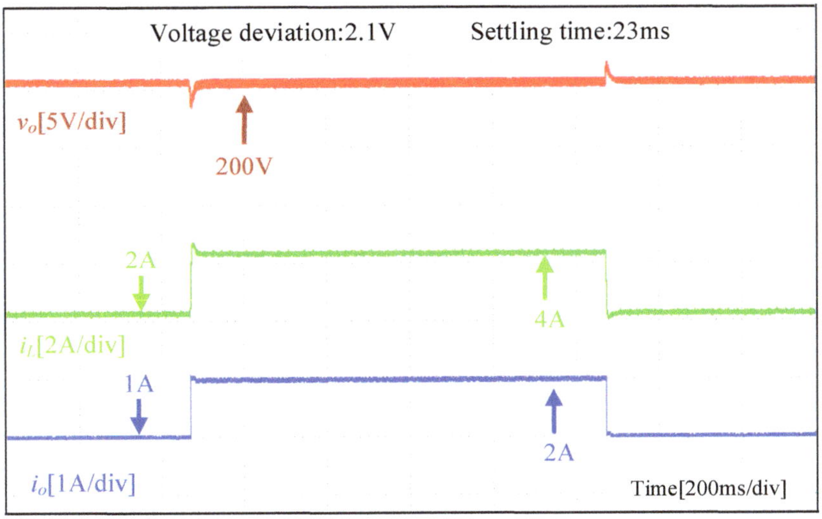

Figure 6. Constant power changes.

6.2. Case 2: Input Voltage Changes

Next, experimental studies were conducted on the proposed control method under input voltage disturbances to evaluate its robustness and stability against source-side variations. The input voltage of the system was varied in the following sequence: 100 V → 80 V → 100 V. Figure 7 illustrates the comparative experimental waveforms corresponding to the input voltage change from 100 V to 80 V and back to 100 V. It can be observed that the proposed control method is able to rapidly stabilize the bus voltage, with a voltage overshoot of 4 V and a settling time of 62 ms.

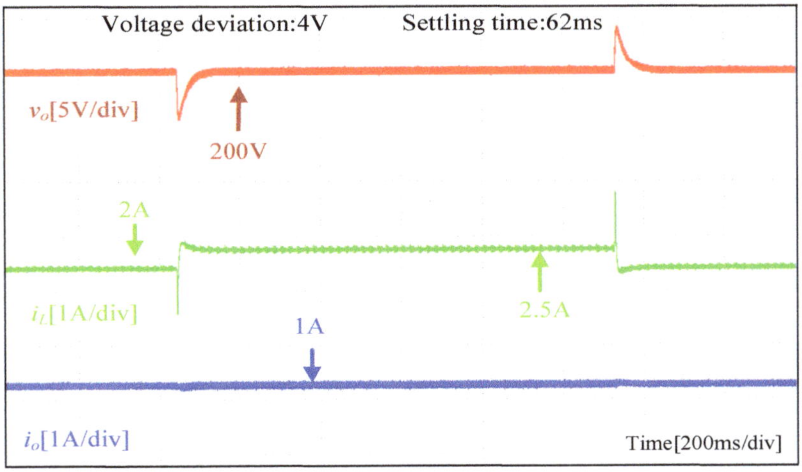

Figure 7. Input voltage changes.

6.3. Case 3: Output Voltage Changes

Finally, the dynamic tracking performance of the proposed control method is evaluated by studying its response to changes in the output voltage reference. In this experiment, the output voltage reference of the converter is stepped from 200 V to 190 V while the load is 200 Ω. As shown in Figure 8, the output voltage rapidly tracks the new reference within

https://doi.org/10.3390/en19030682

30 ms without any voltage overshoot. The peak inductor current during the transient is 1.7 A. These results demonstrate that the proposed control method provides excellent voltage regulation capability.

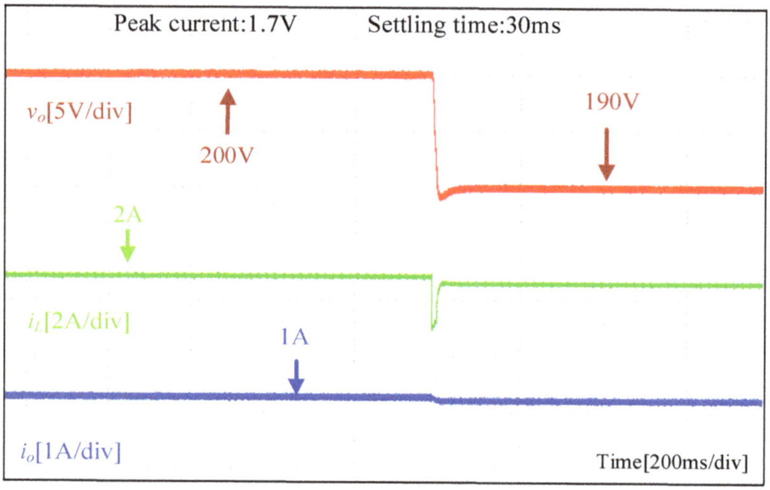

Figure 8. Output voltage changes.

7. Conclusions

This paper has proposed a novel control methodology for DC-DC converters that integrates an Extended Kalman Filter (EKF) with Passivity-Based Control (PBC), effectively enhancing system stability under Constant Power Load (CPL) conditions. Initially, the designed Extended Kalman Filter controller accurately estimates the time-varying load power by leveraging real-time converter measurements. This predicted information is subsequently fed into the proposed passivity-based controller to generate PWM signals for driving the power switches, thereby regulating the boost converter supplying the CPLs.

The presented strategy significantly improves large-signal stability and system robustness, while also providing precise current limiting and rapid voltage regulation. These capabilities collectively ensure reliable and stable operation of DC microgrids under various complex operating scenarios. Furthermore, the nonlinear control loop eliminates the conventional requirements for detailed nominal parameters and complex multi-degree-of-freedom tuning, resulting in a relatively simple and practical control structure. Simulation studies validate the superiority of the proposed approach, demonstrating exceptional control performance across various operating conditions.

Author Contributions: Writing—original draft, Conceptualization, P.W. (Peng Wang); AEKF method, Z.M.; PBC method, J.T. and Z.L. Stability analysis, Y.L. Writing—review and editing, Optimization, P.W. (Panbao Wang); Visualization, Y.Z. All authors have read and agreed to the published version of the manuscript.

Funding: This work was supported by technology cooperation project of State Grid Heilongjiang Electric Power Co., Ltd. under Grants 522422240001.

Data Availability Statement: The original contributions presented in this study are included in the article. Further inquiries can be directed to the corresponding author.

https://doi.org/10.3390/en19030682

Conflicts of Interest: Author P.W., Z.M., J.T., Z.L. and Y.L. were employed by the company Daxinganling Power Supply Company of State Grid Heilongjiang Electric Power Co., Ltd. The remaining authors declare that the research was conducted in the absence of any commercial or financial relationships that could be construed as a potential conflict of interest. The authors declare that this study received funding from technology cooperation project of State Grid Heilongjiang Electric Power Co., Ltd. under Grants (522422240001). The funder was not involved in the study design, collection, analysis, interpretation of data, the writing of this article or the decision to submit it for publication.

References

1. Bidram, A.; Davoudi, A. Hierarchical Structure of Microgrids Control System. *IEEE Trans. Smart Grid* **2012**, *3*, 1963–1976. [CrossRef]
2. Dragičević, T.; Lu, X.; Vasquez, J.C.; Guerrero, J.M. DC Microgrids—Part I: A Review of Control Strategies and Stabilization Techniques. *IEEE Trans. Power Electron.* **2016**, *31*, 4876–4891. [CrossRef]
3. Parhizi, S.; Lotfi, H.; Khodaei, A.; Bahramirad, S. State of the Art in Research on Microgrids: A Review. *IEEE Access* **2015**, *3*, 890–925. [CrossRef]
4. Kumar, D.; Zare, F.; Ghosh, A. DC Microgrid Technology: System Architectures, AC Grid Interfaces, Grounding Schemes, Power Quality, Communication Networks, Applications, and Standardizations Aspects. *IEEE Access* **2017**, *5*, 12230–12256. [CrossRef]
5. Meng, L.; Shafiee, Q.; Trecate, G.F.; Karimi, H.; Fulwani, D.; Lu, X.; Guerrero, J.M. Review on Control of DC Microgrids and Multiple Microgrid Clusters. *IEEE J. Emerg. Sel. Top. Power Electron.* **2017**, *5*, 928–948. [CrossRef]
6. Xu, Q.; Vafamand, N.; Chen, L.; Dragičević, T.; Xie, L.; Blaabjerg, F. Review on Advanced Control Technologies for Bidirectional DC/DC Converters in DC Microgrids. *IEEE J. Emerg. Sel. Top. Power Electron.* **2021**, *9*, 1205–1221. [CrossRef]
7. Lu, X.; Sun, K.; Guerrero, J.M.; Vasquez, J.C.; Huang, L.; Wang, J. Stability Enhancement Based on Virtual Impedance for DC Microgrids with Constant Power Loads. *IEEE Trans. Smart Grid* **2015**, *6*, 2770–2783. [CrossRef]
8. Hassan, M.A.; Su, C.-L.; Pou, J.; Almakhles, D.; Zhan, T.-S.; Lo, K.-Y. Robust Passivity-Based Control for Interleaved Bidirectional DC–DC Power Converter with Constant Power Loads in DC Shipboard Microgrid. *IEEE Trans. Transp. Electrif* **2024**, *10*, 3590–3602. [CrossRef]
9. Sadabadi, M.S.; Shafiee, Q. Scalable Robust Voltage Control of DC Microgrids with Uncertain Constant Power Loads. *IEEE Trans. Power Syst.* **2020**, *35*, 508–515. [CrossRef]
10. Hassan, M.A.; Li, E.-P.; Li, X.; Li, T.; Duan, C.; Chi, S. Adaptive Passivity-Based Control of DC–DC Buck Power Converter with Constant Power Load in DC Microgrid Systems. *IEEE J. Emerg. Sel. Top. Power Electron.* **2019**, *7*, 2029–2040. [CrossRef]
11. Sarrafan, N.; Zarei, J.; Razavi-Far, R.; Saif, M.; Khooban, M.-H. A Novel On-Board DC/DC Converter Controller Feeding Uncertain Constant Power Loads. *IEEE J. Emerg. Sel. Top. Power Electron.* **2021**, *9*, 1233–1240. [CrossRef]
12. Li, X.; Jiang, W.; Wang, J.; Wang, P.; Wu, X. An Autonomous Control Scheme of Global Smooth Transitions for Bidirectional DC-DC Converter in DC Microgrid. *IEEE Trans. Energy Convers.* **2021**, *36*, 950–960. [CrossRef]
13. Zeng, J.; Zhang, Z.; Qiao, W. An Interconnection and Damping Assignment Passivity-Based Controller for a DC–DC Boost Converter with a Constant Power Load. *IEEE Trans. Ind. Appl.* **2014**, *50*, 2314–2322. [CrossRef]
14. Ming, J.; Wang, Y.; Wang, F.; Su, B. Large-Signal Stability Analysis and Shunt Active Damper Compensation for DC Microgrid with Multiple Constant Power Loads. In Proceedings of the 22nd International Conference on Electrical Machines and Systems (ICEMS), Harbin, China, 11–14 August 2019; pp. 1–6. [CrossRef]
15. Ndiwulu, G.W.; De Jaeger, E.; Lusala, A.K.; Matalatala, M. Large-Signal Stability Analysis for Islanded Microgrids Based on Inner Control Loops Approach. In Proceedings of the 2021 IEEE PES/IAS PowerAfrica, Nairobi, Kenya, 23–27 August 2021; pp. 1–5. [CrossRef]
16. Xia, Y.; Lv, Z.; Wei, W.; He, H. Large-Signal Stability Analysis and Control for Small-Scale AC Microgrids with Single Storage. *IEEE J. Emerg. Sel. Top. Power Electron.* **2022**, *10*, 4809–4820. [CrossRef]
17. Peng, D.; Huang, M.; Li, J.; Sun, J.; Zha, X.; Wang, C. Large-Signal Stability Criterion for Parallel-Connected DC–DC Converters with Current Source Equivalence. *IEEE Trans. Circuits Syst. II Express Briefs* **2019**, *66*, 2037–2041. [CrossRef]
18. Xie, W.; Han, M.; Cao, W.; Guerrero, J.M.; Vasquez, J.C. System-Level Large-Signal Stability Analysis of Droop-Controlled DC Microgrids. *IEEE Trans. Power Electron.* **2021**, *36*, 4224–4236. [CrossRef]
19. Leng, M.; Zhou, G.; Xu, G.; Sahoo, S.; Liu, X.; Zhou, Q.; Yin, Y.; Blaabjerg, F. Small-Signal Stability Assessment and Interaction Analysis for Bipolar DC Microgrids. *IEEE Trans. Power Electron.* **2023**, *38*, 5524–5537. [CrossRef]
20. Cespedes, M.; Xing, L.; Sun, J. Constant-Power Load System Stabilization by Passive Damping. *IEEE Trans. Power Electron.* **2011**, *26*, 1832–1836. [CrossRef]
21. Pang, S.; Nahid-Mobarakeh, B.; Pierfederici, S.; Huangfu, Y.; Luo, G.; Gao, F. Large-Signal Stable Nonlinear Control of DC/DC Power Converter with Online Estimation of Uncertainties. *IEEE J. Emerg. Sel. Top. Power Electron.* **2021**, *9*, 7355–7368. [CrossRef]

22. Lei, J.; Zhou, B.; Qin, X.; Wei, J.; Bian, J. Active Damping Control Strategy of Matrix Converter via Modifying Input Reference Currents. *IEEE Trans. Power Electron.* **2015**, *30*, 5260–5271. [CrossRef]
23. Guan, Y.; Xie, Y.; Wang, Y.; Liang, Y.; Wang, X. An Active Damping Strategy for Input Impedance of Bidirectional Dual Active Bridge DC–DC Converter: Modeling, Shaping, Design, and Experiment. *IEEE Trans. Ind. Electron.* **2021**, *68*, 1263–1274. [CrossRef]
24. Hassan, M.A.; Su, C.-L.; Chen, F.-Z.; Lo, K.-Y. Adaptive Passivity-Based Control of a DC–DC Boost Power Converter Supplying Constant Power and Constant Voltage Loads. *IEEE Trans. Ind. Electron.* **2022**, *69*, 6204–6214. [CrossRef]
25. Viswambharan, A.; Errouissi, R.; Debouza, M.; Shareef, H. Experimental Verification of Disturbance Observer-Based Backstepping Control for DC–DC Boost Converter. *IEEE Trans. Circuits Syst. I Regul. Pap.* **2023**, *70*, 5520–5533. [CrossRef]
26. Zhang, Z.; Wang, P.; Zhou, J.; Wang, W.; Xu, D. A Novel Passivity Model Predictive Control for Predefined-Time Feedforward Compensation in DC Microgrids Feeding Constant Power Loads. *IEEE Trans. Power Electron.* **2025**, *40*, 5147–5162. [CrossRef]
27. Pandey, S.K.; Patil, S.L.; Chaskar, U.M.; Phadke, S.B. State and Disturbance Observer-Based Integral Sliding Mode Controlled Boost DC–DC Converters. *IEEE Trans. Circuits Syst. II Express Briefs* **2019**, *66*, 1567–1571. [CrossRef]
28. Martínez-Treviño, B.A.; Aroudi, A.E.; Valderrama-Blavi, H.; Cid-Pastor, A.; Vidal-Idiarte, E.; Martinez-Salamero, L. PWM Nonlinear Control with Load Power Estimation for Output Voltage Regulation of a Boost Converter with Constant Power Load. *IEEE Trans. Power Electron.* **2021**, *36*, 2143–2153. [CrossRef]
29. Song, J.; Fu, C.; Zhang, G.; Duan, B.; Zhang, C. Backstepping Control of High-Frequency Link Matrix Rectifier for Battery Chargers. *IEEE Trans. Power Electron.* **2021**, *36*, 10801–10814. [CrossRef]
30. Xu, Q.; Zhang, C.; Wen, C.; Wang, P. A Novel Composite Nonlinear Controller for Stabilization of Constant Power Load in DC Microgrid. *IEEE Trans. Smart Grid* **2019**, *10*, 752–761. [CrossRef]
31. Middlebrook, R.D.; Cuk, S. A general unified approach to modelling switching-converter power stages. In Proceedings of the IEEE Power Electronics Specialist Conference, Cleveland, OH, USA, 8–10 June 1976; pp. 18–34.
32. Abdolahi, M.; Adabi, J.; Mousavi, S.Y.M. An Adaptive Extended Kalman Filter with Passivity-Based Control for DC-DC Converter in DC Microgrids Supplying Constant Power Loads. *IEEE Trans. Ind. Electron.* **2024**, *71*, 4873–4882. [CrossRef]
33. Komurcugil, H. Improved passivity-based control method and its robustness analysis for single-phase uninterruptible power supply inverters. *IET Power Electron.* **2015**, *8*, 1558–1570. [CrossRef]

https://doi.org/10.3390/en19030682

energies

Article

A Si and SiC Hybrid Arms ANPC Converter Achieving Comprehensive Optimization of Power Quality, Efficiency, and Cost

Tianlun Xia [1], Xinchun Feng [1], Ziyang An [2], Meifang Li [1], Chushan Li [2,*], Haoze Luo [1] and Huan Yang [1]

[1] College of Electrical Engineering, Zhejiang University, Hangzhou 310027, China; xiatianlun@zju.edu.cn (T.X.); xinchun.feng@zju.edu.cn (X.F.); limf2013@aliyun.com (M.L.); haozeluo@zju.edu.cn (H.L.); yanghuan@zju.edu.cn (H.Y.)
[2] Zhejiang University-University of Illinois at Urbana-Champaign Institute, Zhejiang University, Haining 314400, China; ziyang1.23@intl.zju.edu.cn
* Correspondence: chushan@intl.zju.edu.cn

Abstract

Wide bandgap (WBG) power semiconductors such as silicon carbide (SiC) can significantly improve the performance of multilevel converters. However, there are three challenges for large-scale application: high cost, limited power ratings, and reliability issues. In this paper, we propose a Si and SiC hybrid arms active neutral-point-clamped (ANPC) converter, using smaller current rating SiC devices compared to other Si devices in this topology. By employing the hybrid-frequency modulation scheme, the Si devices switch at fundamental frequency (FF) or low frequency (LF), while the SiC devices switch at high frequency (HF). The equivalent circuit of the proposed converter is derived to analyze the principle of LF current ripple compensation. The closed-loop cooperative current control strategy is proposed to realize unequal current sharing in two arms and complete LF current ripple compensation. The Si arm processes major power, while the SiC arm compensates the LF current ripple generated by the Si arm and processes minor power. The proposed topology and control strategy are validated by simulation and experimental results. Compared with the existing typical topologies, the comprehensive optimization of power quality, efficiency, and cost is realized.

Keywords: Si and SiC hybrid; high efficiency; high power quality; low cost; cooperative current control

1. Introduction

Multilevel converters have evolved over many years and are considered a preferred solution for electric energy conversion in high-power applications [1]. Recently, with the rapid development of emerging power conversion applications such as renewable energy generation [2], electric vehicles [3], more electric aircraft (MEA) [4], and medium-voltage motor drives [5], the efficiency, power quality, cost, and power density of multilevel converters have reached a higher level. Due to the inherent material limitations of silicon, the operating temperature and switching frequency of Si-based devices are approaching their theoretical boundaries. As a result, enhancing the performance of high-power multilevel converters based on silicon devices has grown increasingly impossible. WBG power semiconductors represented by SiC MOSFET feature high voltage withstanding capability, low on-state resistance, and high switching frequency [6], which can further improve the performance of multilevel converters. However, the large-scale application of SiC devices still faces three major bottlenecks: high cost, limited power ratings, and reliability issues. Firstly, a SiC MOSFET is four to eight times more expensive than a Si IGBT, particularly in the high current ratings [7]. Secondly, the maximum current rating of an individual SiC MOSFET die lags significantly behind that of a Si IGBT. Thirdly, compared with Si IGBTs, the ruggedness and reliability of SiC MOSFETs are still subjects of widespread concern, particularly due to unresolved issues such as gate oxide degradation [8], threshold voltage drift [9], and body diode reliability [10]. Therefore, the large-scale industrial application of SiC devices in high-power multilevel converters remains significantly constrained.

To overcome these challenges, researchers have introduced hybrid concepts that combine the strengths of both Si and SiC devices. One widely adopted example is the hybrid three-level (3L) ANPC converter, which offers a practical balance between efficiency and device cost in advanced grid-connected inverter and rectifier systems [11,12]. In high-power rectifier applications, hybrid unidirectional rectifiers using SiC MOSFETs and Si diodes in 3L and five-level (5L) topologies have been proposed to reduce system cost in [7,13]. Similarly, three five-level converters with hybrid Si and SiC legs were presented in [14–16]. The above designs employ hybrid-frequency modulation strategies where Si devices switch at fundamental or low frequency, and SiC devices operate at carrier frequency. Thus, these topologies have lower losses, improved harmonic performance, and reduced cost. However, the SiC devices in these topologies handle the full load power, which results in relatively high cost. Moreover, the topologies in [14–16] are just single-phase circuits, which cannot be applied to three-phase high-power scenarios. Therefore, how to configure a three-phase hybrid multilevel converter with lower cost and better performance is still a challenge.

Recently, a novel concept known as WBG fractional power processing (WFPP) has been introduced [17–19]. In these structures, the LF Si two-level converter carries most of the power, while the HF SiC two-level converter is in parallel to cancel the ripple current and process partial power. The equivalent switching frequency of these topologies is the same as that of the SiC devices, maintaining excellent HF performance. Meanwhile, only low current rating WBG devices are required. Based on these topology-specific studies, a WBG and Si hybrid half-bridge (HHB) structure has been proposed in [20], aimed at improving power quality without significantly increasing cost. The HHB configuration has already shown promising results in power factor correction (PFC) application [21]. Further, the efficiency optimization based on the switching frequency and power sharing control has been achieved in [22,23]. Therefore, the WFPP concept provides feasibility for cost reduction in existing hybrid multilevel topologies. However, the implementation of the WFPP concept in three-phase DC-AC hybrid multilevel converters remains unsolved.

In this paper, a three-phase Si and SiC hybrid arms ANPC converter is proposed. The concepts of hybrid-frequency modulation and WFPP are combined to achieve comprehensive optimization of cost, power quality, and efficiency. Compared to the all-SiC and the existing hybrid converters, the number and current rating of SiC devices are further reduced. Meanwhile, from the AC output perspective, the equivalent switching frequency and power quality of the converter are similar to those of an all-SiC solution. The corresponding closed-loop cooperative current control strategy is also proposed to achieve the current tracking and power sharing. The SiC devices process minor power, which are switched under lower current conditions. This significantly reduces the high-field stress events on the SiC gate oxide.

The rest of this paper is organized as follows. Section 2 presents the proposed hybrid arms ANPC topology and operation principle of current ripple compensation. Section 3 discusses the closed-loop cooperative current control strategy. In Section 4, simulation results are presented. In Section 5, experimental results are presented to validate the proposed converter. In Section 6, performance comparison with existing topologies is presented. Finally, Section 7 concludes this paper.

2. Proposed Hybrid Arms ANPC Converter

2.1. Topology and Modulation Scheme

The per-phase configuration of the proposed converter is illustrated in Figure 1a. Each phase comprises a Si-based 3L-ANPC arm paired with a SiC-based half-bridge arm. The Si 3L-ANPC arm integrates six full-power-rating Si IGBTs, where S_1, S_2, S_3, and S_4 switch at the fundamental frequency (FF), while S_5 and S_6 switch at LF. In contrast, the SiC half-bridge arm employs two lower-power-rating SiC MOSFETs (S_7 and S_8), switching at a comparatively HF. The Si arm in each phase connects a filter inductance L_1, and $i_{x1,\, x\, =\, A,\, B,\, C}$ is the Si arm current, which includes a notable LF ripple. The SiC arm in each phase connects a filter inductance L_2, and $i_{x2,\, x\, =\, A,\, B,\, C}$ is the SiC arm current, which includes the compensated LF ripple. The $i_{x,\, x\, =\, A,\, B,\, C}$ is the total output current of the Si and SiC arms.

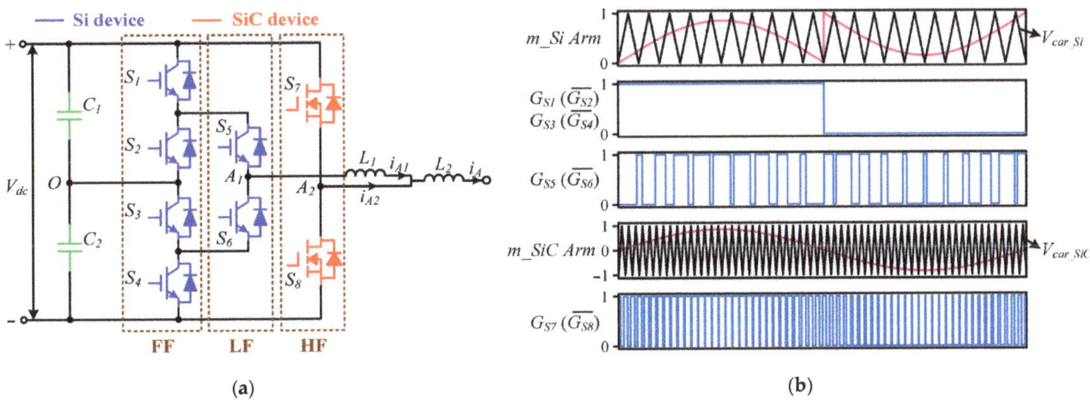

Figure 1. Topology and modulation scheme of the proposed converter. (**a**) The per-phase circuit; (**b**) Waveforms of the modulation scheme.

The corresponding hybrid-frequency sinusoidal pulse width modulation (SPWM) scheme is shown in Figure 1b. There are LF (about 2–5 kHz) and HF (above 50 kHz) two carrier frequencies. In the Si-based 3L-ANPC arm, the FF Si IGBT pairs S_1 and S_3, as well as S_2 and S_4, consistently share identical gating signals. Also, the two FF Si IGBT pairs operate complementarily. The LF Si IGBTs S_5 and S_6 are switched by complementary PWM signals. In the SiC-based half-bridge arm, the HF SiC MOSFETs S_7 and S_8 are switched complementarily. Using the modulation scheme, the switching frequency of Si IGBTs is decreased to reduce the switching loss. Meanwhile, HF switching events are moved to SiC MOSFETs to improve efficiency. The low-conduction-loss advantage of Si IGBT under high current conditions can also be fully utilized.

2.2. Operation Principle of Current Ripple Compensation

Before designing the current control strategy, it is necessary to explore the operation principle of current ripple compensation in this converter. Firstly, taking the grid-connected application as an example, the three-phase equivalent circuit of the proposed converter is derived as shown in Figure 2. In SPWM, the line-to-neutral voltage consists of the fundamental frequency component and the harmonic component by Fourier decomposition.

Moreover, the harmonic component is primarily concentrated around the switching frequency, ignoring the effects of dead-time, device nonlinear characteristics, and other factors.

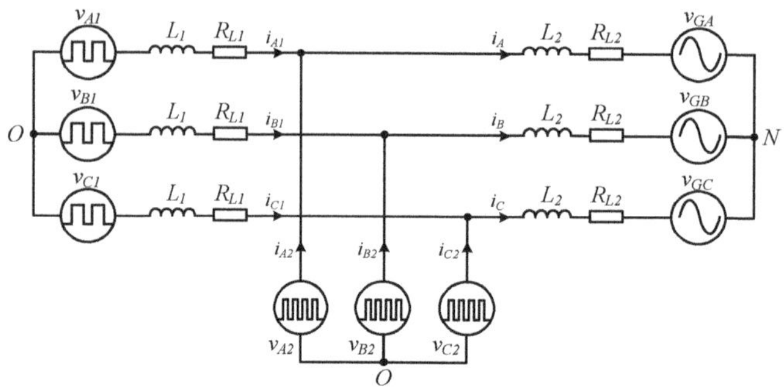

Figure 2. Three-phase equivalent circuit of the proposed converter for grid-connected application.

To improve readability, the equivalent circuits at different frequency ranges are derived as shown in Figure 3, taking phase A as an example. In Figure 3a, for the full frequency range, the line-to-neutral voltages can be expressed.

$$v_{A1N} = v_{A1N,FF} + v_{A1N,LF} \tag{1}$$

$$v_{A2N} = v_{A2N,FF} + v_{A2N,HF} \tag{2}$$

where V_{A1N} represents the line-to-neutral voltage of the Si arm, $V_{A1N,FF}$ represents its fundamental frequency component, and $V_{A1N,LF}$ represents its harmonic component. Similarly, V_{A2N} represents the line-to-neutral voltage of the SiC arm, $V_{A2N,FF}$ represents its fundamental frequency component, and $V_{A2N,HF}$ represents its harmonic component. In practical applications, it is assumed that $\omega L_1 >> R_{L1}$ and $\omega L_2 >> R_{L2}$. Thus, the resistance value of the filter inductors is allowed to be neglected. In Figure 3b, at the fundamental frequency, the Si arm current and grid current can be expressed.

$$i_{A1,FF} = \frac{1}{L_1}\int (v_{A1N,FF} - v_{A2N,FF})dt \tag{3}$$

$$i_{A,FF} = \frac{1}{L_2}\int (v_{A2N,FF} - v_{GA,FF})dt \tag{4}$$

where $v_{GA,FF}$ represents the grid voltage. In Figure 3c, since the switching frequency of the SiC arm is much higher than that of the Si arm, the SiC arm has no harmonic voltage at the LF range. The LF current ripple generated by the Si arm harmonic voltage can be expressed.

$$i_{A1,LF} = \frac{1}{L_1}\int v_{A1N,LF}dt \tag{5}$$

The SiC arm is equivalent to a short circuit for the LF current ripple, preventing it from flowing into the grid. In Figure 3d, the HF harmonic voltage of the Si arm can be disregarded, so the HF harmonic current is only generated by the SiC arm. Moreover, this HF current is insignificant compared to the FF and LF components. Thus, the Si arm current is approximately equal to the main FF current plus LF harmonic current.

$$i_{A1} \doteq i_{A1,FF} + i_{A1,LF} \tag{6}$$

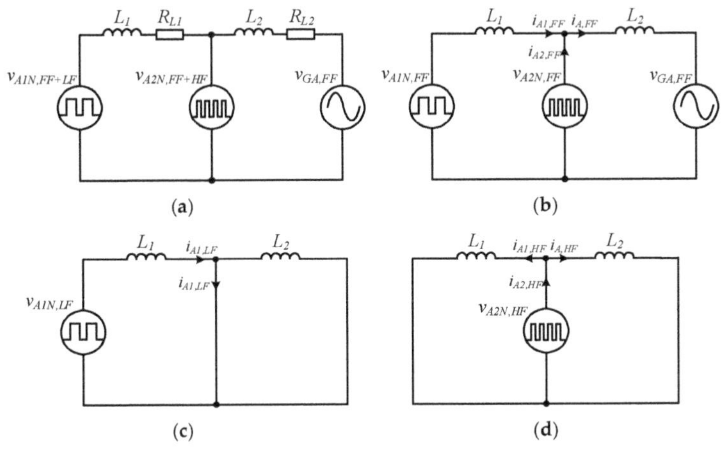

Figure 3. Single-phase equivalent circuit at different frequency ranges. (**a**) Full frequency range; (**b**) Fundamental frequency; (**c**) Si arm switching frequency; (**d**) SiC arm switching frequency.

The SiC arm current is approximately equal to the compensated LF harmonic current plus fractional FF current.

$$i_{A2} \doteq -i_{A1,LF} + i_{A2,FF} \tag{7}$$

The grid current contains the FF component and the HF harmonic component.

$$i_A = i_{A1,FF} + i_{A2,FF} + i_{A,HF} \tag{8}$$

Therefore, the proposed converter has the potential for high power quality.

Based on the above analysis, the schematic diagram of current ripple compensation waveforms is given in Figure 4. It is assumed that the average current value remains constant within one Si arm switching period T_{sw_Si}. In order to reduce cost and volume, the inductance value L_1 is minimized as much as possible. Meanwhile, the switching frequency of the Si arm is decreased to reduce losses, resulting in a large LF current ripple, as shown by the blue solid line in Figure 4. By appropriate current control strategy, the SiC arm operates at HF to compensate the Si arm current ripple and process the minor power, as shown by the yellow solid line in Figure 4. The current ripple of the SiC arm consists of two components. The first is the LF component, which is opposite to that of the Si arm. The second is the HF component, which is generated by the own HF switching events. As shown in the gray solid line in Figure 4, the total current ripple only has the HF component, which is determined only by the SiC arm and has nothing to do with the Si arm, achieving complete LF component compensation. Because of the high switching frequency of the SiC arm, the inductance value L_2 is smaller than L_1. The volume of inductors can be reduced while enhancing power quality.

In Figure 4, the dotted lines represent the average values corresponding to the solid lines. Since the Si and SiC arms share the output power, the average current relationships can be derived as follows.

$$M\_i_{A1} = K * M\_i_A \tag{9}$$

$$M\_i_{A2} = (1 - K) * M\_i_A \tag{10}$$

where M_i_{A1}, M_i_{A2}, and M_i_A represent the Si arm average current, the SiC arm average current, and the total average current, respectively. K ($0 < K < 1$) represents the major power-sharing ratio. The current ripple ratio of the Si arm r_{A1} is defined by the following formula.

$$r_{A1} = \Delta i_{A1} / M\_i_{A1} \quad (11)$$

where Δi_{A1} represents the maximum peak-to-peak current ripple of the Si arm in one FF period. Since the SiC arm completely compensates the LF ripple, the following equation holds true.

$$\Delta i_{A2} = \Delta i_{A1} \quad (12)$$

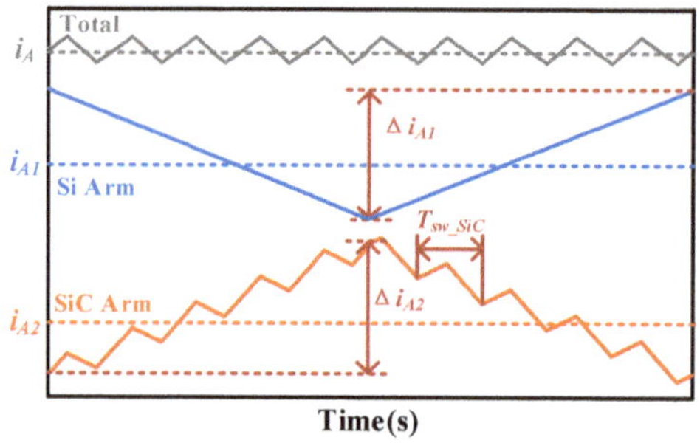

Figure 4. Schematic diagram of current ripple compensation waveforms.

During this LF switching period, the peak current of the Si arm $i_{A1,peak}$ can be expressed.

$$i_{A1,peak} = M\_i_{A1} + \frac{1}{2}\Delta i_{A1} \quad (13)$$

To avoid the circulating current in the SiC arm, $i_{A1,peak}$ should be less than the average value of the total current. Thus, the range of major power-sharing ratio K is limited by the following inequalities.

$$0 < K < \frac{1}{1 + r_{A1}/2} \quad (14)$$

3. Proposed Current Control Strategy

In this section, the closed-loop cooperative current control strategy is proposed to realize unequal current sharing and complete LF current ripple compensation. The block diagram of the control strategy is shown in Figure 5, which consists of two parts: the Si arm control loop (purple box) and the SiC arm control loop (red box). The following provides a detailed explanation of these two parts.

3.1. Si Arm Current Control

According to the control objective, the Si arm processes the main power. The major power sharing ratio K is used to generate the three-phase reference value of the Si arm current.

$$i_{ABC1\_ref} = K * i_{ABC\_ref} \quad (15)$$

where i_{ABC_ref} is the three-phase reference value of the total current, which is usually determined by the target output power. The current distribution in the Si arm can be adjusted by modifying the parameter K, enabling flexible power-sharing capabilities. Considering the reference value is the FF current, the control loop is designed in the dq-frame, rotating with the grid angular frequency ω_0. To enhance the dynamic response and disturbance resistance of the system, the grid voltage v_g is used for feed-forward control. The traditional proportional-integral (PI) controller is used to achieve no-static-error tracking of FF current. The proportional gain K_p is designed to achieve the desired closed-loop bandwidth. Applying the magnitude condition (0 dB gain at the crossover frequency f_c) leads to the dominant relationship $K_p \approx 2\pi f_c L$, ensuring fast dynamic response. The integral gain K_i is designed to shape the phase margin. By setting the zero frequency f_z of the PI controller, we can compensate for the phase lag introduced by the plant's pole, following $K_i = K_p \cdot 2\pi f_z$, guaranteeing stability and robustness. The Si arm controller proportional coefficient K_{p1} and the integral coefficient K_{i1} are designed as follows.

$$K_{p1} = 2\pi f_{c1} L_1 \tag{16}$$

$$K_{i1} = 2\pi f_{z1} K_{p1} \tag{17}$$

where the control bandwidth f_{c1} and integral corner frequency f_{z1} can be designed as follows.

$$f_{c1} \leq \frac{f_{sw\_Si}}{10} \sim \frac{f_{sw\_Si}}{5} \tag{18}$$

$$f_{z1} = \frac{f_{c1}}{10} \sim \frac{f_{c1}}{5} \tag{19}$$

where f_{sw_Si} is the switching frequency of the Si arm. In order to increase the control stability, the control period is equal to the switching period of the SiC arm. Therefore, the updating frequency of the Si arm reference voltage is equal to f_{sw_SiC}, while the carrier frequency is equal to f_{sw_Si}.

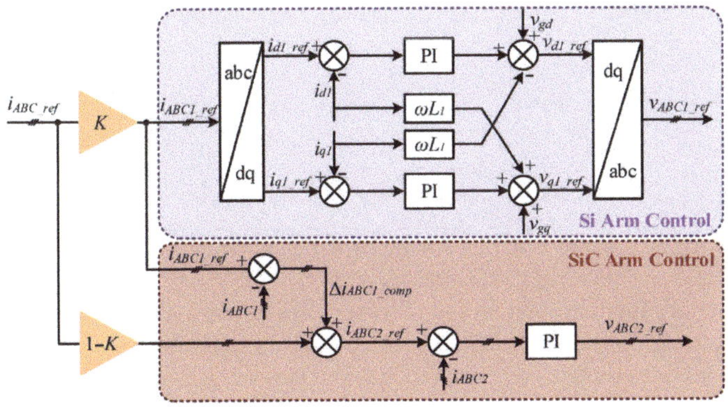

Figure 5. Block diagram of the proposed cooperative current control strategy.

3.2. SiC Arm Current Control

The control objective of the SiC arm is to compensate for the LF ripple current and process the minor power. Thus, the SiC arm three-phase current reference can be divided into two parts. The LF ripple current generated by the Si arm can be expressed.

$$\Delta i_{ABC\_comp} = i_{ABC1\_ref} - i_{ABC1} \tag{20}$$

Thus, the three-phase current reference of the SiC arm can be expressed.

$$i_{ABC2\_ref} = (1 - K) * i_{ABC\_ref} + \Delta i_{ABC\_comp} \tag{21}$$

Considering the reference value consists of the FF and LF current, the control loop is designed in the *abc*-frame. Similarly, the PI controller is selected to track the FF and

LF ripple current. The SiC arm controller proportional coefficient K_{p2} and the integral coefficient K_{i2} are designed as follows.

$$K_{p2} = 2\pi f_{c2} L_2 \tag{22}$$

$$K_{i2} = 2\pi f_{z2} K_{p2} \tag{23}$$

where the control bandwidth f_{c2} and integral corner frequency f_{z2} can be designed as follows.

$$f_{c2} \leq \frac{f_{sw\_SiC}}{10} \sim \frac{f_{sw\_SiC}}{5} \tag{24}$$

$$f_{z2} = \frac{f_{c2}}{10} \sim \frac{f_{c2}}{5} \tag{25}$$

where f_{sw_SiC} is the switching frequency of the SiC arm.

4. Simulation Verification

In this section, a three-phase Si and SiC hybrid arms ANPC converter based on the cooperative current control strategy is simulated in PLECS (4.8.9) software, considering its extremely fast simulation speed and unique thermal analysis function. The parameters of the simulation system are provided in Table 1. The results of performance comparison between the proposed topology and existing typical topologies are also presented.

Table 1. Simulation parameters.

Parameter	Value
DC Bus Voltage V_{dc}	800 V
Fundamental Frequency f_b	50 Hz
Si Arm Switching Frequency f_{sw_Si}	2.5 kHz
SiC Arm Switching Frequency f_{sw_SiC}	50 kHz
Si Arm Filter Inductance L_1	1 mH
SiC Arm Filter Inductance L_2	500 μH
Load Resistor R_{load}	3 Ω

Taking phase-A as an example, the simulation waveforms of the Si arm current, the SiC arm current, and the total current are displayed in Figure 6. The results are consistent with the theoretical analysis. As can be seen in Figure 6a, the Si arm generates a large current ripple because of the low switching frequency, and the total harmonic distortion (THD) is 13.46%. Under the condition $K = 1$, the SiC arm only compensates the LF current ripple. By employing the proposed current control strategy, the THD of the total current is reduced to 1.04%. Figure 6b shows the detailed waveforms of the current ripple compensation. It can be concluded that the SiC arm can completely compensate for the LF ripple current generated by the Si arm, and the total current only includes the FF and HF ripple components. It can be seen from Figure 6c, when K changes from 1 to 0.8, the good power quality can be maintained while the circulating current between the Si and SiC arms can be reduced. The three-phase voltage and current waveforms are shown in Figure 7. At $t = 0.1$ s, the reference value of the three-phase current i_{ABC_ref} changes from 70 to 100. At $t = 0.15$ s, the cosine value of the angle reference changes from 1 to 0.8. It can be concluded that the proposed converter can achieve high control dynamics for the step changes in the current reference and angle reference. A grid-connected operation case where the converter undergoes a step change from outputting pure active power to combined active and reactive power is presented in Figure 8. It can be demonstrated that the proposed control strategy exhibits excellent transient performance with rapid dynamic response, while maintaining superior power quality both before and after the transition period.

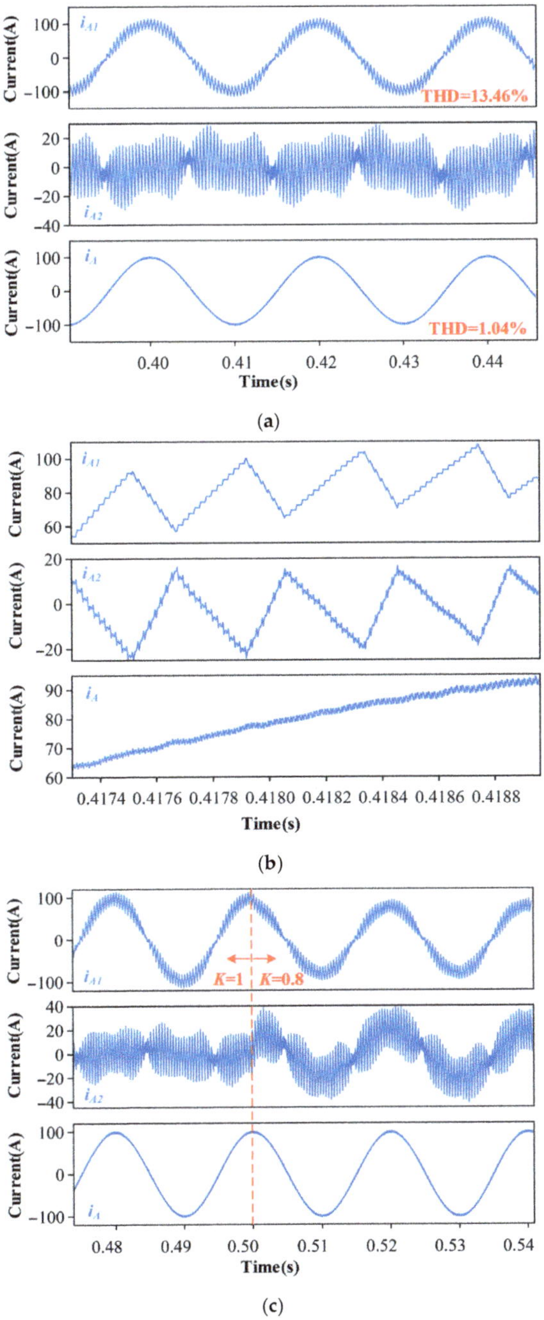

Figure 6. Simulation waveforms of the Si arm current, the SiC arm current, and total current in phase A. (**a**) Steady-state waveforms under K = 1; (**b**) Detailed waveforms of current ripple compensation under K = 1; (**c**) Waveforms of K changing from 1 to 0.8.

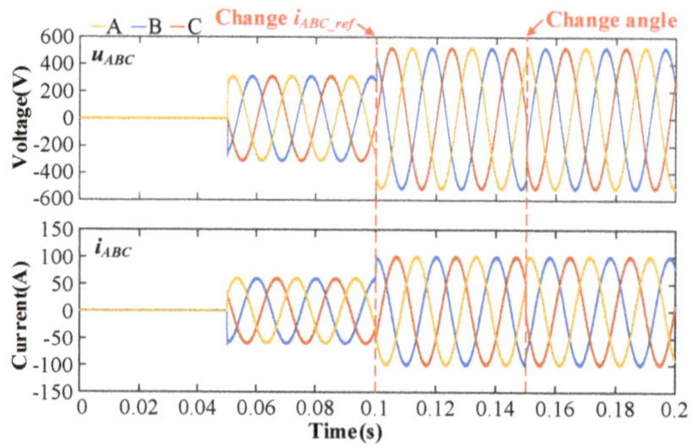

Figure 7. Simulation waveforms of line-to-line voltage and line-to-neutral current for step changes in the current reference and angle reference.

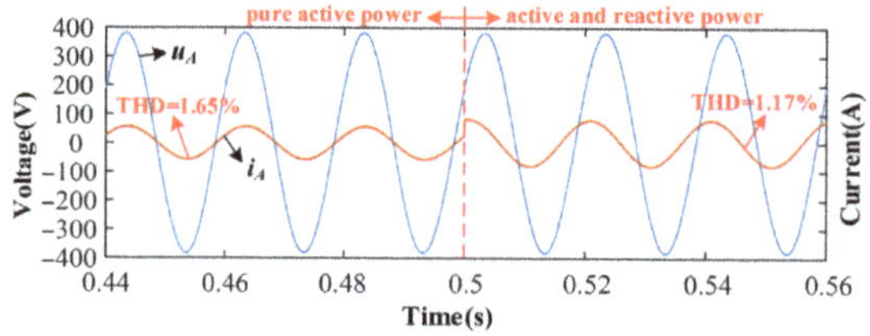

Figure 8. Simulation waveforms of grid-side voltage and current in phase A.

5. Experimental Results

To further verify the effectiveness of the proposed closed-loop cooperative current control strategy, experiments are carried out on a hardware-in-loop platform, as shown in Figure 9. The controller is implemented by a DSP (TMS320F28377D) control system. The power part of the proposed converter is simulated by the RTScale real-time simulator. Critical waveforms were captured and analyzed using a high-bandwidth digital oscilloscope, ensuring accurate measurement of dynamic performance and power quality. Due to the symmetry of the circuit, the waveforms of current in phase A are selected for presentation. The steady-state experimental waveforms of the current under i_{ABC_ref} = 90 A and K = 1 are presented in Figure 10. It can be seen from Figure 10a that the Si arm processes the total load current while generating a considerable ripple current due to the reduced switching frequency. Under this condition, the SiC arm only compensates the LF ripple current. It can be seen from Figure 10b that the LF ripple current is eliminated, but there is a circulating current between the Si and SiC arms, leading to additional power losses. The steady-state experimental waveforms of the current under i_{ABC_ref} = 90 A and K = 0.7 are presented in Figure 11. It can be seen from Figure 11a that the SiC arm compensates the LF ripple current while sharing minor load current. It can be seen from Figure 11b that there is no circulating current between the Si and SiC arms by adjusting the major power sharing ratio K. It can be concluded from Figures 10 and 11 that the flexible power sharing and complete ripple current compensation can be realized by employing the proposed current control strategy. Meanwhile, because of the extremely high equivalent switching frequency, the excellent power quality of the output total current can be achieved. It can also be maintained regardless of changing K.

Furthermore, the dynamic waveforms of the current while K changes from 1 to 0.7 are presented in Figure 12. During the dynamic process of changing the power sharing ratio, there are no fluctuations occurring in the total output current. This provides the possibility of adjusting the optimal K under different operation conditions. Figure 13 shows the dynamic experimental waveforms of the current while i_{ABC_ref} changes from 70 A to 100 A. It can be found that the high control dynamics for the step changes in the current reference can be achieved. When the load power is changed, the steady-state current sharing between the Si and SiC arms can remain unchanged. The dynamic changes in the current sharing between the Si and SiC do not affect the total output current. The dynamic experimental waveforms of the line-to-neutral current while changing the current reference are presented in Figure 14. It can be found that the proposed current control can be applied effectively in the three-phase circuits.

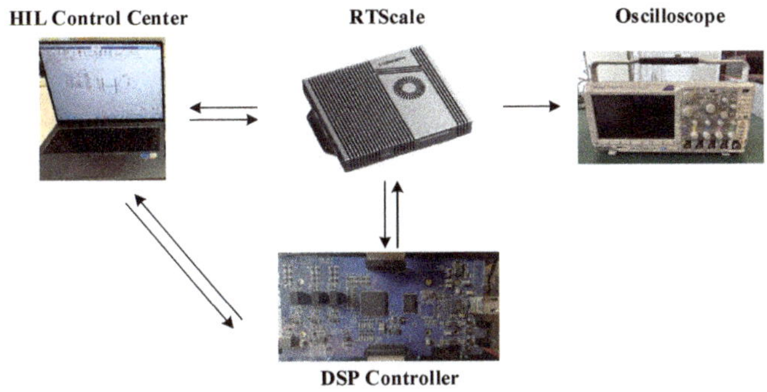

Figure 9. Hardware-in-loop experiment platform.

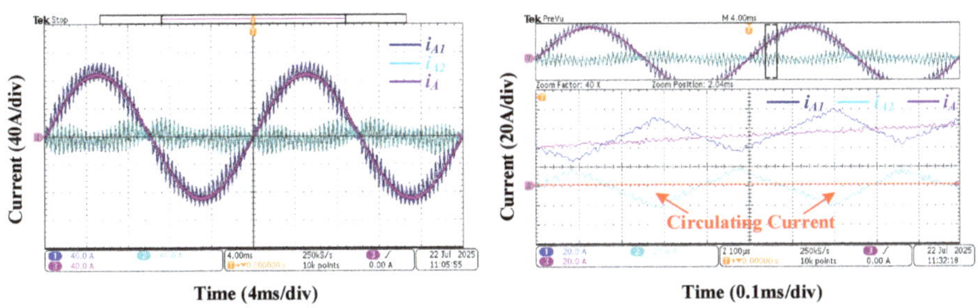

(**a**) (**b**)

Figure 10. Experimental waveforms of the current in phase A under i_{ABC_ref} = 90 A and $K = 1$. (**a**) Steady-state waveforms of the Si arm current, the SiC arm current, and the total current; (**b**) Detailed waveforms of current ripple compensation.

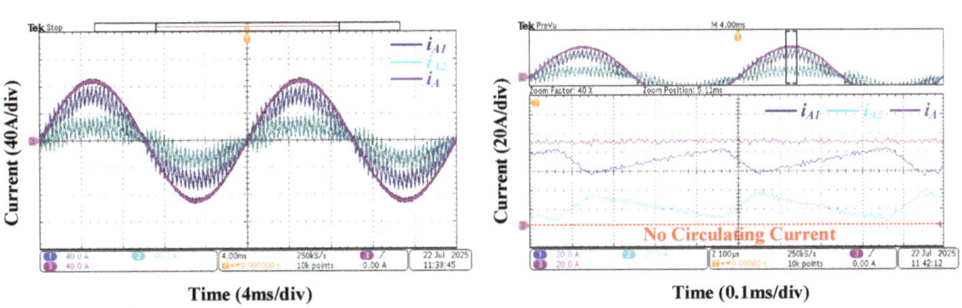

(**a**) (**b**)

Figure 11. Experimental waveforms of the current in phase A under i_{ABC_ref} = 90 A and $K = 0.7$. (**a**) Steady-state waveforms of the Si arm current, the SiC arm current, and the total current; (**b**) Detailed waveforms of current ripple compensation.

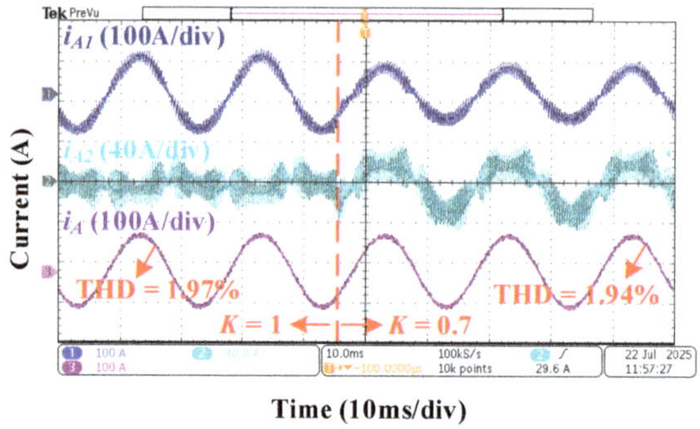

Figure 12. Experimental waveforms of the current in phase A under i_{ABC_ref} = 90 A while K changes from 1 to 0.7.

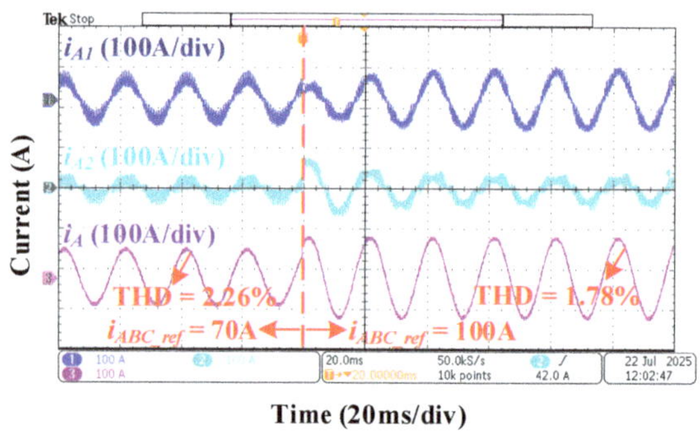

Figure 13. Experimental waveforms of the current in phase A under K = 0.7 while i_{ABC_ref} changes from 70 A to 100 A.

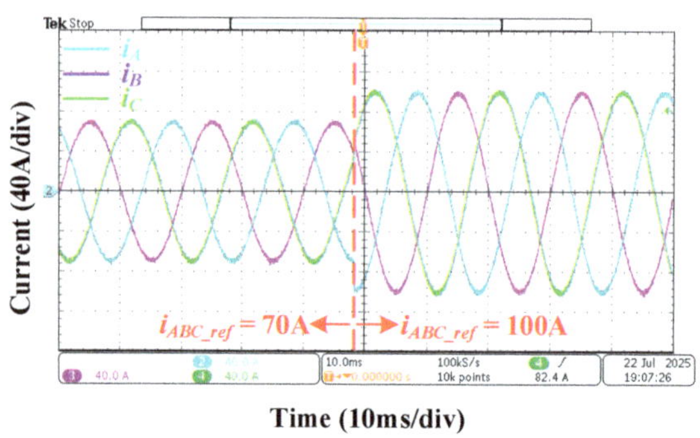

Figure 14. Experimental waveforms of the line-to-neutral current while i_{ABC_ref} changes from 70 A to 100 A.

6. Performance Comparison

6.1. Cost Comparison

To prove the cost-effectiveness of the proposed converter, the all-SiC and all-Si 3L-ANPC topologies are chosen for comparison. Considering the DC bus voltage is 800 V, the 650 V voltage rating Si IGBTs or SiC MOSFETs are selected to configure the all-SiC or all-Si topologies. For the proposed hybrid arms converter, Si IGBTs S_1–S_6 select the 650 V voltage rating devices, while SiC MOSFETs S_7–S_8 select the 1200 V voltage rating and smaller current rating devices. For a fair comparison, the filter inductors of the three topologies are selected to have the same inductance value. It should be considered that the proposed converter uses one more driver chip and current sensor in each phase than the all-SiC and all-Si converters. The chosen components and total price for each phase of three topologies are summarized in Table 2. The unit price of each component is taken from www.mouser.com, accessed on 10 October 2025. It can be found that, compared to the all-SiC converter, the proposed converter can achieve more than 60% total cost reduction.

Table 2. Chosen components and cost comparison.

Components	All-SiC	All-Si	Proposed
SiC MOSFET	C3M0015065K ($35.20 × 6)	/	C3M0040120K1 ($11.19 × 2)
Si IGBT	/	IKZ75N65EL5 ($4.01 × 6)	IKZ75N65EL5 ($4.01 × 6)
Driver IC	UCC21732 ($2.22 × 3)	UCC21732 ($2.22 × 3)	UCC21732 ($2.22 × 4)
Current Sensor	CASR 50-NP ($9.92 × 1)	CASR 50-NP ($9.92 × 1)	CASR 50-NP ($9.92 × 1) CASR 15-NP ($9.06 × 1)
Total Cost	$227.78	$40.64	$74.3

6.2. *Efficiency Comparison*

To evaluate the efficiency performance of the proposed converter, the thermal model was created in the PLECS software environment. The power losses distribution of the proposed converter under i_{ABC_ref} = 90 A and K = 0.8 is presented in Figure 15. Due to the symmetry of the circuit, only half of the devices are presented. According to the results, the FF Si IGBTs S_1 and S_2 have almost no switching losses. The LF Si IGBT S_5 has very little switching loss because of reduced switching frequency. The power losses of the HF SiC MOSFET S_7 are very low due to its excellent switching characteristics. The results are consistent with the designed modulation scheme. Meanwhile, the power-sharing ratio between the Si and SiC arms can seriously affect the efficiency performance. Therefore, the proper selection of the major power sharing ratio K is very important for the efficiency optimization. In Figure 16, the relationship curve between the major power sharing ratio K and power losses under i_{ABC_ref} = 90 A is presented. From this curve, the optimal K under this operating condition can be obtained. In Figure 17, the junction temperature of the devices under different major power sharing ratio K with i_{ABC_ref} = 90 A is presented. It can be found that a key feature of our control strategy is the optimization of temperature equilibrium between the Si and SiC devices via adjusting K. This active balancing prevents localized overheating in any single SiC device, thereby directly countering the temperature-activated mechanisms of gate oxide degradation and threshold voltage drift.

The all-SiC and all-Si 3L-ANPC converters are also chosen for the efficiency comparison. To ensure fairness, the three converters select the same value of filter inductance. Thus, the filter inductance values for all-SiC and all-Si converters are both 1.5 mH. The same modulation scheme is applied in the three converters. Moreover, the switching frequencies of all-SiC and all-Si converters are decided by the THD of the output current, which should be almost equal to that of the proposed converter. Based on the abovementioned conditions, the comparison results are fair and persuasive. In Figure 18, the simulated power losses and efficiency comparisons under different output power are presented. It can be found that the all-Si ANPC converter has the highest power losses. The proposed hybrid arms ANPC converter can achieve 30–40% power loss reduction compared to the all-Si converter in the full power range. In particular, under high-power operation conditions, the efficiency of the proposed converter is higher than that of the all-SiC ANPC converter. The reason is that Si IGBT has a lower conduction voltage than SiC MOSFET under large current conditions, leading to lower conduction losses. Meanwhile, the SiC arm processes the minor load current, which can fully leverage its advantages of low-switching-loss and low-conduction-loss under this condition. Therefore, the proposed topology fully leverages the respective advantages of Si IGBT and SiC MOSFET. It is more attractive than an all-SiC converter in many high-power industrial applications due to higher efficiency and lower cost.

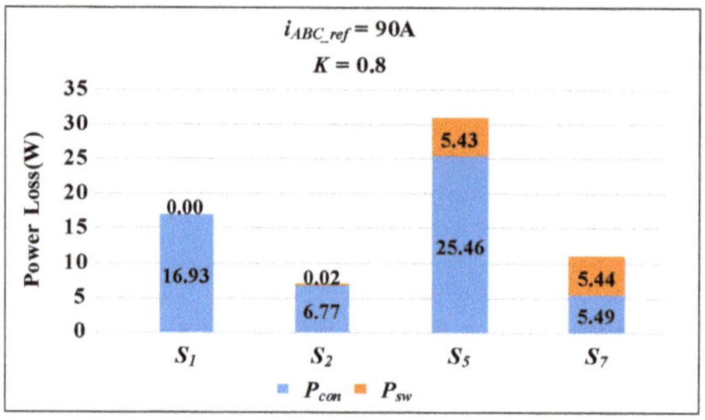

Figure 15. Power losses distribution of the proposed converter under i_{ABC_ref} = 90 A and K = 0.8.

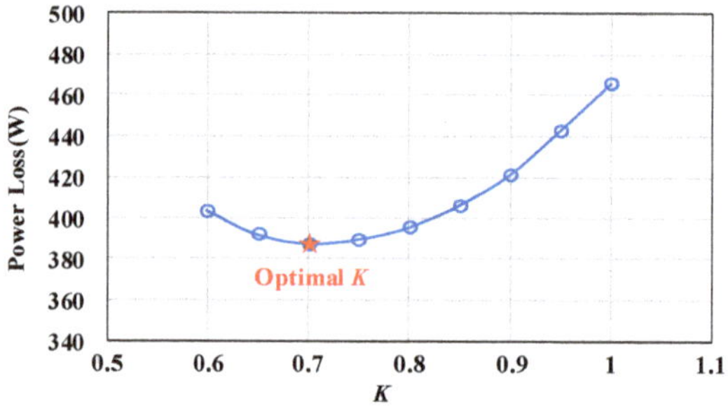

Figure 16. Relationship curve between the major power sharing ratio K and power losses under i_{ABC_ref} = 90 A.

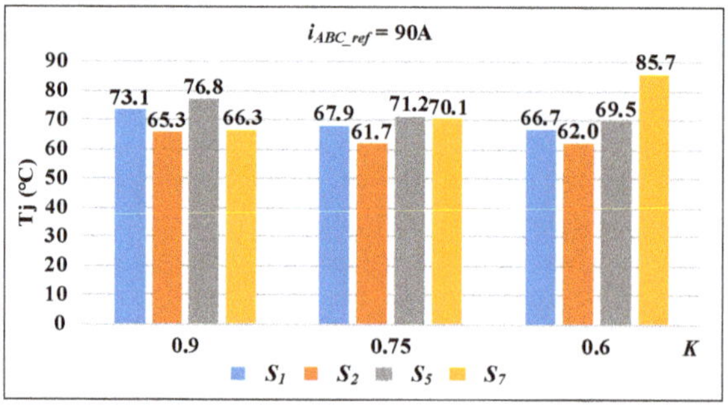

Figure 17. The junction temperature of the devices under different major power sharing ratio K with i_{ABC_ref} = 90 A.

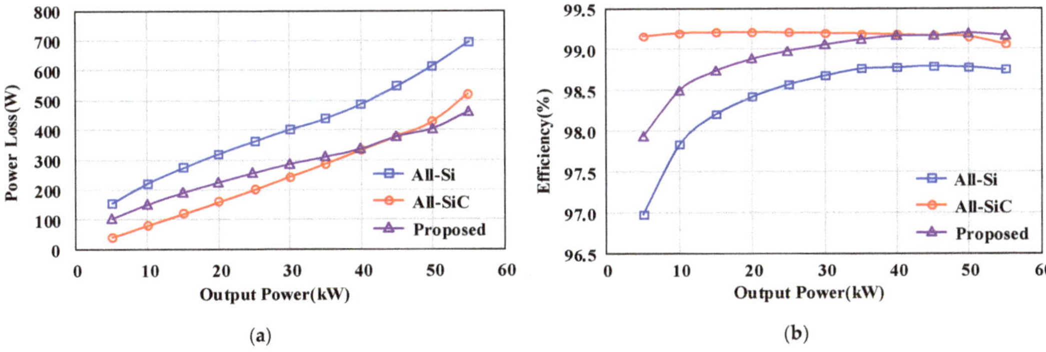

Figure 18. The comparison results of power losses and efficiency under different output power. (**a**) Power losses comparison results; (**b**) Efficiency comparison results.

7. Conclusions

In this paper, a Si and SiC hybrid arms ANPC converter combined with the hybrid-frequency modulation scheme and the WFPP concept was proposed. The advantages of the proposed converter can be concluded as follows.

(1) The proposed converter uses smaller current rating SiC devices, which can achieve more than 60% total cost reduction compared to the all-SiC 3L-ANPC converter.
(2) Although the switching frequency of Si devices is decreased to reduce power losses, the equivalent switching frequency is equal to that of the all-SiC design. Thus, excellent power quality can be achieved.
(3) By employing the hybrid-frequency modulation scheme and proposed current control strategy, the proposed converter can reduce 30–40% power losses compared to the all-Si converter in the full power range. Moreover, the efficiency under high-current conditions is higher than that of the all-SiC design.

The proposed converter fully leverages the respective advantages of Si IGBT and SiC MOSFET, achieving the comprehensive optimization of power quality, efficiency, and cost. It is more attractive than an all-SiC converter in high-power industrial applications due to higher efficiency and lower cost. For extension to medium- and high-voltage applications (e.g., 10-kV grid-connected inverters), the proposed topology offers two clear paths for voltage scaling. On one hand, it can serve as a submodule in cascaded H-bridge (CHB) or the modular multilevel converter (MMC). On the other hand, direct series-connection of devices within each switch presents another viable route.

Author Contributions: Conceptualization, T.X. and C.L.; methodology, T.X.; software, T.X.; validation, T.X., X.F. and Z.A.; formal analysis, T.X. and C.L.; investigation, M.L.; resources, C.L. and H.L.; data curation, T.X. and X.F.; writing—original draft preparation, T.X.; writing—review and editing, X.F., Z.A. and C.L.; visualization, H.L.; supervision, H.Y.; project administration, C.L. and H.Y. All authors have read and agreed to the published version of the manuscript.

Funding: This research was funded by the National Key Research and Development Program of China, grant number 2025YFE0101600, and the Zhejiang Provincial Natural Science Foundation of China, grant number LR24E070001, and the Power Electronics Science and Education Development Program of Delta Group, grant number DREG2024001.

Data Availability Statement: The data presented in this study are available on request from the corresponding author.

Conflicts of Interest: The authors declare no conflicts of interest.

Abbreviations

The following abbreviations are used in this manuscript:

WBG	Wide bandgap
ANPC	Active neutral-point-clamped
FF	Fundamental frequency
LF	Low frequency
HF	High frequency
MEA	More electric aircraft
WFPP	WBG fractional power processing
HHB	Hybrid half-bridge
PFC	Power factor correction
3L	Three-level
5L	Five-level

References

1. Kouro, S.; Malinowski, M.; Gopakumar, K.; Pou, J.; Franquelo, L.G.; Wu, B.; Rodriguez, J.; Pérez, M.A.; Leon, J.I. Recent Advances and Industrial Applications of Multilevel Converters. *IEEE Trans. Ind. Electron.* **2010**, *57*, 2553–2580. [CrossRef]
2. Blaabjerg, F.; Yang, Y.; Kim, K.A.; Rodriguez, J. Power Electronics Technology for Large-Scale Renewable Energy Generation. *Proc. IEEE* **2023**, *111*, 335–355. [CrossRef]
3. Ali, A.; Mousa, H.H.H.; Shaaban, M.F.; Azzouz, M.A.; Awad, A.S.A. A Comprehensive Review on Charging Topologies and Power Electronic Converter Solutions for Electric Vehicles. *J. Mod. Power Syst. Clean Energy* **2024**, *12*, 675–694. [CrossRef]
4. Wang, D.; Hemming, S.; Yang, Y.; Poorfakhraei, A.; Zhou, L.; Liu, C.; Emadi, A. Multilevel Inverters for Electric Aircraft Applications: Current Status and Future Trends. *IEEE Trans. Transp. Electrif.* **2024**, *10*, 3258–3282. [CrossRef]
5. Kumar, R.; Singh, B.; Kant, P. Design and Analysis of a 54-Pulse Converter and 7-Level Hybrid Inverter for Medium Voltage Induction Motor Drive. *IEEE Trans. Ind. Appl.* **2023**, *60*, 560–572. [CrossRef]
6. Millán, J.; Godignon, P.; Perpiñà, X.; Pérez-Tomás, A.; Rebollo, J. A Survey of Wide Bandgap Power Semiconductor Devices. *IEEE Trans. Power Electron.* **2014**, *29*, 2155–2163. [CrossRef]
7. Zhang, Y.; Li, C.; Li, C.; Xin, Z.; Chen, R.; Li, W.; He, X.; Ma, H. A SiC and Si Hybrid Five-Level Unidirectional Rectifier for Medium Voltage Applications. *IEEE Trans. Ind. Electron.* **2022**, *69*, 7537–7548. [CrossRef]
8. Cai, Y.; Sun, P.; Chen, C.; Zhang, Y.; Zhao, Z.; Li, X.; Qi, L.; Chen, Z.; Nee, H.-P. Investigation on Gate Oxide Degradation of SiC MOSFET in Switching Operation. *IEEE Trans. Power Electron.* **2024**, *39*, 9565–9578. [CrossRef]
9. Jiang, H.; Qi, X.; Qiu, G.; Zhong, X.; Tang, L.; Mao, H.; Wu, Z.; Chen, H.; Ran, L. A Physical Explanation of Threshold Voltage Drift of SiC MOSFET Induced by Gate Switching. *IEEE Trans. Power Electron.* **2022**, *37*, 8830–8834. [CrossRef]
10. Kang, M.; Yu, S.; Xing, D.; Liu, T.; Salemi, A.; Booth, K.; Zhu, S.; White, M.H.; Agarwal, A.K. Body Diode Reliability of Commercial SiC Power MOSFETs. In Proceedings of the 2019 IEEE 7th Workshop on Wide Bandgap Power Devices and Applications (WiPDA), Raleigh, NC, USA, 29–31 October 2019; pp. 416–419.
11. Zhang, D.; He, J.; Pan, D. A Megawatt-Scale Medium-Voltage High-Efficiency High Power Density "SiC+Si" Hybrid Three-Level ANPC Inverter for Aircraft Hybrid-Electric Propulsion Systems. *IEEE Trans. Ind. Appl.* **2019**, *55*, 5971–5980. [CrossRef]
12. Lei, Z.; Zhuge, H.; Lou, X.; Zhang, L.; Tao, J. Performance Comparison of 4-SiC and 2-SiC Hybrid Three-Phase Three-Level ANPC Inverters. In Proceedings of the 2021 IEEE Sustainable Power and Energy Conference (iSPEC), Nanjing, China, 23–25 December 2021; pp. 3204–3209.
13. Li, C.; Guan, Q.; Lei, J.; Li, C.; Zhang, Y.; Wang, S.; Xu, D.; Li, W.; Ma, H. An SiC MOSFET and Si Diode Hybrid Three-Phase High-Power Three-Level Rectifier. *IEEE Trans. Power Electron.* **2019**, *34*, 6076–6087. [CrossRef]
14. Zhang, L.; Zheng, Z.; Li, C.; Ju, P.; Wu, F.; Gu, Y.; Chen, G. A Si/SiC Hybrid Five-Level Active NPC Inverter with Improved Modulation Scheme. *IEEE Trans. Power Electron.* **2020**, *35*, 4835–4846. [CrossRef]
15. Zhang, Y.; He, J.; Padmanaban, S.; Ionel, D.M. Transistor-Clamped Multilevel H-Bridge Inverter in Si and SiC Hybrid Configuration for High-Efficiency Photovoltaic Applications. In Proceedings of the 2018 IEEE Energy Conversion Congress and Exposition (ECCE), Portland, OR, USA, 23–27 September 2018; pp. 2536–2542.
16. Belkhode, S.; Shukla, A.; Doolla, S. Five-Level Hybrid Active-NPC H-Bridge Converter with Novel Space-Vector-Modulation Scheme. *IEEE J. Emerg. Sel. Top. Power Electron.* **2022**, *10*, 4146–4159. [CrossRef]
17. Endres, J.; Ackva, A. A Parallel Three-Phase Converter System for Ripple Current Compensation and Passive Filter Reduction. In Proceedings of the 2015 17th European Conference on Power Electronics and Applications (EPE'15 ECCE-Europe), Geneva, Switzerland, 8–10 September 2015; pp. 1–9.

18. Kundu, A.; Na, R.; Amir, A.; Zhou, Y.; Brown, I.P.; John Shen, Z. WBG Fractional Power Processing: A New Si-SiC Hybrid Voltage Source Inverter Design. In Proceedings of the 2020 IEEE Energy Conversion Congress and Exposition (ECCE), Detroit, MI, USA, 11–15 October 2020; pp. 6226–6231.
19. Kundu, A.; Na, R.; Amir, A.; Shen, Z.J. Optimization Strategy of WBG Fractional Power Processing. In Proceedings of the 2021 IEEE Applied Power Electronics Conference and Exposition (APEC), Phoenix, AZ, USA, 14–17 June 2021; pp. 896–901.
20. Zhang, C.; Wang, J.; Qu, K.; Hu, B.; Li, Z.; Yin, X.; Shen, Z.J. WBG and Si Hybrid Half-Bridge Power Processing Toward Optimal Efficiency, Power Quality, and Cost Tradeoff. *IEEE Trans. Power Electron.* **2022**, *37*, 6844–6856. [CrossRef]
21. Zhang, C.; Qu, K.; Hu, B.; Wang, J.; Yin, X.; Shen, Z.J. A High-Frequency Dynamically Coordinated Hybrid Si/SiC Interleaved CCM Totem-Pole Bridgeless PFC Converter. *IEEE J. Emerg. Sel. Top. Power Electron.* **2022**, *10*, 2088–2100. [CrossRef]
22. Zhang, C.; Yuan, X.; Wang, J.; Hu, B.; Yin, X.; Shen, Z.J. Optimization of Power Sharing and Switching Frequency in Si/WBG Hybrid Half-Bridge Converters Using Power Loss Models. *IEEE J. Emerg. Sel. Top. Power Electron.* **2023**, *11*, 2837–2849. [CrossRef]
23. Zhang, C.; Yuan, X.; Wang, J.; Chen, W.; Hu, B.; Shen, Z.J. Adaptive Power Sharing and Switching Frequency Control for Power Loss Optimization in WBG/Si Hybrid Half-Bridge Converters. *IEEE Trans. Power Electron.* **2023**, *38*, 4440–4450. [CrossRef]

energies

MDPI

Article

Design and Optimization of Current-Fed Dual Active Bridge Converter with Dual Coupled-Inductor Structure

Xiaoying Chen [1], Jun Zhang [1], Yizhan Zhuang [1], Yiming Zhang [1] and Jing Guo [2,*]

1 College of Electrical Engineering and Automation, Fuzhou University, Fuzhou 350108, China; fzucxy@fzu.edu.cn (X.C.); 2500127072@fzu.edu.cn (J.Z.); zyz_joe@fzu.edu.cn (Y.Z.); zym@fzu.edu.cn (Y.Z.)

2 School of Automation, Central South University, Changsha 410083, China

* Correspondence: guojingcsu@csu.edu.cn

Abstract

In order to reduce magnetic components for a current-fed dual active bridge converter, this paper proposes a dual coupled-inductor (DCI) structure, which integrates two DC inductors, one high-frequency transformer, and one leakage inductor into two EE cores. By analyzing the principle of the magnetic components, a derivation process is presented to modify the current-fed dual active bridge converter. To simplify the design and enhance efficiency, an equal air gap length optimization method is proposed. And the geometric parameters with the highest efficiency are optimized based on losses. Finally, the feasibility and effectiveness of the above design were verified through a 1 kW test prototype.

Keywords: current-fed dual active bridge; dual coupled-inductor structure; equal air gap; loss optimization design

1. Introduction

Energy storage systems have become an indispensable core component of modern power systems, widely used in key areas [1] such as sustainable energy power systems, microgrids, electric vehicle drive systems, and uninterruptible power supplies (UPS). In these applications, energy storage units such as batteries and supercapacitors typically need to be connected to high-voltage DC buses [2,3] through power electronic conversion devices to achieve efficient, bidirectional flow of energy. Among them, the current-fed dual active bridge (CF-DAB) DC–DC converter [4–7] is widely regarded as one of the ideal choices for achieving this connection function due to its outstanding advantages, such as low input current ripple, wide voltage gain range, and effective support for bidirectional energy flow. It plays a key role in improving the overall efficiency and dynamic response performance of the system.

However, despite the many advantages of CF-DAB, there are still some significant bottlenecks in its practical engineering applications. Firstly, in order to achieve soft switching characteristics and maintain system stability, traditional CF-DAB topology typically requires the configuration of multiple independent magnetic components, such as input-side DC filtering inductors, isolation transformers, and auxiliary inductors for energy regulation [8]. These discrete components not only significantly increase the volume and weight of the system, resulting in a decrease in power density, but also cause additional copper and core losses due to the dispersed layout of magnetic components, limiting further improvement in overall efficiency. Especially in application scenarios such as electric vehicles and portable energy storage that have strict requirements for both volume and efficiency, the above issues become particularly prominent.

Currently, in order to achieve the operational goals of high-power density and high efficiency in the system, researchers generally adopt magnetic component integration technology as the main technical path [9–13]. For example, previous studies have significantly reduced the volume and number of magnetic components by coupling two DC inductors on the low-voltage side and integrating them into a single magnetic core [14,15], or further integrating the DC inductors with a high-frequency transformer into a multiwinding coupled-inductor structure [16,17]. In addition, in order to reduce the manufacturing cost and process complexity of magnetic components, existing solutions usually adopt a flat printed circuit board (PCB) winding, namely planar magnetic technology [18,19]. However, these methods still have a common limitation: they have not achieved internal integration of the leakage inductor and still rely on external series inductors to meet the leakage inductor requirement, which not only increases system volume but also restricts further improvement of power density. To further reduce the number of magnetic components, reference can be made to the leakage inductor integration method adopted in [20–22]. Ref. [20] can achieve controllable leakage inductance by introducing a magnetic shunt, which offers moderate control accuracy. However, the introduction of this additional component increases manufacturing complexity and cost. Ref. [21] can achieve the controllability of leakage inductance by adjusting the spatial distance. Although compared to [20], this method does not add additional magnetic components and has a moderate manufacturing complexity, the calculation is complex, and it is sensitive to mechanical accuracy, resulting in low inductance control accuracy and increased losses as the distance increases. Ref. [22] achieves controllable leakage inductance by adjusting the resistance, which has higher control accuracy compared to [21], but its complex winding structure limits manufacturing feasibility, and the energy consumption of the resistance leads to significant system losses.

In response to the limitations of traditional CF-DAB in terms of power density and efficiency, this paper proposes a dual coupled-inductor (DCI) structure. This structure integrates two DC inductors, one high-frequency transformer, and the required leakage inductor of the circuit into two EE magnetic cores. Compared with existing solutions, the DCI structure reduces the total volume and the occupied space of magnetic components while maintaining functional integrity, providing a new solution for the miniaturization and lightweighting of converters. In terms of parameter control, precise control of DC inductance and leakage inductance can be achieved by designing the number of PCB windings and the length of the magnetic core air gap reasonably. This design not only ensures the flexibility and adjustability of magnetic parameters but also enhances the adaptability to the dynamic performance of the converter. In addition, to further suppress leakage flux, reduce magnetic component losses, and improve the overall efficiency of the converter, this paper proposes a magnetic core optimization method based on the design of equal air gap length for each core. This method not only improves the efficiency of the converter but also simplifies the complexity of designing and manufacturing the planar coupled inductor. On this basis, this article also determines the geometric parameters for achieving optimal system performance through loss optimization.

In this article, Section 2 introduces the derivation process of the DCI structure, improves the traditional CF-DAB based on magnetic component integration, and analyzes the characteristics of the improved topology structure. In Section 3, an equal gap length design scheme is proposed for the optimization design of the planar integrated multi-winding coupled inductor based on the DCI structure, and parameter optimization is carried out for losses. Section 4 validates the proposed optimization scheme based on an experimental prototype. In summary, the DCI integrated structure and its corresponding optimization design method proposed in this article effectively solve the design contradiction between high power density and high efficiency without significantly increasing process complexity, providing a feasible new idea and a technical path with high engineering application value for the development of a new generation of high-performance and compact energy storage interface converters.

2. Current-Fed Dual Active Bridge Converter Modification Based on Magnetic Components Integration

2.1. The Proposed Converter Derivation with DCI Structure

The traditional current-fed dual active bridge (CF-DAB) DC–DC converter includes multiple magnetic components, such as two DC inductors, a high-frequency transformer, and the external leakage inductor, as shown in Figure 1. And their diagrams of individual magnetic components are shown in Figure 2.

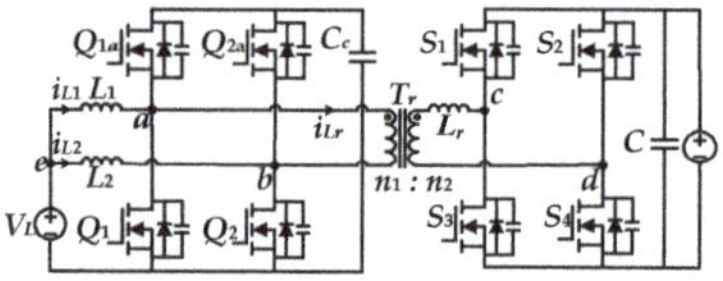

Figure 1. Circuit diagram of CF-DAB converter [23].

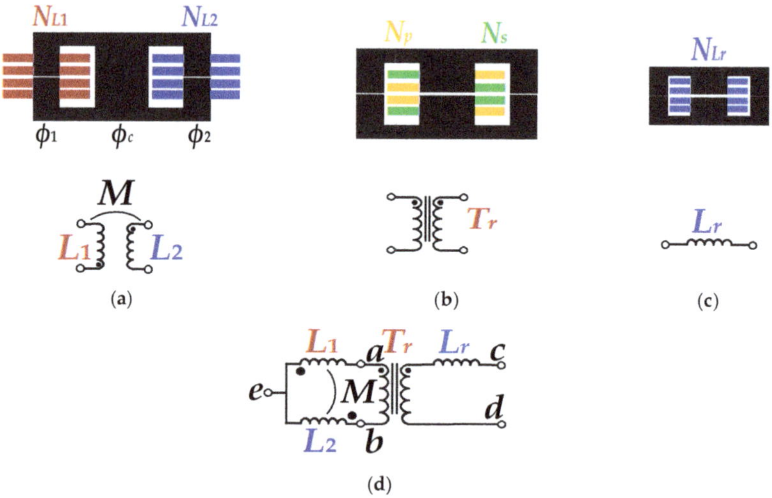

Figure 2. The individual magnetic components of traditional CF-DAB are shown in Figure 1 [23]. (**a**) Decoupled integration of two DC inductors. (**b**) High-frequency transformer. (**c**) External leakage inductor. (**d**) The magnetic components connection circuit.

Among them, Figure 2a shows the integration of two DC inductors, where N_{L1} and N_{L2} represent the number of windings corresponding to L_1 and L_2. N_p and N_s represent the number of windings at the primary and secondary sides, and N_{Lr} represents the number of windings with the external series leakage inductor, which are shown in Figures 2b and 2c, respectively. The magnetic components' equivalent circuit of the traditional CF-DAB converter is shown in Figure 2d. With the traditional individual magnetic structure shown in Figure 2, the number of magnetic components is still large, and the power density cannot meet the requirements.

In order to further realize the integration, consider integrating the high-frequency transformer with the coupled inductor. As shown in Figure 3a, the windings on the two outer columns integrate two DC inductors and a transformer, that is, $N_{L1} = N_{p1}$ and $N_{L2} = N_{p2}$. N_{p1} and N_{p2} are the turns of PCB winding at the primary side of the transformer, and $N_{p1} = N_{p2}$. However, the leakage inductor, shown in Figure 3b, still required an external series inductor. To construct the leakage inductor when all magnetic components are to be integrated together, an air gap needs to be introduced on the middle column based on Figure 3a, as shown in Figure 3c. However, once the air gap is increased, the magnetic flux leakage and loss of the central column will increase, and the DC inductance and leakage inductance are difficult to control. At a certain inductance and core section area, the more windings there are, the larger the air gap will be. To solve this problem, we can reduce the number of PCB layers to achieve a reduction in air gap, thereby reducing the losses and the cost of PCB. Then, to reduce the number of PCB layers, two side columns on the basis of Figure 3c are added. And the primary and secondary windings are separately connected to the two adjacent side columns on both sides of the magnetic core, which is shown in Figure 3d.

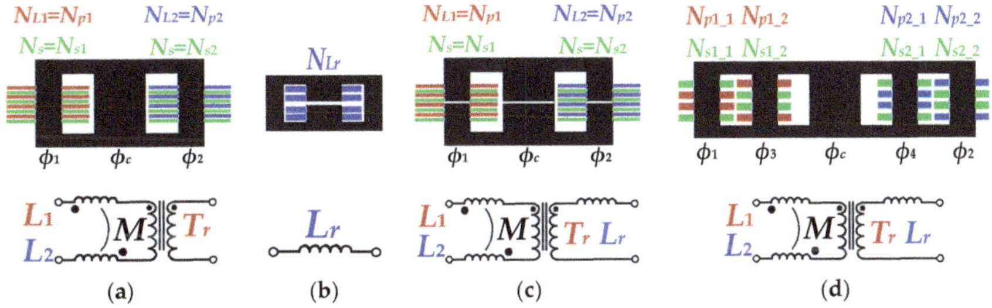

Figure 3. The proposed DCI structure derivation process [23]. (**a**) Magnetic integration of two DC inductors and the transformer. (**b**) External series leakage inductor. (**c**) Three-column structure to integrate the leakage inductor. (**d**) Five magnetic columns structure to integrate the leakage inductor and reduce the required PCB layers.

The magnetic integration shown in Figure 3d is relatively complex. In order to simplify the structure, the flux of the magnetic core can be designed as

$$\phi_1 + \phi_3 = \phi_4 + \phi_2 \tag{1}$$

Then, the magnetic columns in Figure 3d can be separated, as shown in Figure 4a. And a magnetic integrated structure consisting of two EE cores can be obtained, which is called DCI structure in this paper. The circuit model of the DCI structure is shown in Figure 4b. Comparing the above structures in terms of magnetic components, volume, and losses, as shown in Table 1, it can be found that the DCI structure not only reduces the number of magnetic components, increases power density, and reduces the total volume of magnetic components, but also reduces the loss of the center column. And compared with the integrated structure shown in Figure 3c, the DCI structure achieves controllable integration of leakage inductance and DC inductance. In addition, the PCB winding structure and two EE magnetic core designs adopted in this scheme have the advantages of easy implementation and large-scale production. And the DCI structure has good scalability and can be extended to other important topologies. In LLC resonant converters, this structure can achieve high integration of the resonant inductor, magnetizing inductor, and transformer, improving power density. In isolated bidirectional converters such as isolated bidirectional buck-boost, this scheme can effectively integrate a transformer and DC inductor, and optimize soft switching characteristics by utilizing their controllable leakage inductance. This indicates that this study not only can optimize a specific converter but also can provide a universal magnetic integration solution for developing the next generation of high-performance power conversion systems.

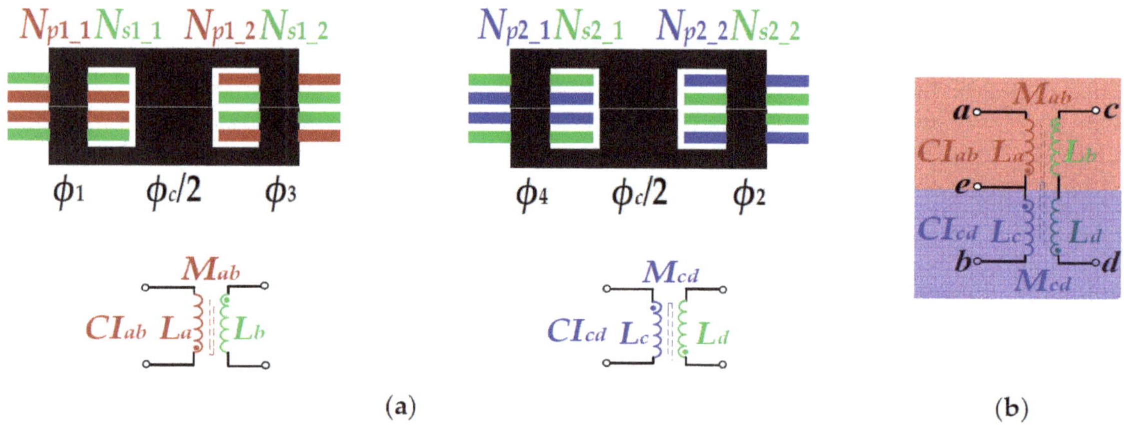

Figure 4. Proposed DCI magnetic integration [23]. (**a**) Magnetic core structure. (**b**) Circuit of the DCI.

Table 1. Structural comparison.

Characteristic	Figure 2	Figure 3a,b	Figure 3c	DCI Structure
Number of components	3	2	1	1
Power density/volume	Maximum	Medium	Minimum	Minimum
Magnetic integration	L_1 and L_2	L_1, L_2 and T_r	L_1, L_2, T_r and L_{rs}	L_1, L_2, T_r and L_{rs}
Losses	/	Medium	Maximum	Minimum

In addition, in this structure, symmetrical design is adopted, that is, $N_{p1} = N_{p1\text{-}1} + N_{p1\text{-}2}$, $N_{p2} = N_{p2\text{-}1} + N_{p2\text{-}2}$, $N_{s1} = N_{s1\text{-}1} + N_{s1\text{-}2}$, and $N_{s2} = N_{s2\text{-}1} + N_{s2\text{-}2}$.

Based on the derived DCI, the topology structure of Figure 1 can be modified to the structure shown in Figure 5, where the red and blue parts represent the DCI structure. For the modified CF-DAB converter, V_L represents the wide input voltage range on the low-voltage side, V_H represents the output voltage, T_{r1} and T_{r2} are two identical transformers, i_a represents the primary current of transformer T_{r1}, i_b represents the secondary current of transformer T_{r2}, L_{rs1} and L_{rs2} represent the internal leakage inductor of the transformer, and L_1 and L_2 represent the magnetizing inductor of the transformer, whose current is equivalent to the DC current in Figure 1, and V_{Cc} represents the clamp voltage of capacitor C_c. In addition, the magnetizing inductor here is equivalent to the DC inductor in Figure 1.

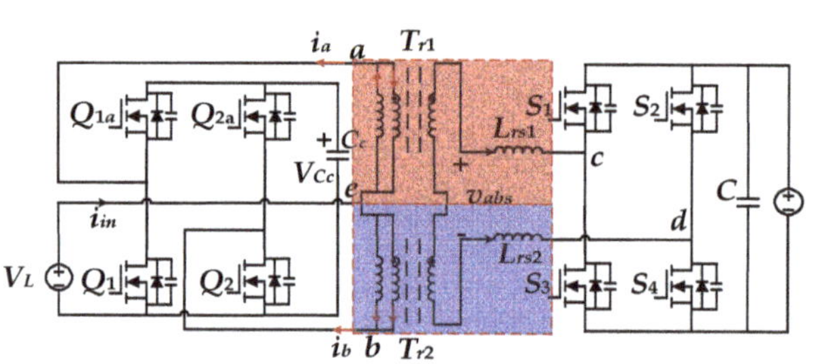

Figure 5. Topology of CF-DAB Converter Based on DCI Structure [24].

2.2. The Principle of the Proposed Converter with DCI Structure

The waveform of this topology under PPS modulation is shown in Figure 6, where PWM is used to maintain the capacitor clamp voltage V_{Cc}, d represents the duty cycle of switches Q_1 and Q_2, and φ represents the phase difference between the rising edges of v_{abs} and v_{cd}. A change in the phase shift angle φ affects the magnitude of the leakage inductor current. According to the division interval of the phase shift angle, it can be divided into two working modes, Mode A($\varphi \in [0, (2d-1)\pi]$) and Mode B($\varphi \in [(2d-1)\pi, \pi]$). The corresponding waveform diagrams are shown in Figure 6a,b, respectively. In addition, the duty cycle of the secondary side switches signal is fixed at 50%, so the midpoint voltage waveform v_{cd} of the secondary side bridge arm is a square wave with a duty cycle of 50%.

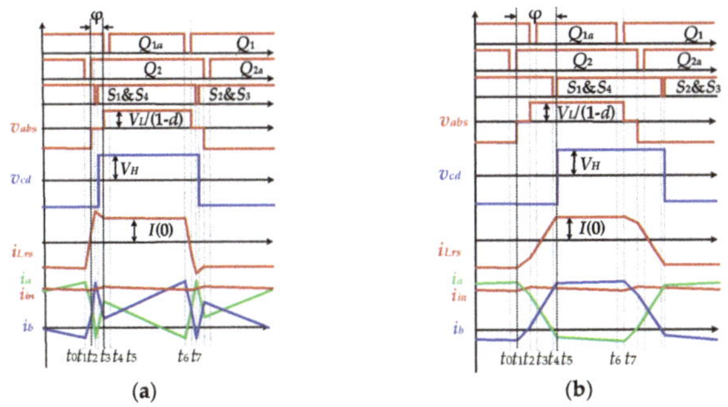

Figure 6. Key waveforms [24]. (**a**) Mode A($\varphi \in [0, (2d-1)\pi]$). (**b**) Mode B($\varphi \in [(2d-1)\pi, \pi]$).

Since the working principles of Mode A and B are similar, in order to simplify the analysis, this article mainly focuses on Mode B. When the converter is working under Mode B, there are seven intervals within half a switching period, as shown in Figure 7. In addition, v_{ae} is defined as the voltage across the primary side of L_1 and T_{r1}, and v_{be} is defined as the voltage across the primary side of L_2 and T_{r2}.

Interval I: Before t_0, the circuit modal diagram is shown in Figure 7a.

Interval II: t_0–t_1, the circuit modal diagram is shown in Figure 7b. At time t_0, Q_{2a} is turned off, while C_{2a} is charged and C_2 is discharged until Q_2 conducts. Among them, the voltage at both ends of C_{2a} and C_2 satisfies (2), and the relationship between the current flowing through them is shown in (3).

$$V_{Ca2} + V_{C2} = V_{Cc} \tag{2}$$

$$C_{a2}\frac{dV_{Ca2}}{dt} - C_2\frac{dV_{C2}}{dt} + i_b = 0 \tag{3}$$

From (2), it can be seen that when V_{Cc} remains constant, the changing trends of V_{Ca2} and V_{C2} are opposite, with the same magnitude. Substituting into (3), the following expression can be obtained,

$$\begin{cases} i_b = 2C_2\frac{dV_{C2}}{dt} = 2i_{C2} \\ i_b = -2C_{2a}\frac{dV_{C2a}}{dt} = 2i_{C2} \end{cases} \tag{4}$$

So, in order to achieve zero voltage turn-on of Q_2, it is necessary to meet the following requirements: $i_b(t_0) < 0$.

Interval III: t_1–t_2, the circuit modal diagram is shown in Figure 7c.

Interval IV: t_2–t_3, the circuit modal diagram is shown in Figure 7d. At time t_2, Q_1 is turned off, while C_1 is charged and C_{1a} is discharged until Q_{1a} conducts. Similarly to Interval II, it can be deduced that the condition for Q_{1a} to implement soft switching is $i_a(t_2) > 0$.

Interval V: t_3–t_4, the circuit modal diagram is shown in Figure 7e.

Interval VI: t_4–t_5, the circuit modal diagram is shown in Figure 7f. At t_4, S_2 and S_3 are turned off, while C_{s2} and C_{s3} charge, and C_{s1} and C_{s4} discharge until S_1 and S_2 conduct. It can be inferred that the condition for S_1 and S_2 to achieve ZVS is $i_{Lrs}(t_4) > 0$.

Interval VII: t_5–t_6, the circuit modal diagram is shown in Figure 7g.

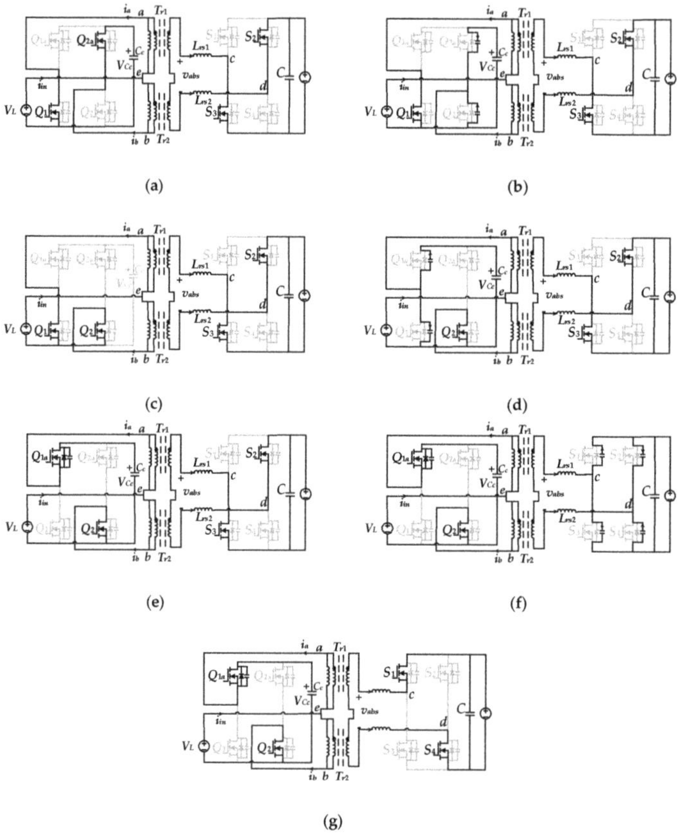

Figure 7. Circuit working intervals in half cycle of the proposed DCI DAB converter [24]. (**a**) Interval I. (**b**) Interval II. (**c**) Interval III. (**d**) Interval IV. (**e**) Interval V. (**f**) Interval VI. (**g**) Interval VII.

Based on the above analysis, the conditions for integrating this circuit to achieve all switch ZVS are

$$\begin{cases} i_b(t_0) < 0 \\ i_a(t_2) > 0 \\ i_{Lrs}(t_4) > 0 \end{cases} \tag{5}$$

Based on the calculation of instantaneous current, the current threshold for implementing Q_1 and Q_2 soft switching can be obtained, which can be referred to [3] and expressed as

$$I_{ZVS\_Q_1Q_2} = i_b(t_0) = \begin{cases} \frac{N_P V_H d(d-1)T_s}{2N_s L_2}, \frac{\varphi}{\pi} \subset [0,(2d-1)] \\ \frac{P}{2V_L} + \frac{N_S V_L^2 T_s}{2N_P V_H L_2} - \frac{V_L T_s}{2L_2} \\ -\frac{N_S V_H T_s}{4N_P L_{rs}} + \frac{N_S}{2N_P L_{rs}} \\ \times\sqrt{(V_H T_S)^2 d(1-d) - 2L_{rs} P T_S}, \frac{\varphi}{\pi} \subset [(2d-1),1] \end{cases} \tag{6}$$

In addition, in order to visually display the ZVS range of all switches and the parameters that affect the ZVS range, this article takes V_L = 18–36 V, V_H = 300 V, f = 150 kHz, and P_N = 1000 W as an example. According to (6), the ZVS range of Q_1 and Q_2 under different d and inductance can be obtained as shown in Figure 8a,b. It can be observed that whether d = 0.5 or d = 0.75, the current $I_{ZVS\_Q1\&Q2}$ are less than zero within the range of [2 μH, 10 μH], as shown in Figure 8a,b, which means that the low-voltage side switches can achieve full load range soft switching under different L_1. Similarly, the ZVS ranges of Q_{1a} and Q_{2a} under different d and inductances are shown in Figure 8c,d, which can also achieve ZVS over the full load range under various inductances. The ZVS range of the high-voltage side switches S_1, S_2, S_3, and S_4 is shown in Figure 8e,f. It can be observed that ZVS cannot be achieved only when V_L = 36 V and P = 0 W, but ZVS can be achieved under different leakage inductance in other load ranges. Additionally, it can be observed that the current threshold for implementing soft switching in low-voltage side switches increases as the leakage inductance decreases.

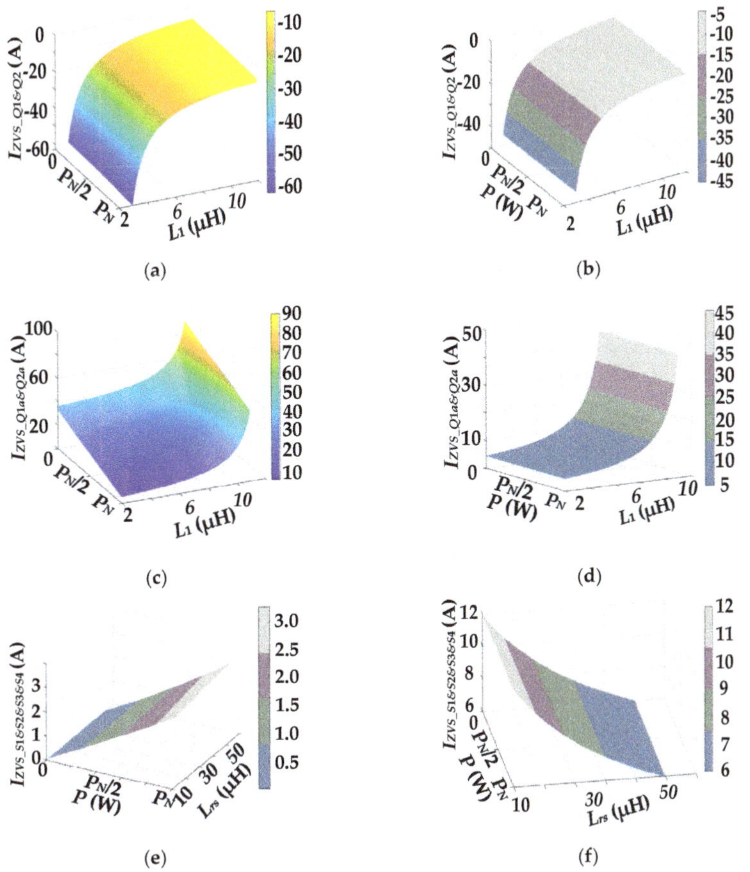

Figure 8. ZVS range of switches [24]. (**a**) Q_1 and Q_2 when V_L = 36 V. (**b**) Q_1 and Q_2 when V_L = 18 V. (**c**) Q_{1a} and Q_{2a} when V_L = 36 V. (**d**) Q_{1a} and Q_{2a} when V_L = 18 V. (**e**) S_1, S_2, S_3, and S_4 when V_L = 36 V. (**f**) S_1, S_2, S_3, and S_4 when V_L = 18 V. (V_L = 18–36 V, V_H = 300 V, P_N = 1 kW).

3. Design and Optimization of Planar Integrated DCI Structure

3.1. Design of Equal Air Gap Length

Because the two coupled inductors CI_{ab} and CI_{cd} shown in Figure 4a are the same, to simplify the analysis, this article takes CI_{ab} as an example and creates its reluctance model as shown in Figure 9. Because the reluctance of the magnetic core is relatively small compared to that of the air gap, it can be ignored. Then, ϕ_1 and ϕ_3 can be obtained as

$$\begin{bmatrix} \phi_1 \\ \phi_3 \end{bmatrix} = \begin{bmatrix} \frac{(N_{P1\_1}R_2+N_{P1\_1}R_c+N_{P1\_2}R_c)i_1-(N_{S1\_1}R_2+N_{S1\_1}R_c+N_{S1\_2}R_c)i_2}{R_1R_2+R_1R_c+R_2R_c} \\ \frac{(N_{S1\_2}R_1+N_{S1\_1}R_c+N_{S1\_2}R_c)i_2-(N_{P1\_2}R_1-N_{P1\_1}R_c-N_{P1\_2}R_c)i_1}{R_1R_2+R_1R_c+R_2R_c} \end{bmatrix} \tag{7}$$

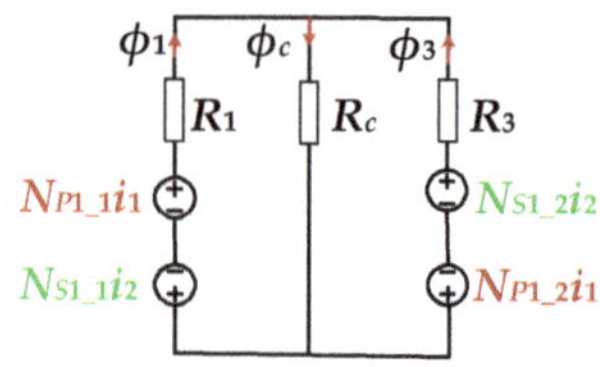

Figure 9. The reluctance model of the coupled inductor CI_{ab} [23].

Therefore, according to Faraday's law, the voltage at both ends of the transformer winding can be obtained as

$$\begin{bmatrix} v_P & v_S \end{bmatrix} = \begin{bmatrix} N_{P1\_1}\frac{d\phi_1}{dt} - N_{P1\_2}\frac{d\phi_3}{dt} & N_{S1\_2}\frac{d\phi_3}{dt} - N_{S1\_1}\frac{d\phi_1}{dt} \end{bmatrix} \tag{8}$$

The inductance matrix of the coupled inductor obtained by (7) and (8) is

$$\begin{cases} L_a = \left[N_{p1\_1}{}^2(R_3+R_c) + N_{p1\_2}{}^2(R_1+R_c) + 2N_{p1\_1}N_{p1\_2}R_c\right]/\Delta R \\ M_{ab} = -\begin{bmatrix} N_{p1\_1}N_{s1\_1}(R_2+R_c) + (N_{p1\_1}N_{s1\_2}+N_{p1\_2}N_{s1\_1})R_c + \\ N_{p1\_2}N_{s1\_2}(R_1+R_c) \end{bmatrix}/\Delta R \\ L_b = \left[N_{s1\_1}{}^2(R_3+R_c) + N_{s1\_2}{}^2(R_1+R_c) + 2N_{s1\_1}N_{s1\_2}R_c\right]/\Delta R \\ \Delta R = R_1R_3 + R_1R_c + R_3R_c \end{cases} \tag{9}$$

The parameters of the transformer are

$$\begin{cases} L_m = L_1 = L_a \\ L_{r\_ab} = L_r/2 = L_b - M_{ab}{}^2/L_a \\ N_a : N_b = n_1 : n_2 = M_{ab}/L_a \end{cases} \tag{10}$$

where L_m represents the equivalent magnetizing inductance, L_{r_ab} represents the leakage inductance, and N_a:N_b represents the transformer turns ratio. From (9) and (10), it can be observed that CI_{ab} is related to the resistance and the number of windings. So, the coupled inductor can be designed based on the number of turns and gap size of the two magnetic cores.

For traditional EE core-based magnetic components, the cross-sectional area of the middle column is twice that of the outer column, and each winding is wound around the middle column. In the model designed in this article, the winding is wound on the outer column, which has a smaller cross-sectional area and leads to a higher magnetic flux density. Therefore, taking V_L = 18–36 V, V_H = 360 V, and P_N = 1000 W as an example, this article proposes a design scheme with equal air gap for each column, which not only facilitates the design of a planar coupled inductor but also effectively reduces leakage of magnetic flux.

The coupled inductor structure based on the EE magnetic core is shown in Figure 10a. Based on this, an Ansys simulation model is established. In this model, the magnetic core material is DMR95, the B-H characteristic is defined according to the DMR95 datasheet, the boundary conditions are set as balloon boundaries to simulate an open field, the excitation current uses periodic current waveforms under full load, and the mesh type adopts a default control strategy based on the maximum element length. The magnetic simulation result is shown in Figure 10b. It can be observed that the outer column has a small cross-sectional area, high magnetic flux density, and high magnetic core loss.

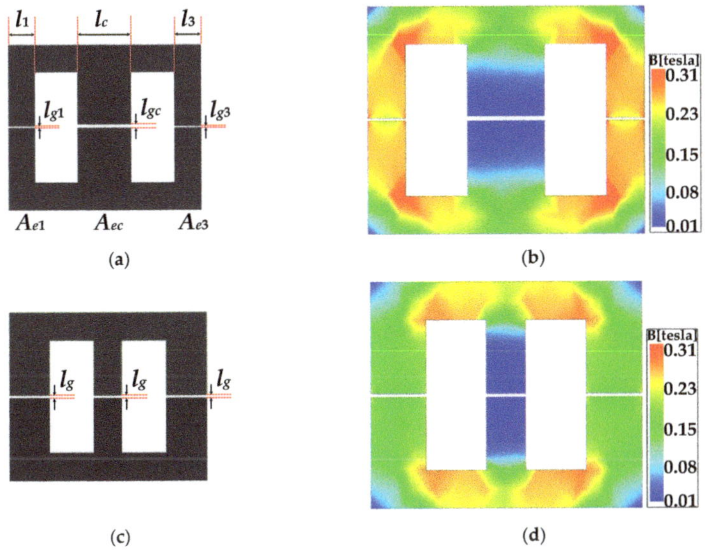

Figure 10. The optimization design of equal air gap length for each magnetic column and the Ansys simulation results. (**a**) Coupled inductor based on EE cores. (**b**) Magnetic flux distribution of EE cores. (**c**) Coupled inductor with the equal air gap length. (**d**) Magnetic flux distribution of equal air gap length design.

And the air gap and cross-sectional area can be expressed as

$$\begin{array}{c} l_c = 2l_1 = 2l_3 \\ A_{ec} = 2A_{e1} = 2A_{e3} \end{array} \tag{11}$$

Combining $R = l/\mu_0 A_e$, it can be concluded that the magnetic reluctance of two outer columns is equal, and it can be found that if the reluctance of the middle column and the outer column are not equal, the air gap of the two are also not equal.

$$R_1 = R_3 \tag{12}$$

$$R_1 \cdot \mu_0 \cdot A_{e1} = l_{g1} \neq l_{gc} = R_c \cdot \mu_0 \cdot A_{ec} \tag{13}$$

Therefore, in order to reduce the magnetic flux density of the outer column, the cross-sectional area of the outer column can be increased, while in order not to change the overall volume of the magnetic core, the cross-sectional area of the middle column can be reduced. The relationship between air gap and cross-sectional area is as follows,

$$A_{e1} = \frac{l_{g1}}{R_1 \cdot \mu_0}, A_{ec} = \frac{l_{gc}}{R_c \cdot \mu_0}, A_{e3} = \frac{l_{g3}}{R_3 \cdot \mu_0} \tag{14}$$

It can be observed that the air gap is directly proportional to the cross-sectional area. By combining Equations (13) and (14), it can be derived that if an equal air gap is designed, the resulting air gap length is

$$l_g = \frac{(A_{e1} + A_{ec} + A_{e3})R_1 R_c R_3 \mu_0}{R_1 R_c + R_c R_3 + R_1 R_3} \tag{15}$$

It can be observed that there must be a positive solution for l_g, which proves the feasibility of equal air gap design. The coupled inductor structure diagram based on equal air gap design is shown in Figure 10c, and its magnetic simulation is shown in Figure 10d. The calculated cross-sectional area of the middle and outer columns under this design is

$$A'_{e1} = \frac{l_g}{R_1 \mu_0}, A'_{ec} = \frac{l_g}{R_c \mu_0}, A'_{e3} = \frac{l_g}{R_3 \mu_0} \tag{16}$$

Comparing Figure 10b,d, it can be found that an equal air gap can reduce the magnetic flux density of the outer column. At the same time, the air gap in the middle column has decreased, and the leakage magnetic flux has also decreased accordingly. And the equal air gap design is also convenient for production practice, so the final choice is the equal air gap scheme.

In addition, the equal air gap scheme largely relies on geometric accuracy, and its small deviations can cause significant fluctuations in leakage inductance L_{rs} due to the inverse proportionality between inductance and air gap length. A small air gap can lead to excessive leakage inductance and increase magnetic core loss, while a large air gap can lead to a small inductance, which cannot effectively suppress ripple current and increase copper loss. Although the large air gap can reduce the magnetic flux density in the magnetic core and decrease core losses, the overall efficiency may decrease due to a significant increase in copper losses. Therefore, it is necessary to determine the specific air gap length based on the optimal efficiency point, and this design indeed requires a certain level of manufacturing precision to ensure its optimized performance.

3.2. Loss Optimization Design

The parameter optimization design is carried out for the loss of DCI structure with equal air gap length under the above working conditions. The specific process is shown in Figure 11.

Combining the formula in Section 3, the parameter design process is shown as follows:

① Transformer turns ratio: under the control of voltage matching PPS, the designed transformer turns ratio is 1:5.

② Equivalent DC inductance: considering ZVS range, ripple current, and effective value of current, L_1 and L_2 are designed to be 6 μH.

③ Leakage inductance L_{rs}: considering the effective value of current, soft switching range, and maximum power capacity, the design of L_{rs} is 1.44 μH.

④ Inductance matrix parameters of coupled inductor: substitute L_1, L_2 and L_{rs} into (10) to obtain self-inductance L_a = 6 μH, L_b = 158 μH, and mutual inductance M_{ab} = 30 μH.

⑤ Coupled inductor turns and reluctance: substitute the inductance matrix parameters into (9), and derive N_{p1} = 3, N_{p2} = 2, N_{s1} = 11, N_{s2} = 15, $R_1 = R_3$ = (2.04 × 10^6) 1/H, and R_c = (2.92 × 10^6) 1/H.

⑥ Cross-sectional area: the maximum magnetic flux of each column is obtained from (7), and finally the cross-sectional area and air gap length are derived. Where, the cross-sectional area is obtained as follows,

$$A_e = \frac{\phi_{\max}}{B_{recommend}} \tag{17}$$

The recommended magnetic flux density is determined according to the working frequency of the converter.

⑦ Air gap: determine the length of the equal air gap according to (15).

Finally, the optimized planar coupled inductor structure can be determined based on ⑤⑥⑦, as shown in Figure 12.

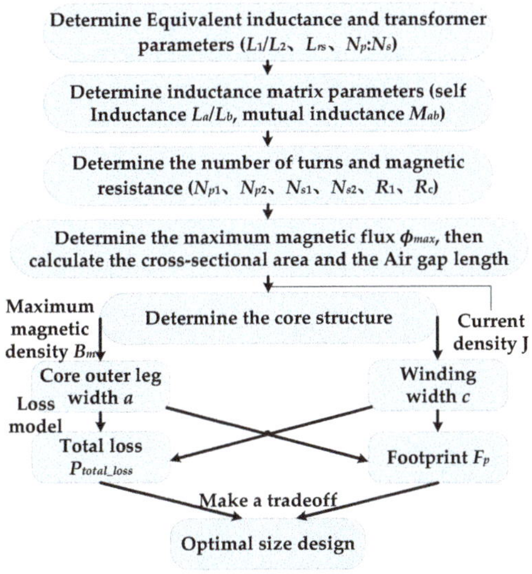

Figure 11. Optimal design flowchart for the planar coupled inductor.

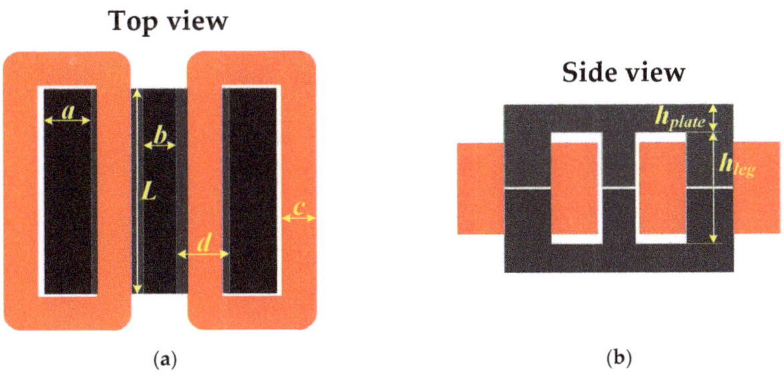

Figure 12. Optimal planar coupled inductor structure. (**a**) Top view. (**b**) Side view.

It can be observed that in order to determine the size, it is necessary to determine the outer column width a and winding width c, where $a = A_{el}/L$, $b = A_{ec}/L$, and $h_{plate} = A_{el}/l$. The height of the magnetic core is determined by the thickness of the PCB board, and the area occupied by this planar coupled inductor is expressed by the following equation,

$$F_p = (2a + 2b + 2c + d)(L + 2c) \tag{18}$$

According to the square waveform modified Steinmetz formula, the iron core loss is expressed as

$$P_{core\_loss} = \frac{8}{\pi^2[4D(1-D)]^{\gamma+1}} k f_s^{\alpha} B_m^{\beta} V_e \tag{19}$$

Among them, D is the duty cycle of the switch, V_e is the volume of the iron core, and γ, k, α and β are the Steinmetz parameters.

It is also known that the DC resistance of each winding is expressed as

$$R_{dc} = \frac{Nl}{ch\sigma_0} \tag{20}$$

Among them, l represents the length of each winding layer, h represents the thickness of the copper skin, and σ_0 represents the conductivity of copper. According to the Dowell one-dimensional AC impedance model, the AC resistance can be calculated as

$$R_{ac} = R_{dc}\left(M' + \frac{(m^2 - 1)D'}{3}\right) \tag{21}$$

$$\begin{cases} M = \lambda h \coth \lambda h \\ D = 2\lambda h \tanh \frac{\lambda h}{2} \\ \lambda = \sqrt{2\pi f_s \mu_0 \sigma_0 \eta} \\ \eta = Nc/b \end{cases} \tag{22}$$

Among them, M is the number of winding layers, M' is the real part of M, D' is the real part of D, and μ_0 is the vacuum magnetic permeability. The winding loss can be obtained from (23).

$$P_{copper\_loss} = I^2_{dcrms\_p} R_{dc\_p} + I^2_{acrms\_p} R_{ac\_p} + I^2_{acrms\_s} R_{ac\_s} \tag{23}$$

By combining (19) and (23), it can be found that both the core loss and winding loss can be expressed as functions of a and c. In this paper, F_p and the total loss of the planar multiwinding coupled inductor are taken as optimization objectives, so the relationship between total loss and F_p can be obtained, as shown in Figure 13, where the red box represents the winding.

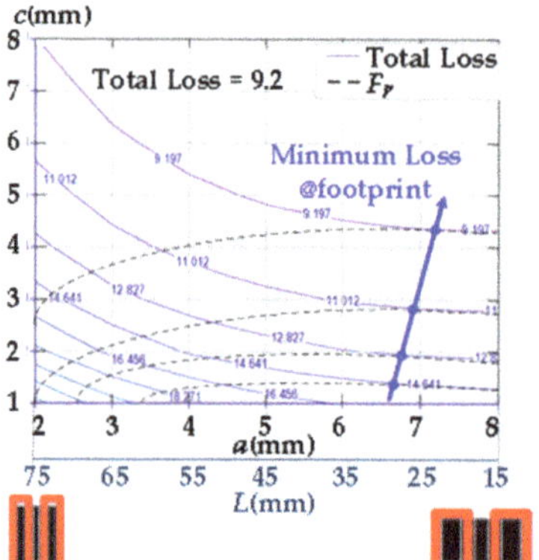

Figure 13. Total power loss calculation with different a and c.

The solid line in the figure represents the waveform of total loss, the dashed line represents F_p at different values of a and c, and the blue tangent point represents the optimal design point, that is, the minimum loss point at a given F_p. It can be found that as F_p increases, the total loss decreases. However, the change in F_p required to reduce the same total loss will increase as the total loss decreases, that is, the degree of decrease in total loss will weaken as F_p continues to increase. Therefore, the final optimization solution is a compromise between F_p and total loss. Based on this, the optimization interval of F_p can be obtained, and the final optimization design point can be selected by combining Figure 13. The final parameter design is determined as a = 6.85 mm, L = 46.5 mm.

4. Experimental Verification

In order to verify the feasibility of the proposed DCI structure with equal air gap planar coupled inductors for the CF-DAB converter and its optimized design, an experimental prototype is designed based on the parameters listed below, as shown in Table 2. And according to the voltage, current stresses and margin that should be considered, device type IPP023N10N5 from Infineon company is selected for the low-voltage side switches, and device type UJ3C065030K3S from United-SiC company is selected for the high-voltage side switches. In addition, the dead time is set to 200 ns to achieve reliable zero voltage switching.

Table 2. Experimental prototype parameters.

Parameter	Value
Power (P/W)	1000
Switching frequency (f/kHz)	150
Input voltage (V_L/V)	18–36
Output voltage (V_H/V)	360
Leakage inductance (L_r/μH)	1.44
DC inductance (L_1, L_2/μH)	6
Transformation ratio (N_p:N_s)	1:5
Clamp capacitor (C_c/μF)	60

This prototype is a CF-DAB converter based on DCI structure with an input voltage range of 18–36 V, an output voltage of 360 V, and a power of 1 kW, as shown in Figure 14a. In addition, Figure 14b shows the dimensions of the designed magnetic core.

(a) (b)

Figure 14. The experimental prototype of the DCI CF-DAB converter. (**a**) Top view of complete converter. (**b**) Optimized core shape.

The steady-state waveforms of the proposed DCI CF-DAB under different loads are shown in Figure 15. Figure 15a,b are the waveforms under light load and full load with an input voltage of 18 V, respectively. Figure 15c,d are the waveforms under light load and full load with an input voltage of 36 V, respectively. It can be seen that the converter can achieve stable operation under this design.

Figure 17 shows the efficiency comparison before and after the equal air gap optimization design under different low-voltage side voltages and loads. It can be found that the optimized low-voltage side voltage of 36 V and load of 550 W has the highest efficiency, reaching 96.4%. It can also be observed that regardless of whether the low-voltage side voltage is 18 V or 36 V and load, the efficiency of traditional CF-DAB is always lower than that of converters using DCI structure. Moreover, after adopting the DCI structure, the efficiency of the optimized design with equal air gap is higher than that without optimization.

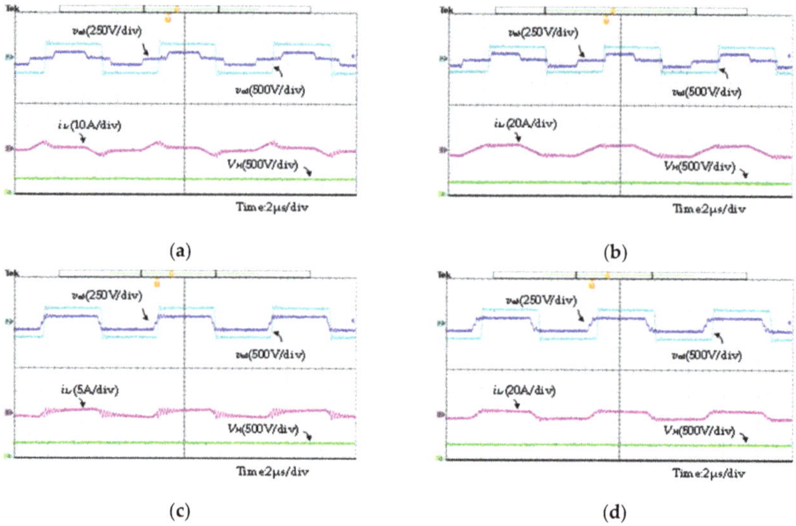

Figure 15. Steady-state waveforms. (**a**) Light load when V_L = 18 V. (**b**) Full load when V_L = 18 V. (**c**) Light load when V_L = 36 V. (**d**) Full load when V_L = 36 V.

Figure 16a,b show the soft switching waveforms of Q_2 and S_1 in light load mode when the input voltage is 36 V. It can be seen that this design can achieve ZVS.

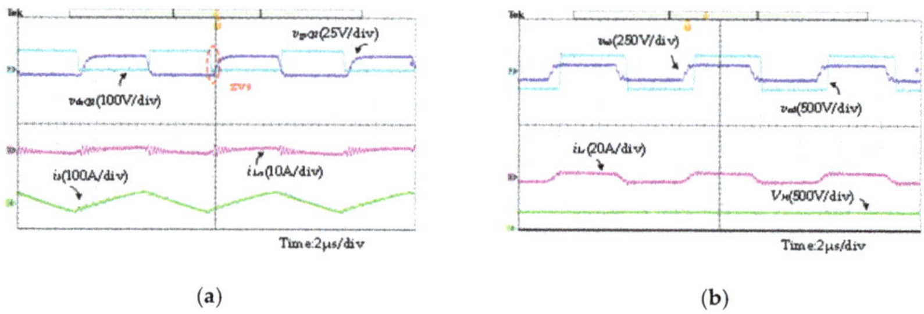

Figure 16. Soft-switching waveforms for light load. (**a**) Q_2 when V_L = 36 V. (**b**) S_1 when V_L = 36 V.

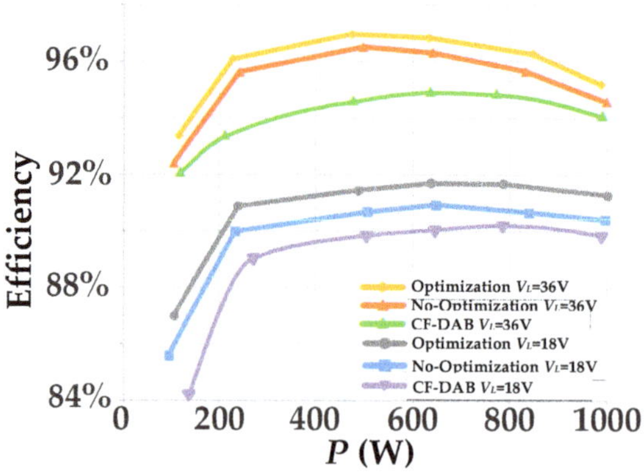

Figure 17. Efficiency comparisons between optimization design, no-optimization design and CF-DAB at different voltages and different loads.

The power analyzer used in this paper has been calibrated for oversampling based on an external meter before measurement, with a voltage measurement accuracy of ±0.1% and a current measurement accuracy of ±1%. In addition, there is a 0.05–0.1% error in the power measurement process due to the length of the measurement probe and other reasons. Under the efficiency improvement shown in Figure 17, this error is acceptable, so the results presented are significant and reliable.

The comparison diagram of losses before and after optimization is shown in Figure 18. Based on this, a loss decomposition table as shown in Table 3 can be obtained. According to the data in Table 3, at V_L = 18 V and rated power P_N, the proposed optimization scheme has the lowest total loss, which is 9.1% lower than before optimization. This advantage mainly stems from the effective optimization of its magnetic core loss. The magnetic core loss of this optimization scheme is much lower than before optimization. Meanwhile, the optimized winding reduces copper loss, reflecting the rationality of its winding layout and the effectiveness of winding width. In summary, the equal air gap optimization scheme using the DCI structure is feasible and effective.

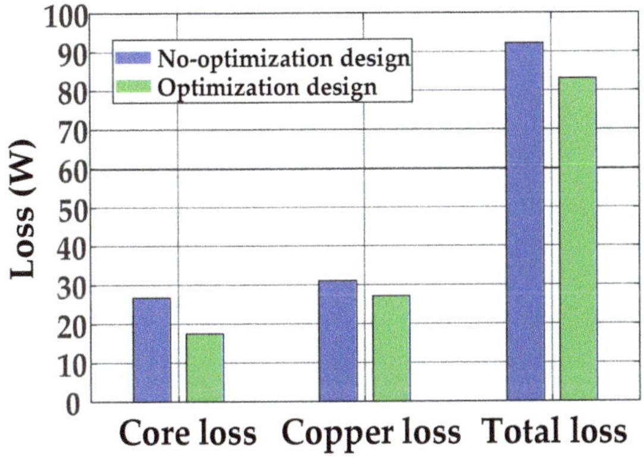

Figure 18. Comparison chart of losses before and after optimization.

Table 3. Loss decomposition table.

Loss Component	No-Optimization Design	Proportion	Optimization Design	Proportion	Decrease
Total loss (W)	92.235	100%	83.086	90.1%	9.1%
Core loss (W)	26.747	29.0%	17.444	18.9%	10.1%
Copper loss (W)	31.091	33.7%	27.106	29.39%	4.31%

5. Conclusions

This article conducted a derivation of the DCI structure, integrating magnetic components to obtain a new CF-DAB topology. Based on the magnetic resistance model, an in-depth analysis was conducted on the inductance matrix parameters of the coupled inductor. The study found that the inductance matrix parameters are closely related to the key structural parameters, such as the number of winding turns and the length of the air gap. In response to the problems of uneven magnetic flux distribution and high losses caused by traditional air gap design, this article proposes an optimized design with equal air gap length, which effectively reduces the magnetic core density of the outer column, simplifies the production process, and provides a practical path for the industrial manufacturing of high-performance magnetic components. The geometric parameters of the planar coupled inductor are optimized, further improving efficiency and reducing losses. Finally, the proposed integration structure and optimization strategy are validated by building an experimental prototype. The experimental results show that the design scheme could improve overall efficiency, fully verifying its feasibility and effectiveness.

Author Contributions: Methodology, J.Z.; validation, Y.Z. (Yizhan Zhuang); investigation, Y.Z. (Yiming Zhang); resources, J.G.; writing—original draft, X.C. All authors have read and agreed to the published version of the manuscript.

Funding: This research was funded by National Natural Science Foundation of China under grant number 52407197.

Data Availability Statement: The original contributions presented in this study are included in the article. Further inquiries can be directed to the corresponding author.

Conflicts of Interest: The authors declare no conflicts of interest.

References

1. Ilyushin, Y.V.; Novozhilov, I.M. Methodology of Inspection of Absolute Stability of Pulse Distributed Control System. In Proceedings of the 2019 XXII International Conference on Soft Computing and Measurements (SCM), St. Petersburg, Russia, 23–25 May 2019.
2. China Society of Automotive Engineers. *Technology Roadmap for Energy Saving and New Energy Vehicles*; China Machine Press: Ningbo, China, 2020; pp. 97–104.
3. Guo, J.; Han, H.; Xu, G.; Cai, Z.; Wang, H.; Sun, Y. Design considerations for PPS controlled current-fed DAB converter to achieve full load range ZVS and low inductor current stress. *IEEE Trans. Ind. Appl.* **2021**, *57*, 6261–6276. [CrossRef]
4. Spiazzi, G.; Rossetto, L.; Mattavelli, P.; Gallo, E.; Cvejic, F. An Isolated Bidirectional Soft-Switching DC-DC Converter for Wide Input/Output Voltage Range. *Energies* **2024**, *17*, 6121. [CrossRef]
5. Carvalho, E.L.; Meneghetti, L.H.; Carati, E.G.; Costa, J.P.D.; Stein, C.M.D.O.; Cardoso, R. Asymmetrical Pulse-Width Modulation Strategy for Current-Fed Dual Active Bridge Bidirectional Isolated Converter Applied to Energy Storage Systems. *Energies* **2020**, *13*, 3475. [CrossRef]
6. Bathala, K.; Kishan, D.; Harischandrappa, N. Soft Switched Current Fed Dual Active Bridge Isolated Bidirectional Series Resonant DC-DC Converter for Energy Storage Applications. *Energies* **2023**, *16*, 258. [CrossRef]
7. Meneghetti, L.H.; Carvalho, E.L.; Carati, E.G.; Denardin, G.W.; da Costa, J.P.; de Oliveira Stein, C.M.; Cardoso, R. Hybrid Inverter and Control Strategy for Enabling the PV Generation Dispatch Using Extra-Low-Voltage Batteries. *Energies* **2022**, *15*, 7539. [CrossRef]
8. Gu, Q.; Xie, B.; Li, R.; Zhu, Z.; Zhang, Y. A ZVS Implementation Method of a Current-fed Dual Active Bridge Converter within the Full Load Range. In Proceedings of the 2024 IEEE Applied Power Electronics Conference and Exposition (APEC), Long Beach, CA, USA, 25–29 February 2024; pp. 2059–2062.
9. Liu, J.; Zheng, Z.; Li, C.; Wang, K.; Li, Y. Automatic current sharing of input-parallel output-parallel dual active bridge converters with coupled inductors. In Proceedings of the IEEE 13th International Conference on Power Electronics and Drive Systems, Toulouse, France, 9–12 July 2019; pp. 1–6.
10. Liu, J.; Li, C.; Zheng, Z.; Wang, K.; Li, Y. Current discrepancy mitigation of input-parallel output-parallel dual active bridge converters using cou- pled inductors. *IEEE Trans. Ind. Electron.* **2021**, *68*, 8182–8192. [CrossRef]
11. Guo, Z.; Sha, D. Dual-active-bridge converter with parallel-connected full bridges in low-voltage side for ZVS by using auxiliary coupling inductor. *IEEE Trans. Ind. Electron.* **2019**, *66*, 6856–6866. [CrossRef]
12. Liu, Y.; Xu, J.; Shuai, Z.; Li, Y.; Peng, Y.; Liang, C.; Cui, G.; Hu, S.; Zhang, M.; Xie, B. A Novel Harmonic Suppression Traction Transformer with Integrated Filtering Inductors for Railway Systems. *Energies* **2020**, *13*, 473. [CrossRef]
13. Yang, J.-W.; Han, S.-K. A Si-FET-Based High Switching Frequency Three-Level LLC Resonant Converter. *Energies* **2019**, *12*, 3082. [CrossRef]
14. Sha, D.; Wang, X.; Liu, K.; Chen, C. A current-fed dual-active-bridgedc–dc converter using extended duty cycle control and magnetic-integrated inductors with optimized voltage mismatching control. *IEEE Trans. Power Electron.* **2019**, *34*, 462–473. [CrossRef]
15. Jiang, C.; Liu, H. A novel interleaved parallel bidirectional dual- active-bridge dc–dc converter with coupled inductor for more-electric aircraft. *IEEE Trans. Ind. Electron.* **2021**, *68*, 1759–1768. [CrossRef]
16. Wang, Z.; Li, C.; Zheng, Z. A high-step-up low-ripple and high- efficiency DC-DC converter for fuel-cell vehicles. *IEEE Trans. Power Electron.* **2022**, *37*, 3555–3569. [CrossRef]
17. Han, H.; Guo, J.; Xu, G.; Xu, J.; Liu, D.; Su, M. Three winding coupled inductor-based dual active bridge DC-DC converter with full load range ZVS under wide voltage range. *IEEE Trans. Ind. Electron.* **2022**, *69*, 6935–6947. [CrossRef]

18. Zhang, Z.; Liu, C.; Wang, M.; Si, Y.; Liu, Y.; Lei, Q. High-efficiency high-power-density CLLC resonant converter with low-stray-capacitance and well-heat-dissipated planar transformer for EV on-board charger. *IEEE Trans. Power Electron.* **2020**, *35*, 10831–10851. [CrossRef]
19. Kolahian, P.; Grimm, F.; Baghdadi, M. A Comprehensive Review on Planar Magnetics and the Structures to Reduce the Parasitic Elements and Improve Efficiency. *Energies* **2023**, *16*, 3254. [CrossRef]
20. Zhang, J.; Ouyang, Z.; Duffy, M.C.; Andesen, M.A.; Hurley, W.G. Leakage inductance calculation for planar transformers with a mag- netic shunt. *IEEE Trans. Ind. Appl.* **2014**, *50*, 4107–4112. [CrossRef]
21. Mu, M.; Xue, L.; Boroyevich, D.; Hughes, B.; Mattavelli, P. Design of integrated transformer and inductor for high frequency dual active bridge GaN charge for PHEV. In Proceedings of the IEEE Applied Power Electronics Conference and Exposition, Charlotte, NC, USA, 15–19 March 2015; pp. 579–585.
22. Li, B.; Li, Q.; Lee, F.C. High-frequency PCB winding transformer with integrated inductors for a bi-directional resonant converter. *IEEE Trans. Power Electron.* **2019**, *34*, 6123–6135. [CrossRef]
23. Guo, J.; Wang, H.; Xu, G.; Li, X.; Sun, Y.; Su, M. Dual Coupled Inductors with Controllable Integrated Leakage Inductance and CM Noise Suppression for CF-DAB Converter. *IEEE Trans. Power Electron.* **2023**, *38*, 8033–8038. [CrossRef]
24. Guo, J.; Wang, H.; Xu, G.; Han, H.; Su, M. Dual-Transformer-Based DAB Converter with Controllable Integrated Inductances. *IEEE Trans. Power Electron.* **2024**, *39*, 6376–6390. [CrossRef]

energies

MDPI

Article

Event-Triggered Control of Grid-Connected Inverters Based on LPV Model Approach

Wensheng Luo *, Zhiwei Zhang, Zejian Shu, Haibin Li and Jianwen Zhang

School of Electrical Engineering and Automation, Harbin Institute of Technology, Harbin 150001, China; 23s136575@stu.hit.edu.cn (Z.Z.); 23s106096@stu.hit.edn.cn (Z.S.); 23s006017@stu.hit.edu.cn (H.L.); jianwen_zhang@stu.hit.edu.cn (J.Z.)

* Correspondence: wensheng.luo@hit.edu.cn

Abstract

This study aims to develop an event-triggered control strategy of grid-connected inverters, based on the linear parameter-varying (LPV) modeling approach. Regarding the changes in grid voltage, filter capacitance and inductance, and random electromagnetic interference, a stochastic LPV model for three-phase two-level inverters is established. To reduce computation burden, an event trigger with a continuous-time form is adopted to derive the state feedback controller for the LPV plant. Unlike the existing common approach to dealing with event-triggered mechanisms, a predesignated event-triggering threshold is used to determine the triggering instant of the event condition. Using parameter-dependent Lyapunov functions, sufficient conditions reliant on parameters are introduced. Based on the derived conditions, the corresponding event-triggered controllers are engineered to ensure uniform ultimate bounded stability for the resulting event-triggered LPV inverter system subject to exogenous disturbance. The simulation results are presented to confirm the efficacy of the proposed methods.

Keywords: grid-connected inverter; linear parameter-varying (LPV) system; stochastic systems; event-triggered control

1. Introduction

Grid-connected inverters are active front ends that connect power sources, such as wind-turbine generators, PV panels, energy storage systems, etc., to the utility grid. Inverter control is normally implemented by a digital microprocessor, which turns on/off the power switches to generate the required active/reactive power. With the increase in control algorithm complexity and switching frequency, the microprocessor faces a greater computational burden; meanwhile, there is more switching loss and the switch span is reduced. To target this problem, researchers have proposed event-triggered control. Event-triggered control implements control action when a certain event occurs, rather than periodically as in traditional digital control systems [1,2]. This feature saves energy, lowers communication, and reduces the computational burden of the control system.

There are some studies on event-triggered control in relation to power converters, but they mainly focus on DC/DC converters. For buck converters, the event-triggered auto-disturbance-rejection control strategy has been proposed, and an extended state observer is adopted to track system state and external interference and update the control signal when the estimation error reaches the limit [3]; in [4], the authors design a proportional–integral observer to deal with the uncertainty of mismatch and external disturbances with continuous derivatives. The above researches have obtained necessary criteria for guaranteeing constrained dynamic stability within the system's theoretical framework, and the validity of the theory has been experimentally verified in buck converters. An event-triggered model predictive control system based on a finite-control-set model prediction method is firstly proposed for buck converters in [5], where the state error modulus is used as a triggering condition. A boost converter is treated as a switching model, and an event-triggered control method is designed to stabilize a linear switching system in [6]; on this basis, an event-triggered control method for a networked linear switching system is further proposed in [7]. Event-triggered sliding-mode control is applied to a buck converter in [8], which greatly reduces the switching frequency of the converter (by about 10 times). Meanwhile, it maintains almost the same control performance as in non-event-triggered control.

Linear parameter-varying (LPV) systems are modeled with temporal change parameters; they have the form of linear systems but includes time-varying uncertainties [9]. Due to this feature, LPV systems are popular in theoretical research and practical engineering, as the well-developed linear system theories can be used. For example, there are theoretical studies that address analysis and synthesis issues, robust filter design formulations, defect diagnosis and localization [10,11], etc. In existing engineering research, an LPV model has been established for lithium-ion batteries to enhance the assessment accuracy of peak-power evaluation under conditions of battery aging [12], an LPV framework has been adopted for a four-wheeled mobile robot to track its trajectory [13], and an uncertain robot system is regulated using an LPV-based control strategy incorporating sliding-mode techniques [14].

In contemporary system dynamics, stochastic systems have emerged as critical tools for analyzing probabilistic behaviors inherent in electromechanical actuators, renewable energy networks, and cellular regulatory pathways. During the last few decades, research has focused on stabilization and filtering problems for stochastic systems. For stabilization problems, stochastic stability analysis and control have been carried out and implemented in secondary frequency regulation for islanded microgrids [15], and in nonlinear systems, the issue of input-to-state stability being subject to stochastic impulsive effects has received systematic investigation [16]. Regarding filtering problems, the recursive estimation issue for specific uncertain random systems incorporating amplify-and-forward relaying techniques were investigated in [17], the decentralized robust filtering problem for randomly switched systems with communication delays under sensor measurement fading conditions was systematically studied in [18]. With the development of networked control, the event-triggering mechanism has been considered in controller design for stochastic systems. For example, an analysis of performance and mean-square exponential stability in discrete-time randomly switched hybrid systems within an event-triggered framework was conducted in [19]; in addition, under stochastic transmission protocols, an event-triggered control challenge in a stochastic nonlinear system was systematically investigated [20].

Regarding grid-connected inverters, they need to deal with grid voltage variations (including magnitude and frequency), the filter capacitance and inductance changes caused by aging and temperature, and the stochastic electromagnetic noise in the environment. To solve these problems, this study develops a variable-threshold event-triggered control approach for grid-connected two-level three-phase inverters using an LPV model. Treated as stochastic excitation terms, stochastic electromagnetic noise processes are incorporated alongside time-variant modeling of filter capacitance, grid voltage, and inductive components. The aims of this paper are to (i) establish a systematic event-triggered control scheme for an LPV system subject to event-triggering conditions; (ii) provide a sufficient condition of stability, namely uniform ultimate bounded stability, fir the LPV system under external disturbance; (iii) grounded in the derived stability conditions, design event-triggered controller gains.

This paper proposes a new modeling method for grid-connected inverters. Unlike the traditional LPV method, this paper includes stochastic perturbations in the model. Moreover, to consider a networked control scenario, which is typical of modern grid control frames, the event-triggered method is introduced to the model, which has a lower computational cost than the continuous control method. This paper is organized as follows: The system overview and inverter modeling are discussed in Section 2. The main results are reported in Section 3, followed by simulation validation in Section 4. Section 5 concludes the paper.

2. Inverter Modeling and System Description

Figure 1 is the grid-connected inverter under investigation, which is of a two-level three-phase topology. To facilitate control design, the inverter is normally described within a synchronously rotating coordinate frame, as elaborated below [21]:

$$L\frac{di_d}{dt} = u_d V_i - \psi_e L i_q - r i_d - v_d, \tag{1a}$$

$$L\frac{di_q}{dt} = u_q V_i + \psi_e L i_d - r i_q - v_q, \tag{1b}$$

where V_i is the DC power supply input, ψ_e represents the frequency of the grid voltage, L signifies the filter inductance and r stands for its series resistance component, and the three-phase *abc* quantities are transformed into dq variables, i.e., the grid voltages v_d, v_q are converted from v_a, v_b, and v_c. i_d, i_q originate from the transformation of grid currents via i_a, i_b, and i_c, while the control-synthesized switching functions are u_d, u_q; subsequently, the switching signals S_a, S_b, and S_c can be derived. To ensure adequate power delivery to the motor or grid, i_d, i_q need regulation to specified setpoints. Figure 2 shows the control scheme of the event-triggered stochastic LPV inverter system, which is divided into four parts: (1) the stochastic LPV dynamical system of the inverter; (2) the event trigger (including the smart sensor and event detector); (3) the event-triggered controller; (4) the zero-order holder (ZOH).

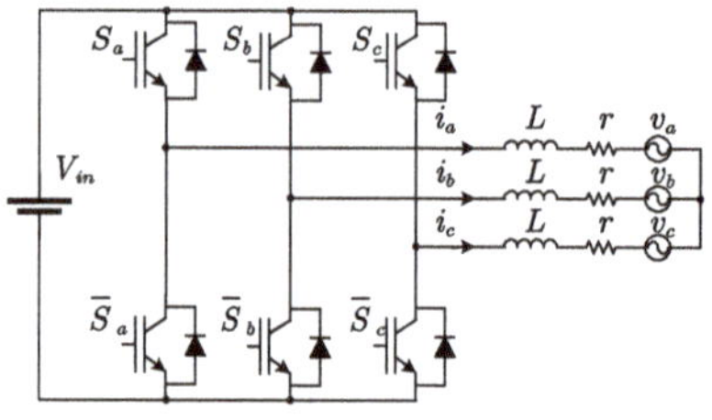

Figure 1. Two-level three-phase inverter.

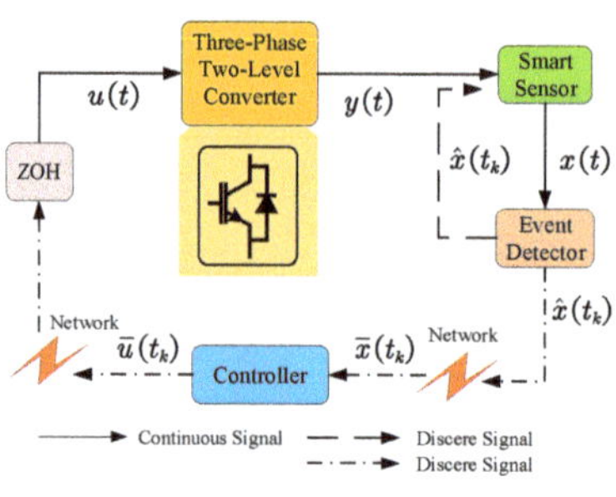

Figure 2. Control scheme of event-triggered stochastic LPV inverter system.

2.1. Dynamical System

The stochastic LPV dynamical system of the inverter can be described in Itô's form as follows:

$$\begin{aligned} dx(t) =&[M(\eta(t))x(t) + N(\eta(t))u(t) + N_\rho(\eta(t))\rho(t)]dt \\ &+ N_\psi(\eta(t))x(t)d\psi(t), \end{aligned} \tag{2a}$$

$$\begin{aligned} dy(t) =&[G(\eta(t))x(t) + H(\eta(t))\rho(t)]dt \\ &+ F(\eta(t))x(t)d\psi(t), \end{aligned} \tag{2b}$$

$$z(t) =C(\eta(t))x(t) + D(\eta(t))\rho(t), \tag{2c}$$

where $x(t) = [i_d; i_q]$ represents the dynamic state of the system; $M(\eta(t))$ represents the dynamic coefficients r, L, and ψ associated with the filtering inductance; $u(t) = [u_d V_i - v_d; u_q V_i - v_q]$ denotes the control input; $N(\eta(t))$ represents the gating matrix associated with L; $\rho(t) \in \mathcal{R}^l$ is a disturbance input with the upper bound $\bar{\rho}$; the controlled output is $z(t) \in \mathcal{R}^q$; $\psi(t)$ is a unidimensional Brownian motion meeting $Z\{d\psi^2(t)\} = dt$ and $Z\{d\psi(t)\} = 0$; and the vector parameter $\eta(t) \in \mathcal{R}^s$ undergoes smooth temporal variation constrained within a bounded convex domain.

The parameter η is unknown but observable in real time. In the rest of this paper, η is used in place of $\eta(t)$ for simplicity. Accordingly, the autonomous form of (2) adopts the following structure:

$$\begin{aligned} dx(t) =&[M(\eta(t))x(t) + N_\rho(\eta(t))\rho(t)]dt \\ &+ N_\psi(\eta(t))x(t)d\psi(t), \end{aligned} \tag{3a}$$

$$y(t) =C(\eta(t))x(t) + D(\eta(t))\rho(t). \tag{3b}$$

To analyze system stability and design controllers, the external disturbance term "$\rho(t)$" within Equation (2) is assumed to be negligible (zero), resulting in

$$\begin{aligned} dx(t) =&[M(\eta(t))x(t) + N(\eta(t))u(t)]dt \\ &+ N_\psi(\eta(t))x(t)d\psi(t), \end{aligned} \tag{4a}$$

$$y(t) =C(\eta(t))x(t). \tag{4b}$$

The autonomous representation of the system (4) is given by

$$dx(t) =M(\eta(t))x(t)dt + N_\psi(\eta(t))x(t)d\psi(t), \tag{5a}$$

$$y(t) =C(\eta(t))x(t). \tag{5b}$$

2.2. State Observer

The state $\hat{x}(t)$ is estimated using the measured output magnitude $y(t)$, based on the following state observer:

$$\begin{aligned} d\hat{x}(t) =&[M(\eta)\hat{x}(t) + N(\eta)u(t) + N_\rho(\eta)\rho(t)]dt \\ &+ N_\psi(\eta)\hat{x}(t)d\psi(t) \\ &+ L(\eta)(dy(t) - [G(\eta)\hat{x}(t) + H(\eta)\rho(t)]dt \\ &- F(\eta)\hat{x}(t)d\psi(t)), \end{aligned} \tag{6}$$

where $\hat{x} \in \mathcal{R}^n$ denotes the observed state, and $L(\eta)$ represents the parameter variable observer gain. Define $\tilde{x} = x(t) - \hat{x}(t)$ as the estimation error; the dynamics of the estimation error are given by

$$\begin{aligned} d\tilde{x}(t) =&[M(\eta) + L(\eta)G(\eta)]\tilde{x}(t)dt \\ &+ [N_\psi(\eta) + L(\eta)F(\eta)]\tilde{x}(t)d\psi(t). \end{aligned} \tag{7}$$

2.3. Event Trigger

Normally, an event trigger includes two functional parts, i.e., an event generator and an event detector. The event detector continuously monitors the state of the dynamic system and checks if a predefined rule is violated, meaning that the event-triggering condition is satisfied. When this condition is achieved, then an event is generated by the event generator and an action is taken.

In this paper, an event is generated when the deviation between the present state and the most recently recorded state exceeds a certain threshold. If there occurs an event, the event generator transfers the latest sampled state to the event-triggered controller.

Let the sequence of event-triggering times be represented by $\{t_k\}_{k=0}^{\infty}$ with $t_k < t_{k+1}$. Without loss of generality, assume the initial triggering time is $t_0 = 0$.

The system state $x(t_k)$ is sampled at each event time t_k, and the subsequent triggering instant t_{k+1} is computed based on

$$t_{k+1} = inf\{t > t_k| \quad \|e(t)\| \geq \sqrt{\varepsilon}\}, \quad \forall k \in \mathcal{N}. \tag{8}$$

where the error $e(t) = x(t) - x(t_k)$, and $\sqrt{\varepsilon}$ denotes the predesignated event-triggering threshold.

2.4. Event-Triggered Controller

The feedback controller of an event-triggered state is employed in this paper. Once an event is generated, the system calculates the control input based on the sampled state

$$\bar{u}(t) = K(\eta)\alpha_{t_k}\hat{x}(t_k), t \in [t_k, t_{k+1}), \quad k \in \mathcal{N}. \tag{9}$$

where $K(\eta)$ refers to parameter-varying gain with suitable dimensions, and $\alpha_{t_k} \in \mathcal{R}$ is a Bernoulli-distributed stochastic variable with

$$\text{Prob}\{\text{ff}_{\text{t}_\text{k}} = 1\} = Z\{\alpha_{t_k}\} := \bar{\alpha}, \tag{10}$$

$$\text{Prob}\{\text{ff}_{\text{t}_\text{k}} = 0\} = 1 - Z\{\alpha_{t_k}\} := 1 - \bar{\alpha}. \tag{11}$$

2.5. Zero-Order Holder

The actuation signal formulated in (9) undergoes parametric updates exclusively at the pre-computed triggering moment $\{t_k\}_{k=0}^{\infty}$; that is, the controller in (9) receives a sampled state $x(t_k)$ which will not change until the next event happens at t_{k+1}. Thus, to keep the continuity of the control signal, a zero-order holder (ZOH) should be embedded:

$$u(t) = \beta_{t_k}\bar{u}(t), \tag{12}$$

where $\beta_{t_k} \in \mathcal{R}$ is a Bernoulli-distributed stochastic variable with

$$\text{Prob}\{\text{fi}_{\text{t}_\text{k}} = 1\} = Z\{\beta_{t_k}\} := \bar{\beta}, \tag{13}$$

$$\text{Prob}\{\text{fi}_{\text{t}_\text{k}} = 0\} = 1 - Z\{\beta_{t_k}\} := 1 - \bar{\beta}. \tag{14}$$

Before delving deeper, several definitions and lemmas are first presented.

Definition 1. *System (5) achieves asymptotic mean-square stability if* $\lim_{t\to\infty} Z\|x(t)\|^2 = 0$ *for arbitrary initial states* $x(0)$.

Definition 2. *The stochastic LPV system (3) denotes robust stability with* γ*-disturbance attenuation in the* $\mathcal{H}_\infty$ *sense when, for every* $\rho(t) \in \mathcal{L}_2[0,\infty]$ *and arbitrary parameter trajectories, the inequality presented below holds:*

$$\|y(t)\|_2 < \gamma\|\rho(t)\|_2.$$

Lemma 1 ([22]). *(Itô's formula) Let $v(t)$, where $t \geq 0$, be an n-dimensional stochastic process satisfying an Itô-type differential equation*

$$dv(t) = f(t)dt + h(t)d\psi(t),$$

where $f(t) \in \mathcal{R}^n$, $h(t) \in \mathcal{R}^{n\times m}$, and $T(v(t),t) \in \mathcal{C}^{2,1}(\mathcal{R}^n \times \mathcal{R}^+;\mathcal{R}^+)$. Accordingly, $T(v(t),t)$ constitutes a real-valued Itô process, and its stochastic differential is formulated as

$$\begin{aligned} dT(v(t),t) =& \mathcal{L}T(v(t),t)dt + T_v(v(t),t)h(t)d\psi(t), \\ \mathcal{L}T(v(t),t) =& T_t(v(t),t) + T_v(v(t),t)f(t) \\ &+ \frac{1}{2}trace(g^T(t)T_{vv}(v(t),t)h(t)), \end{aligned}$$

where $\mathcal{C}^{2,1}(\mathcal{R}^n \times \mathcal{R}^+;\mathcal{R}^+)$ represents the set of all functions with real values $T(v(t),t)$ that possess continuous second derivatives in x and t. Assuming that $T(v(t),t) \in \mathcal{C}^{2,1}(\mathcal{R}^n \times \mathcal{R}^+;\mathcal{R}^+)$, we define

$$\begin{aligned} T_t(v(t),t) &= \frac{\partial T(v(t),t)}{\partial t}, \\ T_v(v(t),t) &= \left(\frac{\partial T(v(t),t)}{\partial v_1},\ldots,\frac{\partial T(v(t),t)}{\partial v_n}\right), \\ T_{vv}(v(t),t) &= \left(\frac{\partial^2 T(v(t),t)}{\partial v_i v_j}\right)_{n\times n}. \end{aligned}$$

Lemma 2. *Consider system (5) and assume the existence of a Lyapunov function $T(x(t),t)$ satisfying*

$$\mathcal{L}T(x(t),t) < 0;$$

then for any non-zero state $x(t) \neq 0$, system (5) exhibits asymptotic mean-square stability according to Definition 1.

3. Main Results

3.1. Event-Triggered Stabilization

Firstly, the designed conventional time-based control procedure is provided for the LPV system (5). The proposition below clarifies the stability condition of LPV system (5) based on Lyapunov stability theory.

Proposition 1. *Provided we have a set of parameter-dependent, continuously differentiable, symmetric positive definite matrices $P(\eta)$, it holds that all parameter trajectories satisfy*

$$\begin{bmatrix} \dot{P}(\eta) + \mathrm{sym}\{P(\eta)M(\eta)\} & N_\psi{}^T(\eta)P(\eta) \\ * & -P(\eta) \end{bmatrix} < 0, \tag{15}$$

Consequently, system (5) is mean-square asymptotically stable.

Proof. Adopt the following candidate Lyapunov function: $T(x,\eta) = x^T(t)P(\eta)x(t)$. In accordance with Itô's formula in Lemma 1,

$$dT(x,\eta) = \mathcal{L}T(x,\eta)dt + 2x^T(t)P(\eta)N_\psi(\eta)x(t)d\psi(t),$$

where

$$\begin{aligned}\mathcal{L}T(x,\eta) =& x^T(t)\dot{P}(\eta)x(t) + 2x^T(t)P(\eta)M(\eta)x(t) \\ &+ x^T(t){N_\psi}^T(\eta)P(\eta)N_\psi(\eta)x(t) \\ =& x^T(t)\left[\dot{P}(\eta) + \mathrm{sym}\{P(\eta)M(\eta)\}\right]x(t) \\ &+ x^T(t)\left[{N_\psi}^T(\eta)P(\eta)N_\psi(\eta)\right]x(t).\end{aligned}$$

Provided that the following inequality stands, then $\mathcal{L}T(x,\eta) < 0$ for any $x(t) \neq 0$:

$$\dot{P}(\eta) + \mathrm{sym}\{P(\eta)M(\eta)\} + {N_\psi}^T(\eta)P(\eta)N_\psi(\eta) < 0, \tag{16}$$

According to Lemma 2, system (5) is mean-square asymptotically stable. By virtue of the Schur complement, inequality (16) is equivalent to inequality (15). The proof is completed. □

Proposition 2. *Given parameter-dependent, continuously differentiable, symmetric positive definite matrices $R(\eta)$ along with matrix $F(\eta)$, the following inequality holds for every parameter trajectory:*

$$\begin{bmatrix} \Pi_{11} & R(\eta){N_\psi}^T(\eta) \\ * & -R(\eta) \end{bmatrix} < 0 \tag{17}$$

where

$$\Pi_{11} \triangleq \mathrm{sym}\{M(\eta)R(\eta) + N(\eta)F(\eta)\} - \sum_{i=1}^{s} \tau_i \frac{\partial R(\eta)}{\eta_i};$$

then under the controller $u(t) = K(\eta)x(t)$ and $K(\eta) = F(\eta)R^{-1}(\eta)$, system (4) is mean-square stable.

Proof. For system (4) under the control law $u(t) = K(\eta)x(t)$, the closed-loop representation of system $K(\eta)$ is given by

$$dx(t) = (M(\eta) + N(\eta)K(\eta))x(t)dt + N_\psi(\eta)x(t)d\psi(t).$$

Multiplying both sides of (16) by $P^{-1}(\eta)$ yields

$$\begin{aligned}&P^{-1}(\eta)\dot{P}(\eta)P^{-1}(\eta) + \mathrm{sym}\left\{P^{-1}(\eta)M(\eta)\right\} \\ &+ P^{-1}(\eta)N_\psi^T(\eta)P(\eta)N_\psi(\eta)P^{-1}(\eta) < 0,\end{aligned} \tag{18}$$

where

$$\dot{P} = \frac{dP}{d\rho} \cdot \frac{d\rho}{dt} = \sum_{i=1}^{s} \tau_i \frac{\partial P(\eta)}{\partial \eta_i}, \quad \tau_i = \frac{d\rho_i}{dt}.$$

Let $R(\eta) = P^{-1}(\eta)$, thus yielding

$$\frac{\partial R(\eta)}{\partial \eta_i} = -R(\eta)\frac{\partial P(\eta)}{\partial \eta_i}R(\eta).$$

Therefore, (18) can be reformulated as

$$\begin{aligned}&-\sum_{i=1}^{s} \tau_i \frac{\partial R(\eta)}{\partial \eta_i} + \mathrm{sym}\{R(\eta)M(\eta)\} \\ &+ R(\eta)N_\psi^T(\eta)P(\eta)N_\psi(\eta)R(\eta) < 0.\end{aligned} \tag{19}$$

Applying the Schur complement , (19) is equivalent to

$$\begin{bmatrix} \mathrm{sym}\{R(\eta)M(\eta)\} - \sum_{i=1}^{s} \tau_i \frac{\partial R(\eta)}{\partial \eta_i} & R(\eta)N_{\psi}^T(\eta) \\ * & -R(\eta) \end{bmatrix} < 0. \tag{20}$$

Considering the closed-loop system governed by the state feedback law $u(t) = K(\eta)x(t)$, (20) is equivalent to

$$\begin{bmatrix} \bar{\Pi}_{11} & R(\eta)N_{\psi}^T(\eta) \\ * & -R(\eta) \end{bmatrix} < 0, \tag{21}$$

where

$$\begin{aligned} \bar{\Pi}_{11} \triangleq & (M(\eta) + N(\eta)K(\eta))R(\eta) \\ & + R(\eta)(N^T(\eta)K^T(\eta) + M^T(\eta)) - \sum_{i=1}^{s} \tau_i \frac{\partial R(\eta)}{\partial \eta_i}. \end{aligned}$$

Let $F(\eta) = K(\eta)R(\eta)$; the result (21), i.e., $\mathcal{L}T(x,\eta) < 0$ for any $x(t) \neq 0$, can be guaranteed by condition (17). The proof is completed. □

Propositions 1 and 2 present the conventional time-based controller design procedure for the considered LPV system without exogenous disturbance. According to the designed procedure, the subsequent theorem presents the event-based control scheme.

Firstly, During the time interval $t \in [t_k, t_{k+1})$, from (8) and (9), the stochastic LPV system in closed-loop form (4) can be equivalently expressed as

$$\begin{aligned} dx(t) = & [M(\eta)x(t) + N(\eta)K(\eta)x(t_k)]dt \\ & + N_{\psi}(\eta)x(t)d\psi(t) \\ = & [(M(\eta) + N(\eta)K(\eta))x(t) - N(\eta)K(\eta)e(t)]dt \\ & + N_{\psi}(\eta)x(t)d\psi(t) \\ = & [\bar{M}(\eta)x(t) + \bar{N}(\eta)e(t)]dt + N_{\psi}x(t)d\psi(t), \end{aligned} \tag{22}$$

where $\bar{M} \triangleq M(\eta) + N(\eta)K(\eta)$ and $\bar{N} \triangleq -N(\eta)K(\eta)$.

On the basis of the resulting control system (22), the subsequent theorem formulates a stability criterion via the Lyapunov functional approach.

Theorem 1. *Consider a closed-loop system (22) with sampling instants determined by (8). For the parameter-varying gain matrices $K(\eta)$ given by (17), assume the existence of a parameter-dependent family of symmetric positive definite matrices $\bar{P}(\eta)$, which satisfy*

$$\begin{aligned} \bar{Q}(\eta) \triangleq & \dot{\bar{P}}(\eta) + \mathrm{sym}\{\bar{P}(\eta)\bar{M}(\eta)\} + N_{\psi}^T(\eta)\bar{P}(\eta)N_{\psi}(\eta) \\ & + \bar{P}(\eta)\bar{N}(\eta)\bar{N}^T(\eta)\bar{P}(\eta) < 0, \end{aligned} \tag{23}$$

consequently, the state expectation of system (22) exhibits global uniform ultimate boundedness and tends to converge within a bounded region.

$$\mathcal{N}(\varepsilon) = \left\{ x(t) \middle| Z\{\|x(t)\|\} \leq \sqrt{\frac{\varepsilon}{\delta \lambda_{\min}(\bar{P}(\eta))}} \right\}, \tag{24}$$

where $\delta = \frac{\lambda_{min}(-\bar{Q}(\eta))}{\lambda_{max}(\bar{P}(\eta))}$.

Proof. Considering the candidate Lyapunov function $T(t) = x^T(t)\bar{P}(\eta)x(t)$, and assuming that the solution to (23) is the parameter-varying symmetric positive definite matrix $\bar{P}(\eta)$, then according to Lemma 1, we determine that

$$
\begin{aligned}
\mathcal{L}T(x,\eta) &= 2x^T(t)\bar{P}(\eta)(\bar{M}(\eta)x(t) + \bar{N}(\eta)e(t)) \\
&\quad + x^T(t)\dot{\bar{P}}(\eta)x(t) + x^T(t)N_\psi^T(\eta)\bar{P}(\eta)N_\psi(\eta)x(t) \\
&= x^T(t)\left(\text{sym}\{\bar{P}(\eta)\bar{M}(\eta)\} + N_\psi^T(\eta)\bar{P}(\eta)N_\psi(\eta)\right)x(t) \\
&\quad + x^T(t)\left(\dot{\bar{P}}(\eta)\right)x(t) + x^T(t)\bar{P}(\eta)\bar{N}(\eta)e(t) \\
&\quad + e^T(t)\bar{N}^T(\eta)\bar{P}(\eta)x(t) \\
&= x^T(t)\left(\text{sym}\{\bar{P}(\eta)\bar{M}(\eta)\} + N_\psi^T(\eta)\bar{P}(\eta)N_\psi(\eta)\right)x(t) \\
&\quad + x^T(t)\left(\dot{\bar{P}}(\eta) + \bar{P}(\eta)\bar{N}(\eta)\bar{N}^T(\eta)\bar{P}(\eta)\right)x(t) \\
&\quad - \left\|e(t) - \bar{N}^T(\eta)\bar{P}(\eta)x(t)\right\|^2 + \|e(t)\|^2 \\
&\leq -\lambda_{\min}(-\bar{Q}(\eta))\|x(t)\|^2 + \|e(t)\|^2.
\end{aligned}
$$

Considering the sampling instants specified in (8), for any $t \in [t_k, t_{k+1})$, $\|e(t)\| \leq \sqrt{\varepsilon}$ holds. Thus, as long as $t \in [t_k, t_{k+1})$, it follows that

$$
\begin{aligned}
\mathcal{L}T(x,\eta) &\leq -\lambda_{\min}(-\bar{Q}(\eta))\|x(t)\|^2 + \varepsilon \\
&\leq -\frac{\lambda_{\min}(-\bar{Q}(\eta))}{\lambda_{\max}(\bar{P}(\eta))}x^T(t)\bar{P}(\eta)x(t) + \varepsilon \\
&= -\delta T(x,\eta) + \varepsilon;
\end{aligned}
$$

thus, it follows that

$$
\begin{aligned}
dT(x,\eta) &= \mathcal{L}T(x,\eta)dt + 2x^T(t)\bar{P}(\eta)N_\psi(\eta)x(t)d\psi(t) \\
&\leq (-\delta T(x,\eta) + \varepsilon)dt + 2x^T(t)\bar{P}(\eta)N_\psi(\eta)x(t)d\psi(t).
\end{aligned}
$$

Consider

$$
\begin{aligned}
de^{\delta t}T(x,\eta) &= \delta e^{\delta t}T(x,\eta)dt + e^{\delta t}dT(x,\eta) \\
&\leq \delta e^{\delta t}T(x,\eta)dt + e^{\delta t}(-\delta T(x,\eta) + \varepsilon)dt \\
&\quad + 2e^{\delta t}x^T(t)N_\psi(\eta)\bar{P}(\eta)x(t)d\psi(t) \\
&= e^{\delta t}\varepsilon dt + 2e^{\delta t}x^T(t)N_\psi(\eta)\bar{P}(\eta)x(t)d\psi(t).
\end{aligned} \tag{25}
$$

Upon performing integration of both sides of (25) between 0 and t, and computing the expectation of the result, the following can be obtained:

$$
Z\left\{e^{\delta t}T(x,\eta) - T(x(0),\eta(0))\right\} \leq \int_0^t e^{\delta s}\varepsilon ds,
$$

which indicates that

$$
\begin{aligned}
Z\{T(x,\eta)\} &\leq e^{-\delta t}(T(x(0),\eta(0)) + \int_0^t e^{\delta s}\varepsilon ds) \\
&= e^{-\delta t}T(x(0),\eta(0)) + \varepsilon\int_0^t e^{-\delta(t-s)}ds \\
&= e^{-\delta t}[T(x(0),\eta(0)) - \frac{\varepsilon}{\delta}] + \frac{\varepsilon}{\delta}.
\end{aligned}
$$

Thus, it can be concluded that, as $t \to \infty$, the closed-loop system (22) is globally uniformly ultimately bounded, as guaranteed by [23], with state trajectories approaching the bounded region specified in (24). The proof is completed. □

Remark 1. *To solve $\bar{P}(\eta)$ in* (23)*, Schur complement transformations should be operated to transform* (23) *into an LMI as follows:*

$$\begin{bmatrix} \Pi_{11} & N_{\psi}^{T}\bar{P}(\eta) & \bar{P}(\eta)\bar{N}(\eta) \\ * & -\bar{P}(\eta) & 0 \\ * & * & -I \end{bmatrix} < 0,$$

where $\Pi_{11} = \sum_{i=1}^{s} \tau_i \frac{\partial \bar{P}(\eta)}{\eta_i} + \mathrm{sym}\{\bar{P}(\eta)\bar{M}(\eta)\}$.

3.2. Event-Triggered $\mathcal{H}_\infty$ Control

Similarly, the designed conventional time-based $\mathcal{H}_\infty$ control procedure is firstly provided for an LPV system (3) in the following propositions for further synthesis of an event-triggered controller.

Proposition 3. *The stochastic LPV system (3) is robustly stabilized despite disturbance attenuation γ. Given the existence of a continuously differentiable, parameter-dependent family of symmetric positive definite matrices $P(\eta)$ for all parameter trajectories, the inequality given below is satisfied:*

$$\begin{bmatrix} \Pi_{11} & P(\eta)N_v(\eta) & C^T(\eta) & N_{\psi}{}^{T}(\eta)P(\eta) \\ * & -\gamma^2 I & D^T(\eta) & 0 \\ * & * & -I & 0 \\ * & * & * & -P(\eta) \end{bmatrix} < 0, \tag{26}$$

where $\Pi_{11} \triangleq P(\eta)M(\eta) + M^T(\eta)P(\eta) + \sum_{i=1}^{s} \tau_i \frac{\partial P(\eta)}{\eta_i}$.

Proof. Set the candidate Lyapunov function to $T(x,\eta) = x^T(t)P(\eta)x(t)$. Without loss of generality, suppose the initial state of the system is zero, i.e., $x(t) = 0$, when $t = 0$. It follows from Itô's formula that

$$Z\{T(x(t),t)\} = Z\Big\{\int_0^t \mathcal{L}T(x(s),s)\Big\}ds,$$

Let

$$\begin{aligned} J(t) &= Z\Big\{\int_0^t \big[y^T(s)y(s) - \gamma^2 v^T(s)v(s)\big]ds\Big\} \\ &= Z\Big\{\int_0^t \big[y^T(s)y(s) - \gamma^2 v^T(s)v(s) + \mathcal{L}T(x(s),s)\big]ds\Big\} \\ &\quad - Z\{T(x(t),t)\} \\ &\leq Z\Big\{\int_0^t \big[y^T(s)y(s) - \gamma^2 v^T(s)v(s) + \mathcal{L}T(x(s),s)\big]ds\Big\} \\ &= Z\left\{\int_0^t \begin{bmatrix} x(s) \\ v(s) \end{bmatrix}^T \bar{\Pi} \begin{bmatrix} x(s) \\ v(s) \end{bmatrix} ds\right\}, \end{aligned}$$

where $\bar{\Pi} \triangleq \begin{bmatrix} \bar{\Pi}_{11} & P(\eta)N_v(\eta) \\ * & -\gamma^2 I \end{bmatrix} + \begin{bmatrix} C^T(\eta) \\ D^T(\eta) \end{bmatrix} \begin{bmatrix} C(\eta) \\ D(\eta) \end{bmatrix}^T$, with $\bar{\Pi}_{11} = \mathrm{sym}\{P(\eta)M(\eta)\} + \sum_{i=1}^{s} \tau_i \frac{\partial P(\eta)}{\eta_i} + N_\psi{}^T(\eta)P(\eta)N_\psi(\eta)$. With the Schur complement, condition (26) guarantees that $\bar{\Pi} < 0$, and consequently, $J(t) < 0$, meaning $y^T(t)y(t) < \gamma^2 v^T(t)v(t)$. Thus, (3) exhibits robust stability according to Definition 2. The proof is completed. □

Proposition 4. *Assume the existence of a parameter-dependent family of symmetric positive definite matrices, $F(\eta)$ and $R(\eta)$; the state feedback controller $K(\eta)$ ensures the robust stabilization of the stochastic LPV system (2) with disturbance attenuation γ, where the following inequality is satisfied under all possible parameter evolutions:*

$$\begin{bmatrix} \Pi_{11} & N_v(\eta) & R(\eta)C^T & R(\eta)N_\psi{}^T(\eta) \\ * & -\gamma^2 I & D^T & 0 \\ * & * & -I & 0 \\ * & * & * & -R(\eta) \end{bmatrix} < 0, \tag{27}$$

where $\Pi_{11} \triangleq \mathrm{sym}\{M(\eta)R(\eta) + N(\eta)F(\eta)\} - \sum_{i=1}^{s} \tau_i \frac{\partial R(\eta)}{\eta_i}$, *and* $K(\eta) = F(\eta)R^{-1}(\eta)$.

Proof. Take system (2) with $u(t) = K(\eta)x(t)$, and replace $M(\eta)$ in (26) with $M(\eta) + N(\eta)K(\eta)$, with the congruence transformation matrix diag $\{P^{-1}(\eta), I, I, P^{-1}(\eta)\}$, and apply it to (26), thus obtaining

$$\begin{bmatrix} \bar{\Pi}_{11} & N_v(\eta) & P^{-1}(\eta)C^T & P^{-1}(\eta)N_\psi{}^T(\eta) \\ * & -\gamma^2 I & D^T & 0 \\ * & * & -I & 0 \\ * & * & * & -P^{-1}(\eta) \end{bmatrix} < 0, \tag{28}$$

where $\bar{\Pi}_{11} \triangleq \mathrm{sym}\{(M(\eta) + N(\eta)K(\eta))P^{-1}(\eta)\} + P^{-1}(\eta)\left(\sum_{i=1}^{s} \tau_i \frac{\partial P(\eta)}{\eta_i}\right)P^{-1}(\eta)$. Let $R(\eta) = P^{-1}(\eta)$ and define $F(\eta) = K(\eta)P^{-1}(\eta)$; substituting these expressions into the transformed inequality yields (27), with $K(\eta) = F(\eta)R^{-1}(\eta)$. The proof is completed. □

To improve the feasibility of the LMIs in Propositions 3 and 4, slack matrices can be introduced, which leads to following proposition.

Proposition 5. *Assume the existence of a parameter-dependent family of symmetric positive definite matrices, $X(\eta)$ and $\bar{P}(\eta)$, and matrices $R(\eta)$ and T, the state feedback controller $K(\eta)$ ensures the robust stabilization of the stochastic LPV system (2) with disturbance attenuation γ, and* (29) *is satisfied for all parameter trajectories.*

$$\begin{bmatrix} \Pi_{11} & \Pi_{12} & N_v(\eta) & 0 & 0 & T \\ * & \Pi_{22} & 0 & T^T C^T(\eta) & T^T N_\psi{}^T(\eta) & 0 \\ * & * & -\gamma^2 I & D^T(\eta) & 0 & 0 \\ * & * & * & -I & 0 & 0 \\ * & * & * & * & -\bar{P}(\eta) & 0 \\ * & * & * & * & * & -X(\eta) \end{bmatrix} < 0, \tag{29}$$

where $\Pi_{11} \triangleq -T - T^T, \Pi_{12} \triangleq X(\eta) + M(\eta)T + N(\eta)R(\eta), \Pi_{22} \triangleq -X(\eta) + \sum_{i=1}^{s} \tau_i \frac{\partial X(\eta)}{\eta_i}$, *and* $K(\eta) = R(\eta)T^{-1}(\eta)$.

Proof. First, the robust $\mathcal{H}_\infty$ stability of the autonomous system (3) is established through a derived sufficient condition. Consider (30):

$$\begin{bmatrix} \hat{\Pi}_{11} & \hat{\Pi}_{12} & W^T N_v(\eta) & 0 & 0 & W^T \\ * & \hat{\Pi}_{22} & 0 & C^T(\eta) & {N_\psi}^T(\eta) & 0 \\ * & * & -\gamma^2 I & D^T(\eta) & 0 & 0 \\ * & * & * & -I & 0 & 0 \\ * & * & * & * & -P^{-1}(\eta) & 0 \\ * & * & * & * & * & -P(\eta) \end{bmatrix} < 0, \tag{30}$$

where $\hat{\Pi}_{11} \triangleq -W - W^T, \hat{\Pi}_{12} \triangleq P(\eta) + W^T M(\eta), \hat{\Pi}_{22} \triangleq -P(\eta) + \sum_{i=1}^{s} \tau_i \frac{\partial P(\eta)}{\eta_i}$. $P(\eta)$ is a continuous differentiable, parameter-dependent, symmetric positive matrix, and W is a general matrix of proper dimensions. We will demonstrate that if $P(\eta)$ and W are feasible, under disturbance attenuation γ, system (3) is robustly stable.

Inequality (30) is equivalent to (31):

$$\begin{bmatrix} 0 & P(\eta) & 0 & 0 & 0 & 0 & 0 \\ * & \Pi_{22} & 0 & 0 & C^T(\eta) & {N_\psi}^T(\eta) & 0 \\ * & * & \Pi_{33} & 0 & C_h^T(\eta) & 0 & 0 \\ * & * & * & -\gamma^2 I & D^T(\eta) & 0 & 0 \\ * & * & * & * & -I & 0 & 0 \\ * & * & * & * & * & -P^{-1}(\eta) & 0 \\ * & * & * & * & * & * & -P(\eta) \end{bmatrix} + \begin{bmatrix} -I \\ M^T(\eta) \\ M_h^T(\eta) \\ N_v^T(\eta) \\ 0 \\ 0 \\ I \end{bmatrix} W(\eta) \begin{bmatrix} I \\ 0 \\ 0 \\ 0 \\ 0 \\ 0 \\ 0 \end{bmatrix}^T + \begin{bmatrix} -I \\ M^T(\eta) \\ M_h^T(\eta) \\ N_v^T(\eta) \\ 0 \\ 0 \\ I \end{bmatrix}^T W^T(\eta) \begin{bmatrix} I \\ 0 \\ 0 \\ 0 \\ 0 \\ 0 \\ 0 \end{bmatrix} < 0. \tag{31}$$

for which $\begin{bmatrix} -I & M(\eta) & N_v(\eta) & 0 & 0 & I \end{bmatrix}$ and $\begin{bmatrix} I & 0 & 0 & 0 & 0 & 0 \end{bmatrix}$, and the null spaces are, respectively,

$$\begin{bmatrix} M(\eta) & N_v(\eta) & 0 & 0 & I \\ I & 0 & 0 & 0 & 0 \\ 0 & I & 0 & 0 & 0 \\ 0 & 0 & I & 0 & 0 \\ 0 & 0 & 0 & I & 0 \\ 0 & 0 & 0 & 0 & I \end{bmatrix} \text{and} \begin{bmatrix} 0 & 0 & 0 & 0 & 0 \\ I & 0 & 0 & 0 & 0 \\ 0 & I & 0 & 0 & 0 \\ 0 & 0 & I & 0 & 0 \\ 0 & 0 & 0 & I & 0 \\ 0 & 0 & 0 & 0 & I \end{bmatrix}.$$

Based on the Projection Lemma [24], it can be determined that

$$\begin{bmatrix} \check{\Pi}_{11} & P(\eta) N_v(\eta) & C^T(\eta) & {N_\psi}^T(\eta) & P(\eta) \\ * & -\gamma^2 I & D^T(\eta) & 0 & 0 \\ * & * & -I & 0 & 0 \\ * & * & * & -P^{-1}(\eta) & 0 \\ * & * & * & * & -P(\eta) \end{bmatrix} < 0, \tag{32}$$

where $\check{\Pi}_{11} \triangleq \sum_{i=1}^{s} \tau_i \frac{\partial P(\eta)}{\eta_i} + P(\eta)M(\eta) + M^T(\eta)P(\eta) - P(\eta)$. Inequality (32) can be transformed into (26) via Schur complement transformation, thereby demonstrating that (30) guarantees the robust stability of system (3).

Now, take (2) with $u(t) = K(\eta)x(t)$, and replace $M(\eta)$ in (30) with $M(\eta) + N(\eta)K(\eta)$, followed by a congruence transformation via diag$\{W^{-1}, W^{-1}, I, I, I, W^{-1}\}$; then, (33) can be obtained:

$$\begin{bmatrix} \bar{\Pi}_{11} & \bar{\Pi}_{12} & N_v(\eta) & 0 & 0 & W^{-1} \\ * & \bar{\Pi}_{22} & 0 & W^{-T}C^T(\eta) & 0 & 0 \\ * & * & -\gamma^2 I & D^T(\eta) & 0 & 0 \\ * & * & * & -I & 0 & 0 \\ * & * & * & * & -P^{-1}(\eta) & 0 \\ * & * & * & * & * & \bar{\Pi}_{66} \end{bmatrix} < 0, \tag{33}$$

where $\bar{\Pi}_{11} \triangleq -W^{-1} - W^{-T}$, $\bar{\Pi}_{22} \triangleq W^{-T}\left(-P(\eta) + \sum_{i=1}^{s} \tau_i \frac{\partial P(\eta)}{\eta_i}\right)W^{-1}$, $\bar{\Pi}_{12} \triangleq W^{-T}P(\eta)W^{-1} + (M(\eta) + N(\eta)K(\eta))W^{-1}$, $\bar{\Pi}_{66} \triangleq -W^{-T}P(\eta)W^{-1}$. Suppose matrices $X(\eta) = W^{-T}P(\eta)W^{-1}$, $\bar{P}(\eta) = P^{-1}(\eta)$, $R(\eta) = K(\eta)W^{-1}$, and $T = W^{-1}$; then, (29) is obtained, and $K(\eta) = R(\eta)T^{-1}$. The proof is completed. □

Similarly to the modeled system (22), during the time span $t \in [t_k, t_{k+1})$, based on (8) and (9), the stochastic LPV system in closed-loop form (2) can be rewritten as

$$\begin{aligned} dx(t) =& [\bar{M}(\eta)x(t) + \bar{N}(\eta)e(t) + N_\rho(\eta)\rho(t)]dt \\ &+ N_\psi x(t)d\psi(t), \end{aligned} \tag{34}$$

where $\bar{M} \triangleq M(\eta) + N(\eta)K(\eta)$, $\bar{N} \triangleq -N(\eta)K(\eta)$.

Referencing the design procedure, the following theorem provides a uniformly ultimately bounded stability criterion for the resulting event-triggered control system (34). It provides a bounded region describing convergence for the overall event-based control system.

Theorem 2. *Take the closed-loop system (34) with sampling instants specified by (8). For the parameter-varying gain matrices $K(\eta)$ given by (29), assuming the existence of a family of parameter-varying symmetric positive definite matrices $\bar{P}(\eta)$ that satisfy*

$$\begin{aligned} \bar{Q}(\eta) \triangleq& \dot{\bar{P}}(\eta) + \bar{M}^T(\eta)\bar{P}(\eta) + \bar{P}(\eta)\bar{M}(\eta) \\ &+ N_\psi^T(\eta)\bar{P}(\eta)N_\psi(\eta) + \bar{P}(\eta)\bar{N}(\eta)N^T(\eta)\bar{P}(\eta) \\ &+ \bar{P}(\eta)N_\rho(\eta)N_\rho^T(\eta)\bar{P}(\eta) < 0, \end{aligned} \tag{35}$$

the LPV stochastic system (34) exhibits robust stability with disturbance attenuation γ; furthermore, the state expectation satisfies global uniform ultimate boundedness and converges to the bounded region

$$\mathcal{N}(\varepsilon) = \left\{ x(t) \middle| Z\{\|x(t)\|\} \leq \sqrt{\frac{\varepsilon + \bar{\rho}^2}{\delta \lambda_{\min}(\bar{P}(\eta))}} \right\}, \tag{36}$$

where $\delta = \frac{\lambda_{min}(-\bar{Q}(\eta))}{\lambda_{max}(\bar{P}(\eta))}$, and $\bar{\rho}$ is the upper bound of external disturbance.

Proof. With $K(\eta)$ given by (29), the $\mathcal{H}_\infty$ stability of the dynamical system is achieved. Next, we shall prove that the system states will converge to a bounded region. Considering the candidate Lyapunov function $T(t) = x^T(t)\bar{P}(\eta)x(t)$, it follows from Lemma 1 that

$$
\begin{aligned}
&\mathcal{L}T(x,\eta)\\
&= x^T(t)\dot{\bar{P}}(\eta)x(t) + x^T(t)N_\psi^T(\eta)\bar{P}(\eta)N_\psi(\eta)x(t)\\
&+ 2x^T(t)\bar{P}(\eta)(\bar{M}(\eta)x(t) + \bar{N}(\eta)e(t) + N_\rho(\eta)\rho(t))\\
&= x^T(t)\Big(\mathrm{sym}\{\bar{P}(\eta)\bar{M}(\eta)\} + N_\psi^T(\eta)\bar{P}(\eta)N_\psi(\eta)\Big)x(t)\\
&x^T(t)\Big(\dot{\bar{P}}(\eta)\Big)x(t) + \mathrm{sym}\Big\{x^T(t)\bar{P}(\eta)\bar{N}(\eta)e(t)\Big\}\\
&+ \mathrm{sym}\Big\{x^T(t)\bar{P}(\eta)N_\rho(\eta)\rho(t)\Big\}\\
&= x^T(t)\bar{Q}(\eta)x(t) - \Big\|e(t) - \bar{N}^T(\eta)\bar{P}(\eta)x(t)\Big\|^2\\
&+ \|e(t)\|^2 - \Big\|\rho(t) - \bar{N}_\rho^T(\eta)\bar{P}(\eta)x(t)\Big\|^2 + \|\rho(t)\|^2\\
&\leq -\lambda_{\min}(-\bar{Q}(\eta))\|x(t)\|^2 + \|e(t)\|^2 + \|\rho(t)\|^2.
\end{aligned}
$$

For every $t \in [t_k, t_{k+1})$, $\|e(t)\| \leq \sqrt{\varepsilon}$ holds; thus, it follows that

$$
\begin{aligned}
\mathcal{L}T(x,\eta) &\leq -\lambda_{\min}(-\bar{Q}(\eta))\|x(t)\|^2 + \varepsilon + \bar{\rho}^2\\
&\leq -\frac{\lambda_{\min}(-\bar{Q}(\eta))}{\lambda_{\max}(P(\eta))}\chi^T(t)\bar{P}(\eta)\chi(t) + \varepsilon + \bar{\rho}^2\\
&= -\delta T(\chi,\eta) + \varepsilon + \bar{\rho}^2;
\end{aligned}
$$

thus, it can be determined that

$$
\begin{aligned}
dT(x,\eta) &= \mathcal{L}T(x,\eta)dt + 2x^T(t)\bar{P}(\eta)N_\psi(\eta)x(t)d\psi(t)\\
&\leq (-\delta T(x,\eta) + \varepsilon + \bar{\rho}^2)dt\\
&\quad + 2x^T(t)\bar{P}(\eta)N_\psi(\eta)x(t)d\psi(t).
\end{aligned}
$$

Consider

$$
\begin{aligned}
de^{\delta t}T(x,\eta) &= \delta e^{\delta t}T(x,\eta)dt + e^{\delta t}dT(x,\eta)\\
&\leq \delta e^{\delta t}T(x,\eta)dt + e^{\delta t}(-\delta T(x,\eta) + \varepsilon + \bar{\rho}^2)dt\\
&\quad + 2e^{\delta t}x^T(t)\bar{P}(\eta)N_\psi(\eta)x(t)d\psi(t)\\
&= e^{\delta t}\varepsilon dt + e^{\delta t}\bar{\rho}^2 dt\\
&\quad + 2e^{\delta t}x^T(t)\bar{P}(\eta)N_\psi(\eta)x(t)d\psi(t);
\end{aligned}
$$

thus,

$$
Z\Big\{e^{\delta t}T(x,\eta) - T(x(0),\eta(0))\Big\} \leq \int_0^t e^{\delta s}(\varepsilon + \bar{\rho}^2)ds,
$$

which indicates that

$$
\begin{aligned}
Z\{T(x,\eta)\} &\leq e^{-\delta t}\Big(\int_0^t e^{\delta s}(\varepsilon + \bar{\rho}^2)ds + T(x(0),\eta(0))\Big)\\
&= (\varepsilon + \bar{\rho}^2)\int_0^t e^{-\delta(t-s)}ds + e^{-\delta t}T(x(0),\eta(0))\\
&= e^{-\delta t}\Big[T(x(0),\eta(0)) - \frac{(\varepsilon + \bar{\rho}^2)}{\delta}\Big] + \frac{(\varepsilon + \bar{\rho}^2)}{\delta}.
\end{aligned}
$$

Thus, it can be seen that, as $t \to \infty$, the closed-loop system (34) is shown to be globally uniformly ultimately bounded. Additionally, the states converge to the bounded region specified in (36). The proof is completed. □

Remark 2. To solve $\bar{P}(\eta)$ in (35), Schur complement transformations should be operated to transform (35) to an LMI as follows:

$$\begin{bmatrix} \Pi_{11} & N_{\psi}^T\bar{P}(\eta) & \bar{P}(\eta)\bar{N}(\eta) & \bar{P}(\eta)N_{\rho}(\eta) \\ * & -\bar{P}(\eta) & 0 & 0 \\ * & * & -I & 0 \\ * & * & * & -I \end{bmatrix} < 0,$$

where $\Pi_{11} = \sum_{i=1}^{s} \tau_i \frac{\partial \bar{P}(\eta)}{\eta_i} + \text{sym}\{\bar{P}(\eta)\bar{M}(\eta)\}$.

4. Simulation Results

The simulation employs the grid and inverter parameters $v_d = 400$ V, $v_d = 0$ V, $V_i = 750$ V, $L = 0.002$ H, $r = 0.004\ \Omega$, and $\psi = 314$ rad/s, which are typical values for grid-connected operation. The LPV inverter model (2) is characterized by the following matrices:

$$\begin{aligned} M(\eta) &= \begin{bmatrix} -\frac{r}{L} + 0.1\rho_1(t) & -\psi + 0.2\rho_2(t) \\ \psi - 0.2\rho_2(t) & -\frac{r}{L} + 0.1\rho_1(t) \end{bmatrix}, \\ N(\eta) &= \begin{bmatrix} \frac{1}{L} + 0.1\rho_1(t) \\ \frac{1}{L} + 0.1\rho_1(t) \end{bmatrix}, N_v(\eta) = \begin{bmatrix} 0.2 + 0.1\rho_1(t) \\ 0.1 + 0.1\rho_2(t) \end{bmatrix}, \\ N_{\psi}(\eta) &= \begin{bmatrix} 10 + 0.1\rho_1(t) & 2 + 0.2\rho_1(t) \\ 3 & 5 \end{bmatrix}, \\ C(\eta) &= \begin{bmatrix} 1.4 + 0.2\rho_1(t) + 0.1\rho_2(t) \\ 1.2 \end{bmatrix}^T, \\ D(\eta) &= 1.5 + 0.3\rho_1(t) + 0.3\rho_2(t), \end{aligned}$$

where the time-varying parameters $\eta_1(t) = \sin(t)$ and $\eta_2(t) = |\cos 5t|$ satisfy $\eta_1(t) \in [-1, 1]$, $\eta_2(t) \in [0, 1]$, $\tau_1 \in [-1, 1]$, and $\tau_2 \in [-5, 5]$.

To carry out simulations, the basis functions $f_1(\eta) = 1, f_2(\eta) = \eta_1(t), f_3(\eta) = \eta_2(t)$ are used to approximate the parameter-dependent matrices; therefore, $\mathcal{X}(\eta) = \sum_{j=1}^{3} f_j(\eta)\mathcal{X}_j$, where $\mathcal{X}(\eta)$ represents the parameter-dependent matrices appearing in this paper. Therefore, to solve a matrix variable $\mathcal{X}(\eta)$ is to solve $\mathcal{X}_j (j = 1, 2, 3)$. The event-triggering threshold is set at $\varepsilon = 0.05$.

Case i: Event-triggered Stabilization

Design an event-triggered controller $K(\eta)$ that guarantees system (4) is asymptotically mean-square stable. Based on Theorem 1, it can be determined that

$$\begin{aligned} R_1 &= 10^{-9} \begin{bmatrix} 0.0605 & -0.0270 \\ -0.0270 & 0.3260 \end{bmatrix}, F_1 = 10^{-10} \begin{bmatrix} -0.0624 \\ -0.1241 \end{bmatrix}^T, \\ R_2 &= 10^{-10} \begin{bmatrix} 0.0079 & 0.0003 \\ 0.0003 & -0.2444 \end{bmatrix}, F_2 = 10^{-12} \begin{bmatrix} -0.0300 \\ 0.3625 \end{bmatrix}^T, \\ R_3 &= 10^{-11} \begin{bmatrix} -0.0892 & -0.1508 \\ -0.1508 & -0.0635 \end{bmatrix}, F_3 = 10^{-13} \begin{bmatrix} 0.8140 \\ 0.2516 \end{bmatrix}^T, \end{aligned}$$

The obtained controller is as follows:

$$K_1 = \begin{bmatrix} -0.1246 \\ -0.0484 \end{bmatrix}^T, K_2 = \begin{bmatrix} -0.0373 \\ -0.0149 \end{bmatrix}^T, K_3 = \begin{bmatrix} 0.0081 \\ -0.0588 \end{bmatrix}^T.$$

Throughout the time period $t \in [t_k, t_{k+1})$, the control signal is

$$u(t) = \begin{bmatrix} -0.1246 - 0.0373\eta_1(t) + 0.0081\eta_2(t) \\ -0.0484 - 0.0149\eta_1(t) - 0.0588\eta_2(t) \end{bmatrix}^T x(t_k).$$

Set the initial state to $x(t_0) = [-1 \quad 1]^T$. The state response of the closed-loop system is illustrated in Figure 3, from which it can be observed that the state converges to a bounded region. This means the system is stabilized with the event-triggered controller. The control input signal is shown in Figure 4, the triggering error norm is presented in Figure 5, and the intervals between consecutive events are shown in Figure 6.

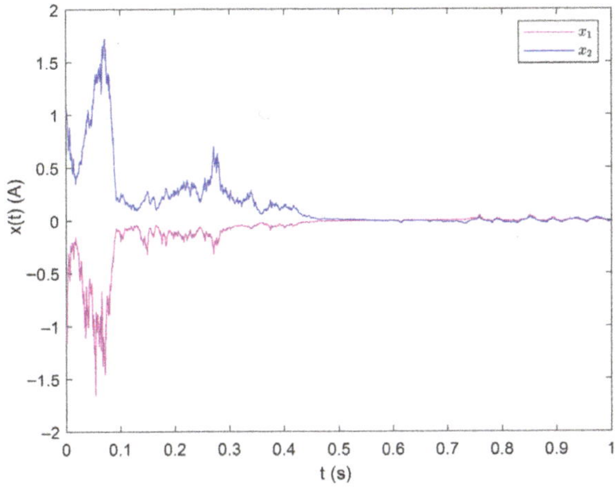

Figure 3. State response $x(t)$ under Case i.

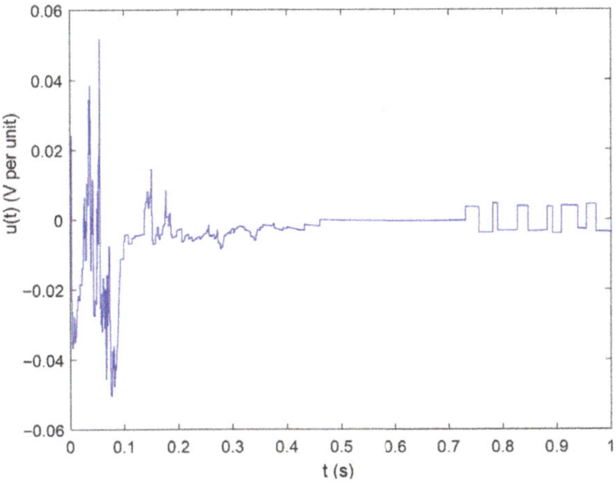

Figure 4. Control input $u(t)$ under Case i.

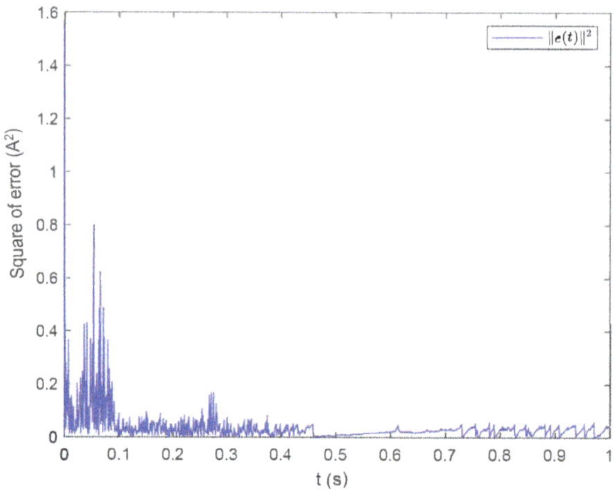

Figure 5. Trajectory of $\|e(t)\|^2$ under Case i.

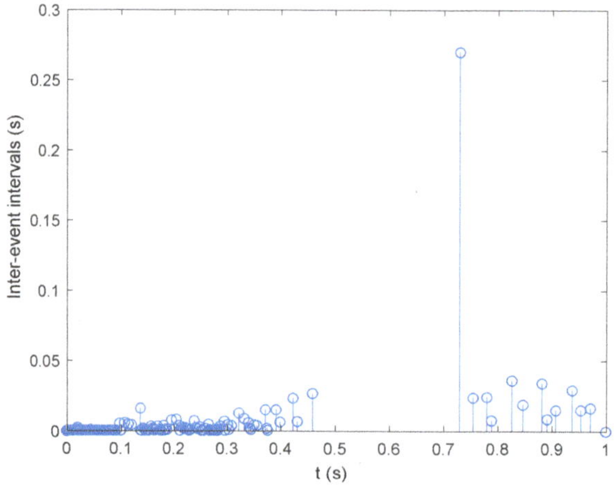

Figure 6. Event intervals under Case i.

Case ii: Event-triggered $\mathcal{H}_\infty$ Control

The disturbance signal $\rho(t) = 1/1 + 0.5t^2$ is introduced.The design objective is to construct an event-triggered controller $K(\eta)$ to stabilize system (2) and meet the disturbance attenuation criterion γ. According to Theorem 2, it can be determined that

$$T = \begin{bmatrix} 43.9114 & -22.8416 \\ -25.6409 & 57.2889 \end{bmatrix}, R_1 = \begin{bmatrix} -8.4817 \\ -10.3793 \end{bmatrix}^T,$$
$$R_2 = \begin{bmatrix} -0.3749 \\ -0.3681 \end{bmatrix}^T, R_3 = \begin{bmatrix} 0.0066 \\ -0.0226 \end{bmatrix}^T, K_1 = \begin{bmatrix} -0.3897 \\ -0.3365 \end{bmatrix}^T,$$
$$K_2 = \begin{bmatrix} -0.0160 \\ -0.0128 \end{bmatrix}^T, K_3 = 10^{-3} \begin{bmatrix} -0.1032 \\ -0.4348 \end{bmatrix}^T;$$

thus, over the period $t \in [t_k, t_{k+1})$, the control signal is

$$u(t) = \begin{bmatrix} -0.3897 - 0.0160\eta_1(t) - 10^{-3} \cdot 0.1032\eta_2(t) \\ -0.3365 - 0.0128\eta_1(t) - 10^{-3} \cdot 0.4348\eta_2(t) \end{bmatrix}^T x(t_k).$$

Set the initial state at $x(t_0) = [-1 \quad 1]^T$. With this controller, the system is stabilized, as demonstrated in Figure 7. Figure 8 shows the control input signal. while Figures 9 and 10 display the error norm and event intervals, respectively.

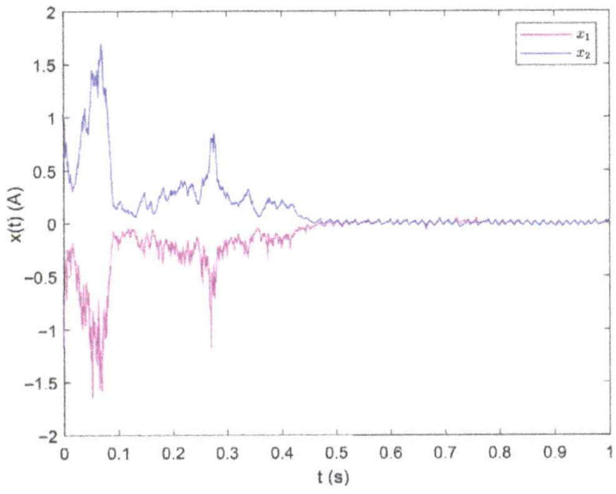

Figure 7. State response $x(t)$ under Case ii.

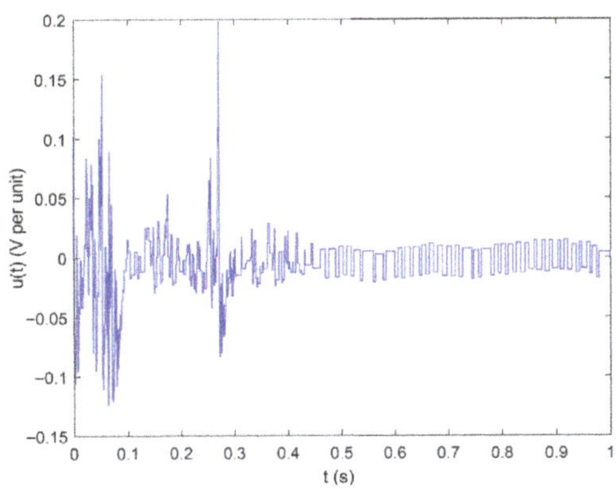

Figure 8. Control input $u(t)$ under Case ii.

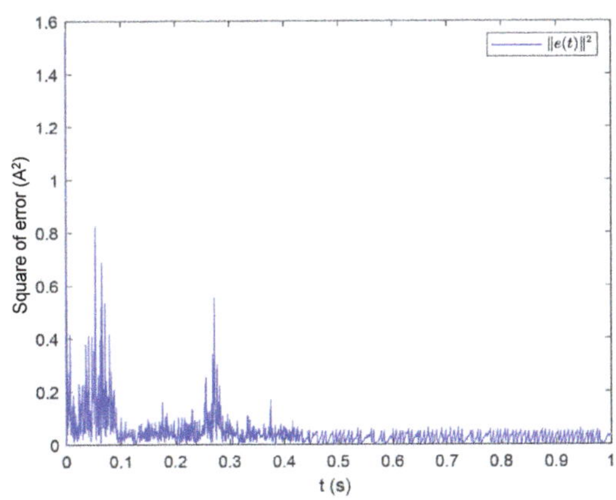

Figure 9. Trajectory of $\|e(t)\|^2$ under Case ii.

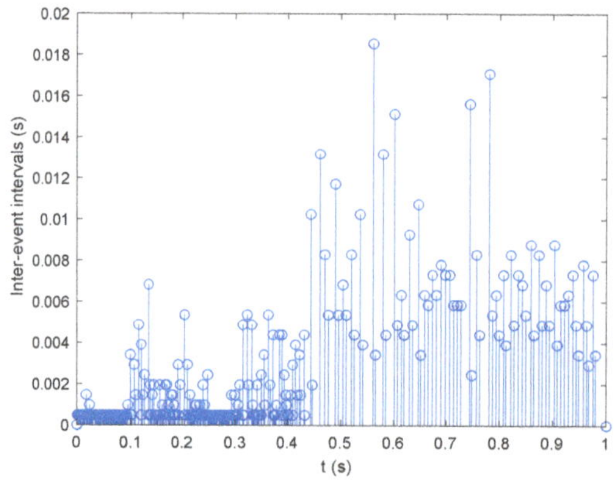

Figure 10. Event intervals under Case ii.

5. Discussion

The aim of the simulation for Case i is to verify the effectiveness of the proposed event-triggered controller and the correctness of the theoretical deduction. The inverter states converge to equilibrium under the proposed controller, and the control signal is a discrete one, which complies with the event-triggering mechanism. Moreover, the average control interval is 0.04 s, which saves 400 control cycles for the 10 kHz digital control cycle. The aim of the simulation for Case ii is to further test the robustness of the system when disturbed by a time-varying signal. The simulation results show that the system stability is maintained, and the control output also follows a event-triggering pattern. In this case, the average control interval is 0.006 s, which saves 60 control cycles for the 10 kHz digital control cycle. Moreover, due to the disturbance, the event intervals are narrower.

6. Conclusions

An event-triggered controller for grid-connected inverters is designed within the frame of a linear parameter-varying (LPV) system, which takes into consideration the time-varying feature of inverter parameters and the stochasticity of electromagnetic noise. A stochastic LPV model for three-phase two-level inverters is established, for which an event trigger with a continuous-time form is designed with a predesignated event-triggering threshold. This mechanism is typical for modern grid control frames, and has a lower computational cost than the continuous control method. The sufficient conditions of the uniform ultimate bounded stability are obtained for the resulting event-triggered LPV system subject to exogenous disturbance. The simulation results show that the inverter system is stabilized, with minor performance degradation, while the computational burden and communication density are reduced.

Author Contributions: Conceptualization, W.L.; methodology, W.L.; validation, W.L., Z.Z., and Z.S.; formal analysis, J.Z.; investigation, W.L.; data curation, Z.S. and H.L.; writing—original draft preparation, W.L., Z.Z., and Z.S.; writing—review and editing, W.L. and Z.Z.; supervision, W.L. All authors have read and agreed to the published version of the manuscript.

Funding: This work is supported in part by the National Natural Science Foundation of China, under grant 62003114; the Provincial Natural Science Foundation of Heilongjiang, under grant LH2023F019; the China Postdoctoral Science Foundation, under grant 2020M681097; the Heilongjiang Postdoctoral Fund, under grant LBH-Z20134; the Fundamental Research Funds for the Central Universities, under grant HIT.NSFJG202207; and the National Key Laboratory of Laser Spatial Information Foundation, under grant LSI2025WDZC07.

Data Availability Statement: The original contributions presented in this study are included in the article. Further inquiries can be directed to the corresponding author.

Conflicts of Interest: The authors declare no conflicts of interest.

References

1. Zhang, Y.; Wu, Z.-G.; Shi, P. Resilient Event-/Self-Triggering Leader-Following Consensus Control of Multiagent Systems Against DoS Attacks. *IEEE Trans. Ind. Inform.* **2023**, *19*, 5925–5934. [CrossRef]
2. Zhou, J.; Xu, Y.; Sun, H.; Li, Y.; Chow, M.-Y. Distributed Power Management for Networked AC–DC Microgrids with Unbalanced Microgrids. *IEEE Trans. Ind. Inform.* **2020**, *16*, 1655–1667. [CrossRef]
3. Sun, J.; Yang, J.; Li, S.; Zheng, W.X. Sampled-Data-Based Event-Triggered Active Disturbance Rejection Control for Disturbed Systems in Networked Environment. *IEEE Trans. Cybern.* **2017**, *49*, 556–566. [CrossRef] [PubMed]
4. Yang, J.; Sun, J.; Zheng, W.X.; Li, S. Periodic Event-Triggered Robust Output Feedback Control for Nonlinear Uncertain Systems with Time-Varying Disturbance. *Automatica* **2018**, *94*, 324–333. [CrossRef]
5. Wang, B.; Huang, J.; Wen, C.; Rodriguez, J.; Garcia, C.; Gooi, H.B.; Zeng, Z. Event-Triggered Model Predictive Control for Power Converters. *IEEE Trans. Ind. Electron.* **2020**, *68*, 715–720. [CrossRef]
6. Ma, G.; Qin, L.; Liu, X.; Wu, G. Event-Triggered Output-Feedback Control for Switched Linear Systems with Applications to a Boost Converter. In Proceedings of the 2016 IEEE 11th Conference on Industrial Electronics and Applications (ICIEA), Hefei, China, 5–7 June 2016; pp. 2431–2436.
7. Ren, H.; Zong, G.; Li, T. Event-Triggered Finite-Time Control for Networked Switched Linear Systems with Asynchronous Switching. *IEEE Trans. Ind. Inform.* **2018**, *48*, 1874–1884. [CrossRef]
8. Rathore, N.; Fulwani, D. Event Triggered Control Scheme for Power Converters. In Proceedings of the IECON 2016—42nd Annual Conference of the IEEE Industrial Electronics Society, Florence, Italy, 23–26 October 2016; pp. 1342–1347.
9. Yang, X.; Yin, S. Robust Global Identification and Output Estimation for LPV Dual-Rate Systems Subjected to Random Output Time-Delays. *IEEE Trans. Ind. Inform.* **2017**, *13*, 2876–2885. [CrossRef]
10. Han, K.; Feng, J.; Zhao, Q. Robust Estimator-Based Dual-Mode Predictive Fault-Tolerant Control for Constrained Linear Parameter Varying Systems. *IEEE Trans. Ind. Inform.* **2021**, *17*, 4469–4479. [CrossRef]
11. de Oliveira, M.S.; Pereira, R.L. LMI-Based Filter Design Conditions for Discrete-Time LPV Systems with Bounded Parameter Variation. *IEEE Trans. Autom. Control* **2021**, *66*, 910–915. [CrossRef]
12. Hu, X.; Xiong, R.; Egardt, B. Model-Based Dynamic Power Assessment of Lithium-Ion Batteries Considering Different Operating Conditions. *IEEE Trans. Ind. Inform.* **2014**, *10*, 1948–1959. [CrossRef]
13. Rotondo, D.; Puig, V.; Nejjari, F.; Romera, J. A Fault-Hiding Approach for the Switching Quasi-LPV Fault-Tolerant Control of a Four-Wheeled Omnidirectional Mobile Robot. *IEEE Trans. Ind. Electron.* **2015**, *62*, 3932–3944. [CrossRef]
14. Incremona, G.P.; Ferrara, A.; Utkin, V.I. Sliding Mode Optimization in Robot Dynamics with LPV Controller Design. *IEEE Control Syst. Lett.* **2022**, *6*, 1760–1765. [CrossRef]
15. Liu, S.; Hu, Z.; Wang, X.; Wu, L. Stochastic Stability Analysis and Control of Secondary Frequency Regulation for Islanded Microgrids Under Random Denial of Service Attacks. *IEEE Trans. Ind. Inform.* **2019**, *15*, 4066–4075. [CrossRef]
16. Tang, Y.; Wu, X.; Shi, P.; Qian, F. Input-to-State Stability for Nonlinear Systems with Stochastic Impulses. *Automatica* **2020**, *113*, 108766. [CrossRef]
17. Tan, H.; Shen, B.; Peng, K.; Liu, H. Robust Recursive Filtering for Uncertain Stochastic Systems with Amplify-and-Forward Relays. *Int. J. Syst. Sci.* **2020**, *51*, 1188–1199. [CrossRef]
18. Chen, Y.; Wang, Z.; Yuan, Y.; Date, P. Distributed H_∞ Filtering for Switched Stochastic Delayed Systems Over Sensor Networks with Fading Measurements. *IEEE Trans. Cybern.* **2020**, *50*, 2–14. [CrossRef]
19. Liu, J.; Wu, L.; Wu, C.; Luo, W.; Franquelo, L.G. Event-Triggering Dissipative Control of Switched Stochastic Systems via Sliding Mode. *Automatica* **2019**, *103*, 261–273. [CrossRef]
20. Liu, K.-Z.; Teel, A.R.; Sun, X.-M. Event-Triggered Nonlinear Systems with Stochastic Dynamics, Transmission Times, and Protocols. *IEEE Trans. Autom. Control* **2022**, *67*, 1973–1979. [CrossRef]
21. Luo, W.; Vazquez, S.; Liu, J.; Gordillo, F.; Franquelo, L.G.; Wu, L. Control System Design of a Three-Phase Active Front End Using a Sliding-Mode Observer. *IEEE Trans. Syst. Man Cybern. Syst.* **2022**, *52*, 739–748. [CrossRef]
22. Mao, X.; Yuan, C. *Stochastic Differential Equations with Markovian Switching*; Imperial College Press: London, UK, 2006.

23. Corless, M.; Leitmann, G. Continuous State Feedback Guaranteeing Uniform Ultimate Boundedness for Uncertain Dynamic Systems. *IEEE Trans. Autom. Control* **1981**, *26*, 1139–1144. [CrossRef]
24. Gahinet, P.; Apkarian, P. A Linear Matrix Inequality Approach to H_∞ Control. *Int. J. Robust Nonlinear Control* **1994**, *4*, 421–448. [CrossRef]

energies

MDPI

Article

Full-Bridge T-Type Three-Level LLC Resonant Converter with Wide Output Voltage Range

Kangjia Zhang [1], Kun Zhao [1], Xiaoxiao Yang [1], Muyang Liu [1] and Zhigang Yao [2,*]

1 National Rail Transit Electrification and Automation Engineering Technology Research Center, Southwest Jiao Tong University, Chengdu 611756, China; zkj0825@my.swjtu.edu.cn (K.Z.); zhaokun@my.swjtu.edu.cn (K.Z.); xiaoxyang@my.swjtu.edu.cn (X.Y.); 2021200442@my.swjtu.edu.cn (M.L.)
2 Energy Research Institute, Nanyang Technological University, Singapore 639798, Singapore
* Correspondence: zhigang.yao@ntu.edu.sg; Tel.: +86-13368363187

Abstract

Traditional LLC resonant converters face significant challenges in wide-output-voltage-applications, such as limited voltage gain, efficiency degradation under wide-gain range, and increased complexity in magnetic component design. For example, in electric vehicle charging power modules, achieving wide output voltage typically relies on changing the transformer turns ratio or switching the series-parallel circuit configuration via relays, which prevents real-time dynamic adjustment. To overcome these limitations, this paper proposes a wide-gain-range control method based on a full-bridge T-type three-level LLC resonant converter, capable of achieving a voltage gain range exceeding six times. By integrating a T-type three-level bridge arm with PWM modulation and employing a variable-topology and variable-frequency control strategy, the proposed method achieves synergistic optimization for wide-output-voltage-applications. PWM modulation enables wide-range voltage output by dynamically adjusting both the converter topology and switching frequency. Finally, the proposed method is validated through circuit simulations and experimental results based on a full-bridge T-type three-level LLC converter prototype, demonstrating its effectiveness and feasibility.

Keywords: LLC resonant converter; T-type three-level; wide gain range

1. Introduction

As a highly efficient power conversion scheme, the LLC resonant converter is widely used in various types of power electronic devices, particularly in medium- and high-power applications that require high efficiency, low loss, and low electromagnetic interference (EMI). Its most significant advantage lies in the implementation of soft-switching operation, which enables the converter to maintain zero-voltage-switching (ZVS) or zero-current-switching (ZCS) states during both conduction and turn-off of the switching devices. This capability effectively reduces switching losses and electromagnetic interference [1–3]. The soft-switching feature allows the LLC resonant converter to maintain high efficiency even at high frequencies, making it especially suitable for high-frequency, high-power-density applications such as high-efficiency power supplies and electric vehicle charging systems [4,5].

However, the frequency regulation range of LLC resonant converters is typically narrow. When operating far from the resonant frequency, conventional adjustments to the duty cycle or phase shift angle may fail to maintain soft-switching, leading to increased switching currents in the power devices and a significant drop in efficiency [6,7]. This limitation poses challenges for achieving wide output voltage ranges, especially under large load variations, where system efficiency can degrade substantially [8].

Many existing wide gain control methods primarily rely on adjusting the gain near the resonance point [9,10], but this approach limits the converter's output voltage range to some extent. To address this issue, researchers have proposed various topology design schemes, with full-bridge LLC and half-bridge LLC being the most common structures. These topologies achieve a wider output voltage range within a certain range by adjusting the frequency, duty cycle, and phase shift angle [11–15]. However, they still have a limited gain range and pose significant design challenges under efficient operating conditions [16,17].

To address this issue, multilevel topologies have gained increasing attention in recent years. By increasing the number of levels, multilevel topologies can significantly reduce the voltage stress on the switching devices and provide a stable output across a wider voltage range [18–20]. For instance, converters with three or more levels can not only extend the gain range but also enhance both the efficiency and power density of the system [21–23]. However, the adoption of multilevel topologies often comes with more complex control strategies and higher cost burdens, particularly in high-power applications [24,25]. Additionally, adjusting the transformer turns ratio is a common method to extend the gain range. By altering the turns ratio or modifying the topology, the voltage gain can be adjusted within a specific range, thus enabling a wider output voltage [26–30]. However, this approach is constrained by factors such as transformer size, cost, and magnetic materials, making it unsuitable for all application scenarios [31].

Another three-level topology is the T-type three-level LLC converter. This topology typically includes four switching tubes, though only three are required when the T-type bridge is replaced by a bidirectional switch, which helps achieve low conduction losses and reduced switching stress [32]. The T-type bridge has also evolved into a hybrid bridge topology, which meets the wide gain requirement through frequency shifting [33]. Furthermore, recent research on the T-type three-level LLC converter has focused on widening the gain range of the T-type bridge LLC converter by achieving a 0.25-fold gain through alternating modulation of the topology [34–36]. Reference [34] indicates that the T-type bridge structure exhibits lower on-resistance and voltage drop compared to stacked half-bridge and NPC three-level LLC converters [37,38]. As illustrated in Figure 1, the switch network of the T-type bridge demonstrates reduced conduction resistance and voltage drop. Specifically, for power semiconductor devices with the same current rating and technology generation, the conduction resistance of a 1200 V SiC MOSFET is lower than the combined resistance of two 650 V SiC MOSFETs connected in series. Likewise, the conduction voltage drop of a 1200 V IGBT is less than that of two 650 V IGBTs in series. These characteristics indicate that the T-type three-level LLC converter achieves lower conduction losses, thereby improving overall system efficiency [39–41]. Reference [42] indicate that compared with T-type LLC resonant converters, dual-resonant cavity structures are more prone to current imbalance. In addition, the use of multiple transformers further increases the complexity of design and debugging. Additionally, the T-type three-level LLC converter is more cost-effective when using three independent drive supplies. The T-type bridge can also be replaced with bidirectional switches, further reducing the overall cost.

To expand the gain of the LLC converter while utilizing the advantages of the T-type bridge, this paper proposed a wide-range control method based on a full-bridge T-type three-level LLC resonant converter. The T-type three-level bridge arm reduces the voltage stress on the switching devices and extends the voltage gain range through multilevel modulation. Meanwhile, the wide-output-voltage design, which incorporates variable-frequency and variable-topology control strategies, ensures soft-switching characteristics across the entire load range.

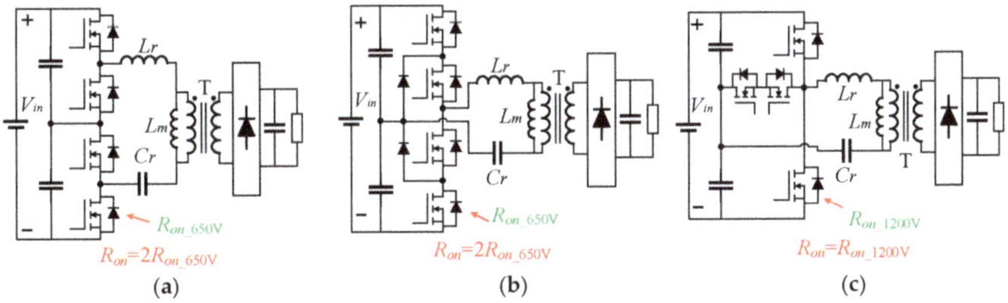

Figure 1. Conduction resistances at 0.5× normalized gain for common three-level topologies, with Vin = 800 V as an example. (**a**) Stacked half-bridge LLC converter; (**b**) NPC three-level LLC converter; (**c**) T-type three-level LLC converter.

The main contributions of this paper are as follows:

(1) This paper proposes a wide-range control method based on a full-bridge T-type three-level LLC resonant converter and analyzes the working principle and wide output voltage characteristic.
(2) A variable-topology strategy is employed to tackle the challenge of achieving both a wide output voltage range.
(3) A PWM modulation method is proposed to achieve a wide output voltage range through variable-topology and variable-frequency control strategies.

The structure of this paper is as follows: Section 2 describes in detail the conventional control methods to realize wide output voltage and their defects and shortcomings; Section 3 presents the variable-topology control strategy; Section 4 presents experimental results and comparative analyses; and Section 5 concludes the paper.

2. Conventional Control Method for Full-Bridge T-Type Three-Level LLC Resonant Converter

2.1. Operating Principle of the Full-Bridge T-Type Three-Level LLC Converter

As shown in Figure 2, the topology full-bridge T-type LLC resonant converter utilizes two interleaved half-bridge T-type three-level converters operating with a 180° phase shift between phases. Each phase contains four power switches (S_{a1}–S_{a4}), where S_{a1}/S_{a3} and S_{a2}/S_{a4} operate as complementary pairs with a 50% duty cycle. The voltage divider capacitors (C_{dc1}, C_{dc2}) ensure balanced voltage distribution ($Vc_{dc1} = Vc_{dc2}$), while the LLC resonant tank—comprising the resonant inductor (L_r), resonant capacitor (C_r), and the transformer's magnetizing inductance (L_m)—supports soft-switching operation. The secondary-side full-bridge rectifier (D_1–D_4), along with the output capacitor (C_o), provides a regulated DC output.

2.2. Analysis of Conventional Control Methods

Conventional control methods often achieve a wide voltage output range by varying the duty cycle at frequencies higher than the resonant frequency. For example, Figure 3 shows the modulated waveform of a wide voltage output achieved by varying the duty cycle at frequencies greater than the resonant frequency.

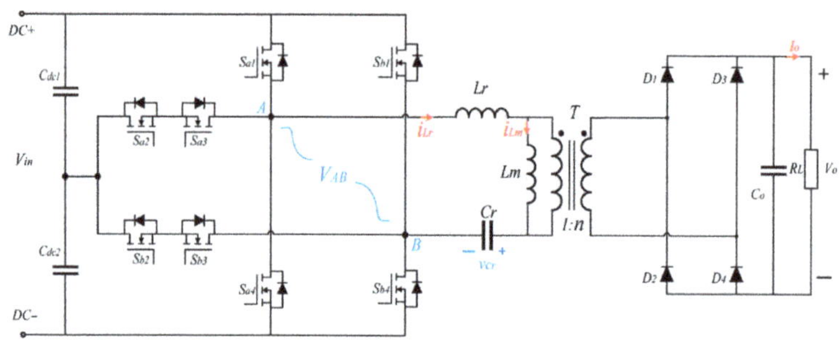

Figure 2. Topology of full-bridge T-type three-level LLC resonant converter.

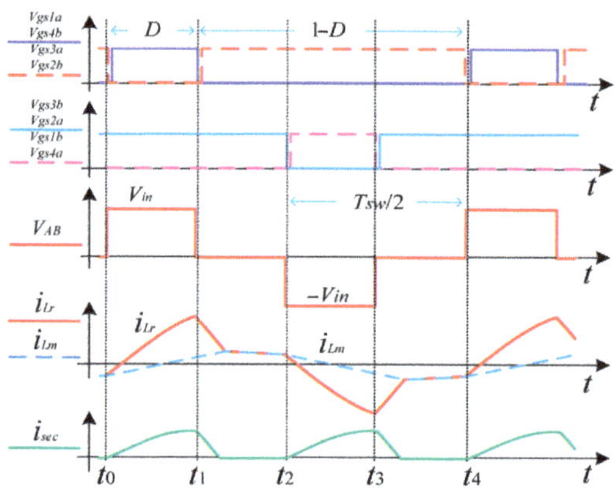

Figure 3. Waveform using variable duty cycle modulation when the switching frequency is less than the resonant frequency.

Phase [t_0–t_1]: During this phase, S_{a1} and S_{a2} of phase A conduct, while S_{b3} and S_{b4} of phase B conduct. The input voltage to the resonant cavity is equal to the input voltage of the system, and the resonant current is greater than the magnetizing current. The magnetizing inductance is clamped by the output voltage, causing the magnetizing current to rise linearly.

Phase [t_1–t_2]: At time t_1, S_{a1} and S_{b4} turn off. At this moment, the resonant current is greater than zero and is sustained by the body diodes of S_{a3} and S_{b2} (i.e., S_{a2} and S_{b3}). The input voltage to the resonant cavity becomes zero, allowing for ZVS when S_{a3} and S_{b2} turn on. Once S_{a3} and S_{b2} are on, the resonant current decreases, while the magnetizing current continues to rise linearly. When the magnetizing current equals the resonant current, the magnetizing inductance and resonant capacitance jointly resonate with the resonant inductor and capacitor. At this point, the primary side of the transformer is decoupled from the secondary side, and the magnetizing inductance resonates with the resonant capacitor and resonant inductor.

Phase [t_2–t_4]: The converter enters the next half-cycle. When the voltages across C_{dc1} and C_{dc2} are equal, the waveform of the converter in the [t_2–t_4] phase becomes symmetrical to that of the [t_0–t_2] phase.

A wide voltage output range is often achieved by varying the switching frequency below the resonant frequency, as shown in Figure 4, which illustrates the modulation waveform at frequencies below the resonant frequency.

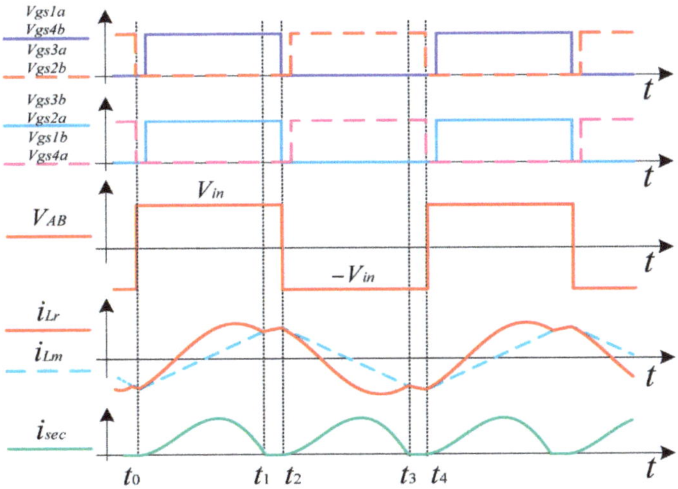

Figure 4. Waveform using variable frequency modulation when the switching frequency is less than the resonant frequency.

Phase [t_0–t_1]: At time t_0, the switching tubes S_{a3} and S_{b1} are turned off. Since the excitation inductance L_m is relatively large and the duration of this mode is short, it can be approximated that the resonant inductance current remains unchanged. At this point, the resonance and excitation currents are equal. When the body diodes of S_{a1} and S_{b4} conduct, these transistors can be turned on at zero voltage. To ensure the zero-voltage turn-on of S_{a1} and S_{b4}, they must be switched on before the current i_{Lr} changes from negative to positive. As the source-drain voltages of S_{a1} and S_{b4} drop to zero, zero-voltage turn-on is achieved. At this point, the voltage between points A and B is V_{in}, and the rectifier diode D_1 conducts. The excitation inductance is then clamped at V_o/n, ceasing to participate in the resonance, while the excitation inductance current increases linearly. Meanwhile, the resonant network consisting of L_r and C_r begins to resonate, with the voltage across it being $V_{in} - V_o/n$.

Phase [t_1–t_2]: At time t_1, the current in the resonant inductor equals the excitation inductor current. At this moment, the current on the transformer secondary side is zero, effectively disconnecting the load from the resonant network. The rectifier diode D_1 is turned off, ensuring ZCS without any reverse recovery issues. The voltage across the excitation inductor L_m is no longer clamped at V_o/n, and the resonant components L_r, L_m, and C_r continue to resonate and operate.

Phase [t_2–t_4]: During this phase, the operating principle of the converter is similar to the second half of the previous operating cycle, so it will not be repeated here.

When the full-bridge T-type three-level LLC resonant converter operates at $f_s > f_r$, a wide voltage output range is achieved by varying the duty cycle. Additionally, when operating below the resonant frequency, a wide voltage output range is typically achieved using Pulse Frequency Modulation (PFM) converter control. The voltage gain function of the converter can be derived through fundamental approximation equivalent analysis:

$$G = \frac{\sin D\pi}{\sqrt{\left(1 + \frac{1}{k}\left(1 - \frac{1}{fn^2}\right)\right)^2 + \left(Q^2\left(fn - \frac{1}{fn}\right)^2\right)}} \tag{1}$$

where $k = L_m/L_r$, $f_n = f_s/f_r$, f_s is the switching frequency, the resonant frequency is $fr = 1/2\pi\sqrt{L_r C_r}$, the quality factor is $Q = \sqrt{L_r C_r}/R_{eq}$, and n is the transformer transformation ratio.

As shown in Figure 5, the traditional method achieves a wide voltage gain range by controlling the converter at frequencies both below and above the resonant frequency, primarily by varying the duty cycle. Although theoretical analysis indicates that a near-zero output voltage can be attained by reducing the duty cycle, this results in certain operating ranges where ZVS soft switching is not achievable. In these conditions, the turn-off current increases significantly, which leads to lower efficiency and greater EMI. Therefore, new methods for wide output regulation need to be explored. The 0.75× and 0.25× voltage gain modulation methods proposed in this paper offer a balance between wide voltage regulation and efficient operation. The proposed variable topology control strategy defines three discrete gain ratios (0.75, 0.5, and 0.25) at the resonant frequency. Along with the 1× fundamental ratio, this results in three distinct operating modes. Pulse Frequency Modulation (PFM) further enhances voltage regulation, and together these techniques enable a wide output voltage range.

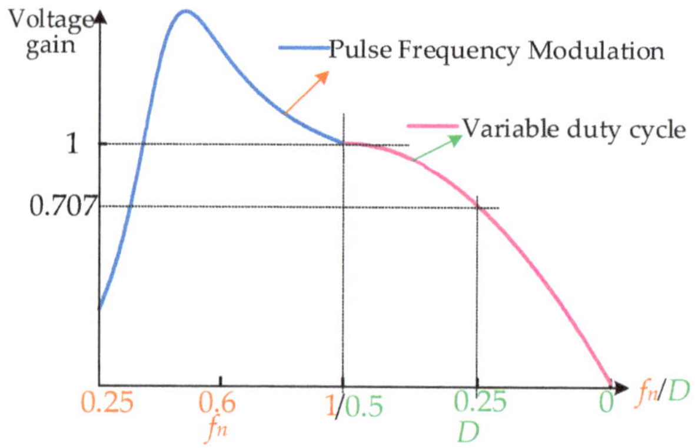

Figure 5. Normalized gain curves under conventional control methods with variable switching frequency or duty cycle.

3. Proposed Modulation and Control Method for Wide Gain

3.1. Proposed 0.75 Gain Operating Mode

The modulation method proposed in Section 2 achieves a wide output voltage range through duty ratio variation. However, at small duty ratios, excessive resonant current spikes occur, leading to efficiency degradation. To achieve a wide gain range, the full-bridge T-type LLC converter employs variable-topology modulation, enabling multiple normalized gain points at the resonant frequency. The first part of this chapter presents the variable-topology modulation scheme that achieves a 0.75-times normalized gain, where the original reference normalization is unity gain. This introduces a 0.75-times voltage gain mode for the full-bridge T-type LLC converter operating at the resonant frequency. The 0.75-times gain mode consists of two frequency-controlled sub-modes (Mode 1 and Mode 2). Assuming that the voltages across C_{dc1} and C_{dc2} remain balanced at 0.5 V_{in} with sufficient capacitance, Figure 6a,b show the equivalent circuit and operational waveforms for Mode 1, respectively.

When the converter operates in Mode 1 of the 0.75-times gain mode, S_{a2} remains on, while S_{a4}, S_{b2}, and S_{b3} stay off. S_{a1} and S_{a3} operate complementarily with a 50% duty cycle, and S_{b1} and S_{b4} operate complementarily with a 50% duty cycle, with a 180° phase difference between S_{b1} and S_{a1}. The converter exhibits the following operational phases within one cycle:

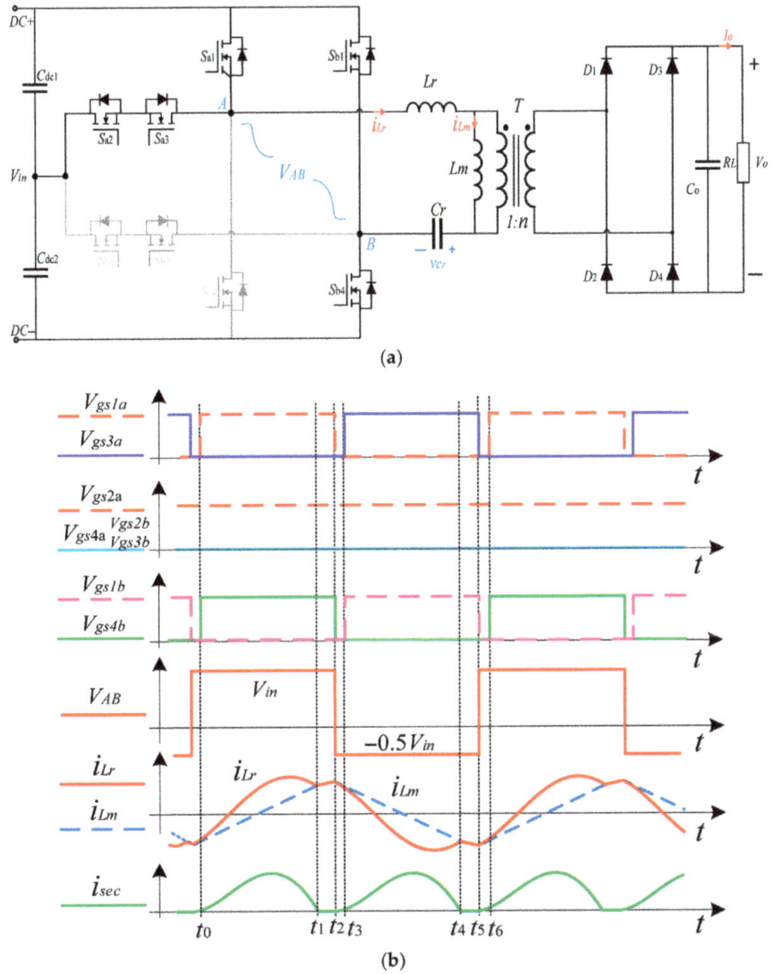

Figure 6. Circuit diagram and operating waveforms of 0.75 gain Mode I. (**a**) Equivalent circuit of Mode I; (**b**) waveforms of the Mode I circuit.

Phase [t_0–t_1]: During this phase, S_{a1} and S_{b4} conduct. The resonant tank input voltage V_{AB} equals V_{in}, and the resonant inductor and capacitor resonate in series. The magnetizing inductance is clamped to V_o/n, causing its current to increase linearly.

Phase [t_1–t_2]: When the magnetizing current equals the resonant current, the primary and secondary sides of the transformer decouple, allowing the magnetizing inductance to resonate with the resonant inductor and capacitor.

Phase [t_2–t_3] (Dead Time): At t_2, S_{a1} and S_{b4} turn off. Since the resonant current remains positive, the circulating current flows through the body diodes of S_{a3} and S_{b1}, changing the resonant tank input voltage to $-0.5\ V_{in}$. With the body diodes of S_{a3} and S_{b1} conducting, ZVS turn-on is achieved at t_3.

Phase [t_3–t_4]: With S_{a3} and S_{b1} conducting, the resonant tank input voltage is $-0.5\ V_{in}$. Since the resonant current is lower than the magnetizing current, the resonant inductor and capacitor continue to resonate in series. The magnetizing inductance is clamped to $-V_o/n$, causing its current to decrease linearly.

Phase $[t_4–t_5]$: When the resonant and magnetizing currents equalize, the transformer decouples again, allowing the magnetizing inductance to resonate with the resonant components.

Phase $[t_5–t_6]$ (Dead Time): At t_5, S_{a3} and S_{b1} turn off. With negative resonant current, the current flows through the body diodes of S_{a1} and S_{b4}, changing V_{AB} to V_{in}. ZVS is achieved when S_{a1} and S_{b4} turn on after the dead time.

Mode 2 of the 0.75-times gain mode is symmetrical to Mode 1: S_{a1}, S_{b2}, and S_{b3} remain off, while S_{a3} stays on. S_{a2} and S_{a4} operate complementarily at a 50% duty cycle, as do S_{b1} and S_{b4}. The corresponding waveforms and equivalent circuits are shown in Figures 6b and 7a, respectively. Given the high symmetry between the modes, where only C_{dc2} replaces C_{dc1} as the unilateral power source, the operational process of Mode 2 is omitted here.

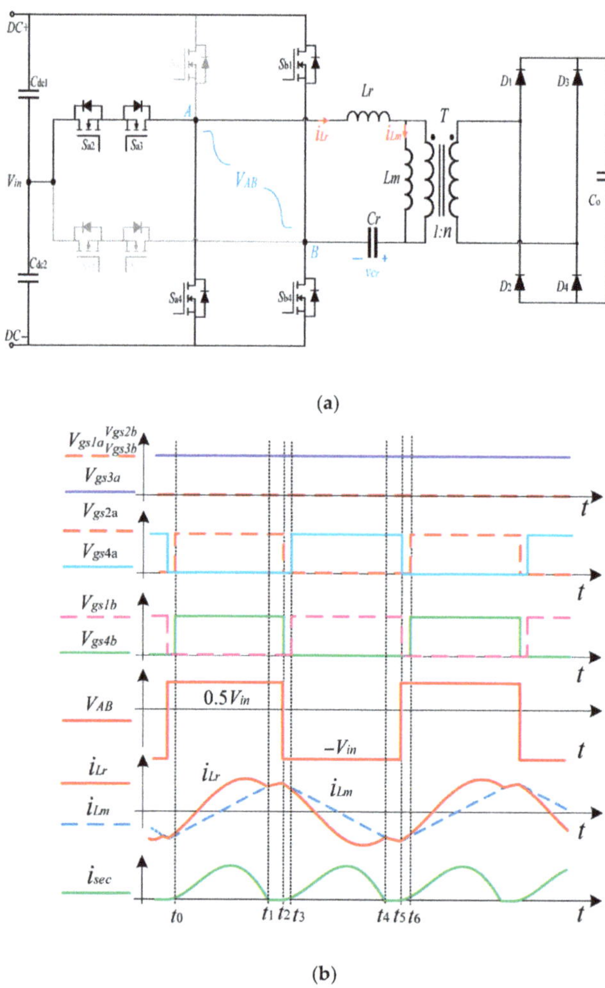

Figure 7. Circuit diagram and operating waveforms of 0.75 gain Mode II. (**a**) Equivalent circuit of Mode II; (**b**) waveforms of the Mode II circuit.

Due to the presence of negative-voltage half-cycle intervals in 0.75-times gain Mode 1, the resonant network is powered solely by the upper input capacitor, C_{dc1}, during these periods. Similarly, in 0.75-times gain Mode 2, there is a duration where only the lower input capacitor, C_{dc2}, supplies power.

To maintain voltage balance between the input capacitors, the converter should alternate between both modes while operating in 0.75-times gain mode. This alternation prevents excessive voltage droop caused by prolonged operation of a single capacitor. Ideally, the converter operates in Mode 1 for a fixed duration (typically several dozen switching cycles in high-frequency operation), then switches to Mode 2 for an equal duration, repeating this sequence.

Thanks to the four-diode rectifier bridge on the secondary side, mode transitions have minimal impact on the output load. Consequently, the input voltage V_{AB} of the resonant network can be approximated as a constant 0.75 V_{in} during this process. In summary, the voltage-normalized gain formula for the full-bridge T-type LLC converter in 0.75-times gain mode is derived as follows:

$$G = \frac{0.75}{\sqrt{\left(1+\frac{1}{k}\left(1-\frac{1}{fn^2}\right)\right)^2+\left(Q^2\left(fn-\frac{1}{fn}\right)^2\right)}} \tag{2}$$

3.2. Proposed 0.25 Gain Operating Mode

This section proposes a mode of operation for the full-bridge T-type LLC resonant converter that achieves a voltage gain of 0.25 at the resonant frequency, referred to as the 0.25-fold gain mode. This mode consists of two sub-modes, each controlled by frequency conversion. Assuming that the voltages across C_{dc1} and C_{dc2} are both 0.5 V_{in} and that the capacitance is sufficiently large, the control method for Mode 1 of the 0.25-fold gain mode is as follows: switches S_{a1}, S_{b1}, S_{b2}, and S_{b3} remain off, while switches S_{a3} and S_{b4} remain on, and switches S_{a2} and S_{a4} operate complementarily with a 50% duty cycle. The corresponding circuit waveforms and equivalent circuits are shown in Figures 8a and 8b, respectively.

Phase [t_0–t_1]: During this stage, S_{a2} is conducting, and the resonant cavity input voltage (V_{AB}) is 0.5 V_{in}. Initially, the resonant inductor and resonant capacitor resonate in series, and the excitation inductance is clamped to V_o/n, with the current increasing linearly. Since $f_s < f_r$, the excitation inductance and resonant inductance resonate together towards the end of this stage when the excitation inductance equals the resonant inductance.

Phase [t_1–t_2]: The time interval between t_1 and t_2 is the dead time. At t_1, S_{a2} turns off, and the resonant current becomes greater than zero. The resonant current loop is refreshed through the body diode of S_{a4}, and the resonant cavity input voltage becomes 0 V. After the dead time, at t_2, S_{a4} can achieve ZVS turn-on.

Phase [t_2–t_3]: During this stage, S_{a4} turns on. The resonant cavity input voltage remains at 0 V, and the resonant current is smaller than the excitation current. The resonant inductor and resonant capacitor continue to resonate in series. The excitation inductance is clamped to $-V_o/n$, causing the excitation current to decrease linearly. When the resonant current equals the excitation current, the transformer's primary and secondary sides are decoupled. At this point, the excitation inductance resonates with the resonant inductor and capacitor.

Phase [t_3–t_4] Stage: At t_3, S_{a4} turns off, and the resonant current becomes negative. The resonant current is restored by the body diode of S_{a2} and capacitor C_{dc2}, and S_{a2} reverses conduction. The resonant cavity input voltage becomes V_{in}, after which S_{a2} can achieve ZVS turn-on.

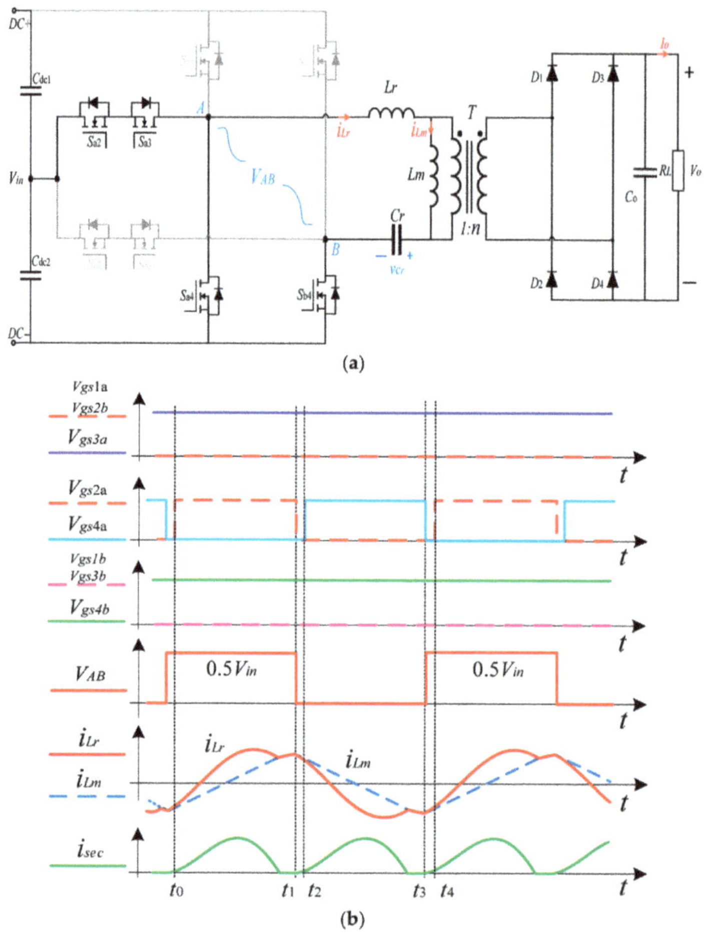

Figure 8. Circuit diagram and operating waveforms of 0.25 gain Mode I. (**a**) Equivalent circuit of Mode I; (**b**) waveforms of the Mode I circuit.

Mode II of the 0.25-times gain mode of the full-bridge T-type LLC converter is symmetrical to Mode I. When operating in Mode II, switches S_{a2} and S_{b1} remain continuously on, while switches S_{a4}, S_{b2}, S_{b3}, and S_{b4} remain off. Switches S_{a1} and S_{a3} operate complementarily with a 50% duty cycle. The corresponding waveforms and equivalent circuits for this mode are shown in Figures 9a and 9b, respectively.

Using the fundamental harmonic approximation (FHA) method, it can be shown that the normalized DC voltage gain function of the full-bridge T-type LLC converter in both 0.25-times gain modes can be expressed by Equation (3). From Equation (3), it is evident that the voltage gain of the converter in both 0.25-times gain modes is 0.25 when the switching frequency equals the resonant frequency, representing only 0.25 times the voltage gain of the full-bridge mode with identical circuit parameters and switching frequency.

Since the initial voltages of C_{dc1} and C_{dc2} may not be balanced in practice, and since operating in a single mode would cause the capacitor voltage to drift, Mode 1 and Mode 2 of the 0.25-times gain mode should operate alternately. This alternation ensures that the voltage across the dividing capacitors remains balanced within the desired range by controlling the duration of each mode.

$$G = \frac{0.25}{\sqrt{\left(1+\frac{1}{k}\left(1-\frac{1}{fn^2}\right)\right)^2+\left(Q^2\left(fn-\frac{1}{fn}\right)^2\right)}} \quad (3)$$

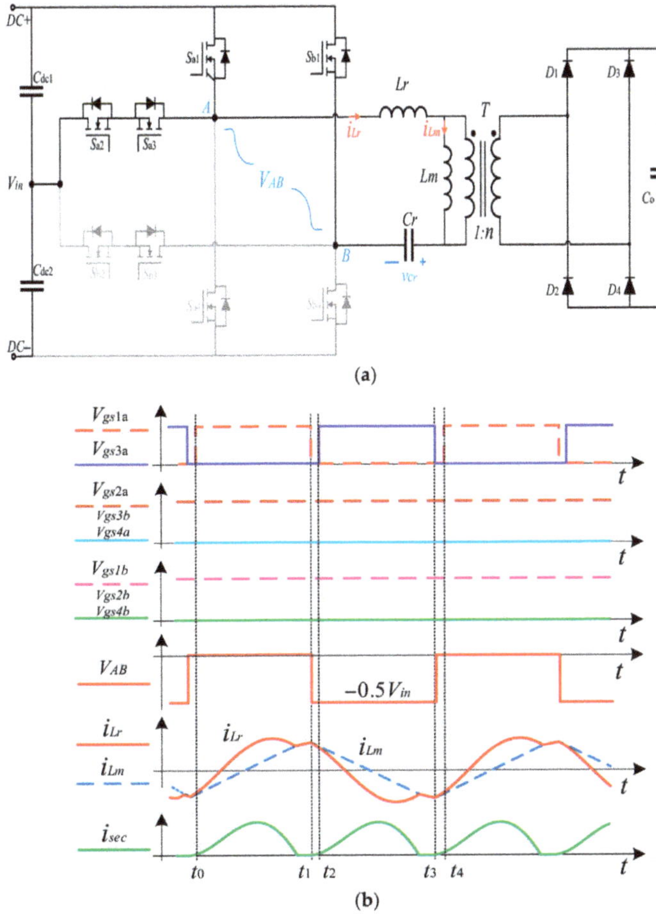

Figure 9. Circuit diagram and operating waveforms of 0.25 gain Mode II. (**a**) Equivalent circuit of Mode II; (**b**) waveforms of the Mode II circuit.

3.3. *Conventional 1 and 0.5 Gain Operating Mode*

The conventional 1-times gain modulation method, introduced in Section 2, utilizes variable duty cycle and frequency modulation to achieve one times the normalized gain. This method is not repeated here. The full-bridge T-type three-level LLC converter, shown in Figure 10, operates in half-bridge mode, where S_{a2} and S_{a3} remain on, S_{a1}, S_{a4}, S_{b2}, and S_{b3} remain off, and S_{b1} and S_{b4} operate with a 50% duty cycle.

The full-bridge T-type three-level LLC resonant converter operating in half-bridge mode undergoes the following stages during one cycle:

[t_0–t_1]: During this stage, S_{b4} conducts, and the resonant cavity input voltage is 0.5 V_{in}.

[t_1–t_2]: At t_1, S_{b4} is turned off. Since the resonant current is greater than zero, it forces the body diode of S_{b1} to reverse conduction, renewing the current. After the current is renewed, the resonant cavity input voltage becomes $-0.5\ V_{in}$. The time interval from t_1 to t_2 is the dead time. At t_2, S_{b1} turns on, and ZVS is achieved due to the reverse conduction during the turn-on.

[t_2–t_4]: The converter then enters the next half-cycle of operation, which is symmetric to the previous one (from t_0 to t_2). ZVS is also realized for S_{b4} at t_4.

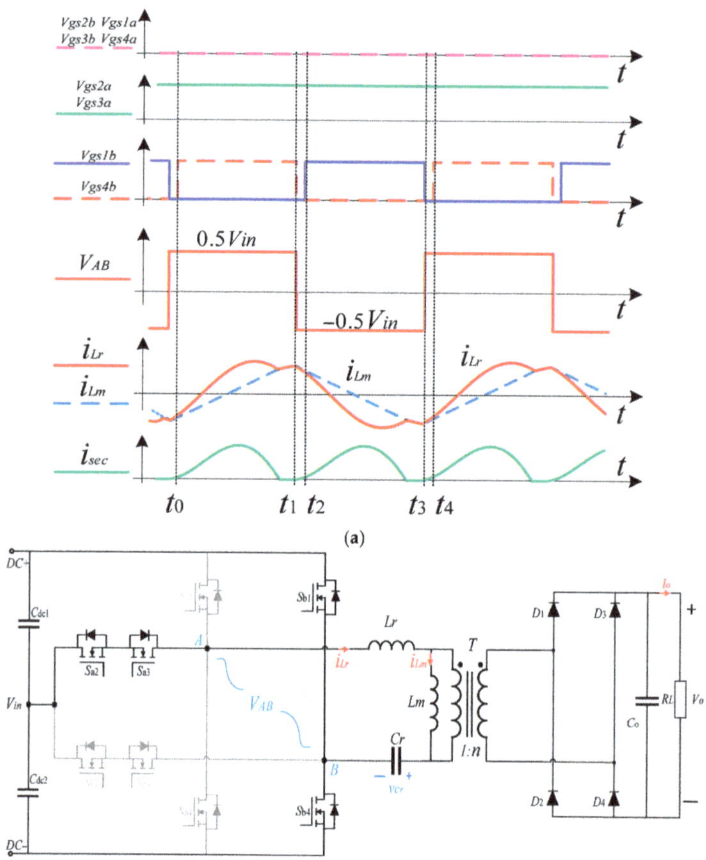

Figure 10. Waveforms and equivalent circuits at 0.5 gain half-bridge mode. (**a**) 0.5 gain operating waveform; (**b**) 0.5 gain operating equivalent plot.

Through the analysis of the 0.5-times gain mode, it can be seen that the alternate conduction of the upper and lower bridges not only reduces the voltage stress on the switching devices by half but also helps balance the input capacitors ($Vc_{dc1} = Vc_{dc2}$), mitigating the effects of uneven voltage distribution. The normalized voltage gain in this modulation mode is as follows:

$$G = \frac{0.5}{\sqrt{\left(1 + \frac{1}{k}\left(1 - \frac{1}{fn^2}\right)\right)^2 + \left(Q^2\left(fn - \frac{1}{fn}\right)^2\right)}} \tag{4}$$

3.4. Control Strategy of Full-Bridge T-Type Three-Level LLC Resonant Converter

In summary, based on the gain function of the full-bridge T-type three-level LLC converter operating in different modes, the corresponding gain curves can be plotted, as shown in Figure 11. As seen in Figure 11, compared to the original operating modes, the proposed 0.75-times and 0.25-times gain modes enable the full-bridge T-type LLC converter to achieve an extended voltage gain range with minimal switching frequency variation, thereby enabling a wide output voltage range.

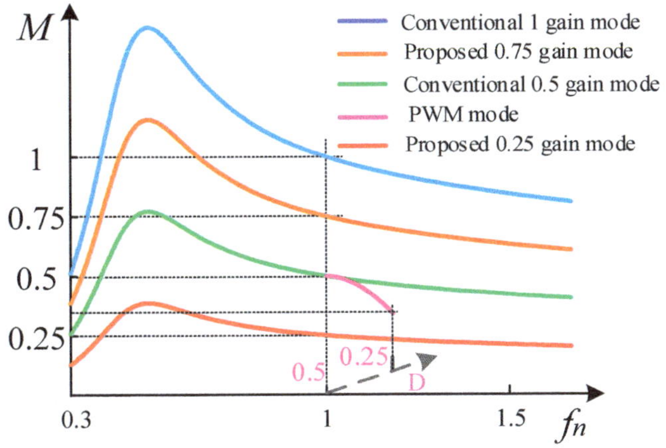

Figure 11. Normalized gain curves under the proposed method with variable circuit topology and switching frequency.

The full-bridge T-type three-level LLC resonant converter operates in four distinct modes, each of which is selected based on the gain range, as shown in Figure 12. The PI controller, G_V, adjusts the switching frequency according to the rated output voltage, V_{ref}. In this figure, V_{ref} represents the reference output voltage, V_O denotes the actual output voltage, and V_{err} indicates the voltage error. Additionally, G_{pre} refers to the predicted voltage gain, and f_s represents the output frequency of the primary side inverter bridge section.

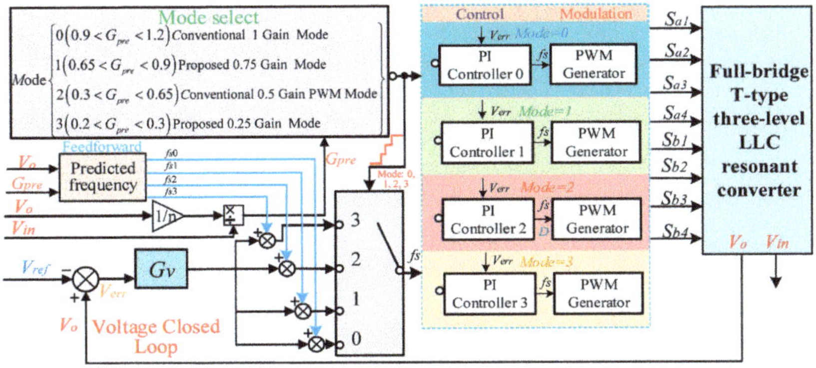

Figure 12. Proposed converter control block diagram based on quadruple normalized gain mode selection.

To enhance the accuracy of the converter's gain prediction, the mode prediction is incorporated into the mode selection process. Based on the output voltage, the required gain of the converter is determined, and the appropriate operation mode is selected according to the voltage gain. The relationship between voltage gain and the mode of operation is described by a specific relational equation, as shown in the mode selection diagram in Figure 12. This approach enables the converter to achieve a wide output voltage range through mode selection and variable frequency control, resulting in a voltage gain of more than six times the normalized voltage.

4. Simulation and Experimental Results

4.1. Simulation of Different Modulation Modes for Full-Bridge T-Type Three-Level LLC Converter

To verify the feasibility of the proposed modulation method to achieve the goal of wide voltage output, a complete simulation model of a full-bridge T-type three-level LLC resonant converter is constructed in this section based on the MATLAB/Simulink simulation platform. The main parameters of the simulation are shown in Table 1.

Table 1. Parameter table for simulation experiment of full-bridge T-type three-level LLC resonant converter.

Component	Value	Component	Value
V_{in}	200 V	L_r	7.8 μH
V_{out}	100 V–400 V	C_r	900 nF
f_r	60 kHz	L_m	31.2 μH
n	2	R_{load}	60 Ω
C_{dc1}/C_{dc2}	800 μF		

For the modulation method proposed in Section 3, simulation experiments were conducted to verify its effectiveness. As shown in Figure 13, with an input voltage of 200 V and a transformer turns ratio of 1:2 on both the primary and secondary sides, the system operates at a resonant frequency of 60 kHz. The output voltage was switched between 400 V and 100 V three times, as shown in the simulation waveforms. Figure 13 clearly demonstrates the feasibility of this modulation method, which meets the requirement for a four-fold output voltage ratio under resonant frequency conditions. Moreover, during the switching process, no overvoltage or overcurrent conditions occurred, preventing significant damage to the converter. The input filter capacitor voltage remained stable at approximately 100 V, with only a small deviation of about 0.5 V. This deviation can be attributed to the converter's balancing of capacitance charging and discharging during the switching between sub-modes.

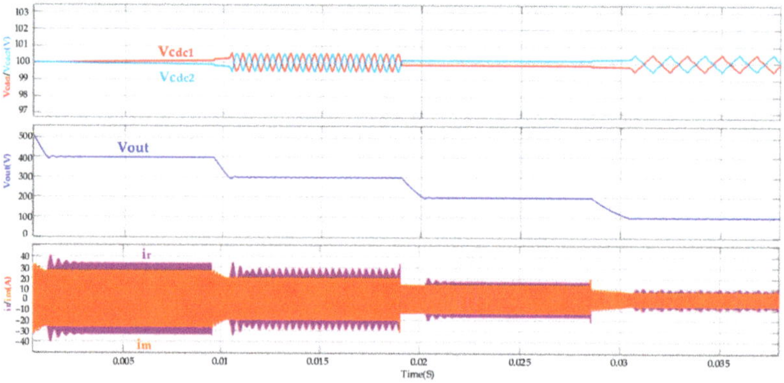

Figure 13. Dynamic simulation waveforms of output voltage change from 400 V to 100 V and waveforms of input capacitor voltage and resonant current and excitation current.

Figure 14 shows the simulated waveforms of the converter in the unity gain mode, with an input voltage of 200 V and operating at the resonant frequency. When a complementary drive signal is detected at unit gain and a dead time is applied, the reverse-connected T-type devices remain in the off state. At this point, the output voltage stabilizes around 400 V with a ripple of approximately 0.25%. Meanwhile, the secondary current follows a continuous half-sine waveform, enabling continuous energy transfer from the primary side to the secondary side.

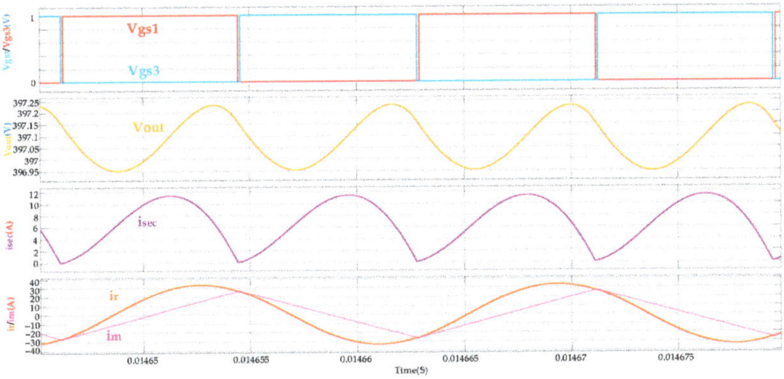

Figure 14. Steady-state waveforms in unity gain mode.

As shown in Figure 15, the simulated experimental waveform of the transition from sub-mode 1 to sub-mode 2 in 0.75-times gain mode is presented. It can be observed that the gate drive signal V_{gs1} changes from a 50% duty cycle to a constant off state, while V_{gs3} changes from a 50% duty cycle to a constant on state. With an input voltage of 200 V, the output voltage fluctuates around 300 V with a ripple rate not exceeding 1.67%. The switching process lasts 1 to 2 cycles, with minimal energy loss, no overcurrent in the resonant and excitation currents, and very little current jitter. This effectively achieves the transition between sub-modes while maintaining the voltage balance of the input capacitor. A 0.75-times modulation method is implemented at the resonant frequency, resulting in an output voltage that is 0.75 times the normalized gain.

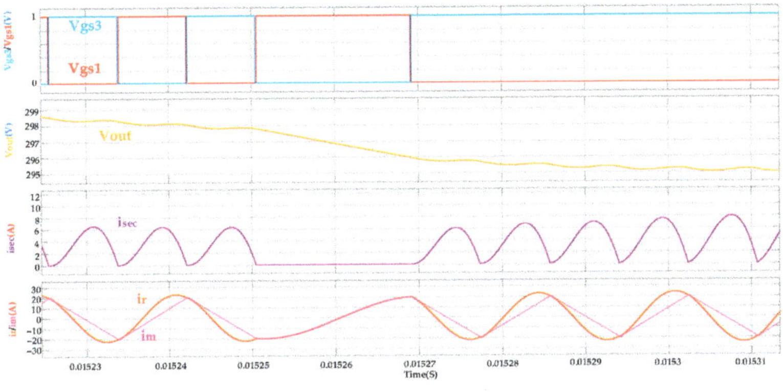

Figure 15. Dynamic transition waveform between two sub-circuit modes at a normalized voltage gain of G = 0.75.

As shown in Figure 16, the steady-state waveform of the bridge arm operation in 0.5× gain mode is presented. In this mode, the switching tubes S_{a2} and S_{a4} operate complementarily at the resonant frequency. The output voltage is half of the output voltage in the doubled gain mode, stabilizing around 200 V, with a ripple rate not exceeding 1.5%. The secondary-side current follows a continuous half-sinusoidal waveform, and energy is continuously transferred from the primary to the secondary side. Both the resonant current and excitation current remain within the permissible current range, ensuring that the converter achieves an output voltage half of the normalized gain at the resonant frequency.

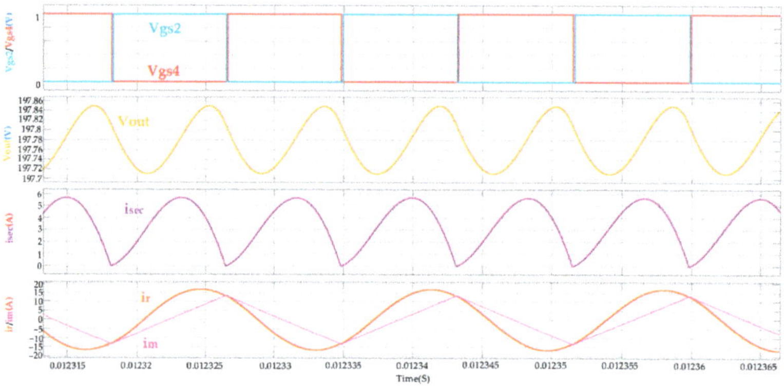

Figure 16. Steady-state waveform at a normalized voltage gain of G = 0.5.

As shown in Figure 17, the steady-state waveform of the switching process from sub-mode 2 to sub-mode 1 in the 0.25-times gain mode is presented. It can be observed that the gate drive signal V_{gs1} changes from a 50% duty cycle to a constant off state, while V_{gs3} changes from a 50% duty cycle to a constant on state. The input voltage is 200 V, and the output voltage fluctuates around 100 V, with a ripple rate not exceeding 3%. The switching process lasts for 1 to 2 cycles, with minimal energy loss, no overcurrent in the resonant and excitation currents, and minimal current jitter. This effectively enables switching between sub-modes while maintaining the voltage balance of the input capacitor. The resonant frequency is modulated by a factor of 0.25, resulting in an output voltage that is 0.25 times the normalized gain.

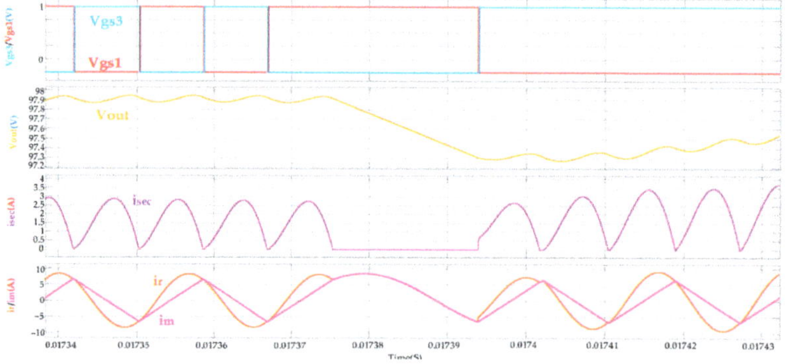

Figure 17. Dynamic transition waveforms between two sub-circuit modes at a normalized voltage gain of G = 0.25.

4.2. Experiments on Different Modulation Modes of Full-Bridge T-Type Three-Level LLC Converter

To verify the feasibility of the output voltage in the proposed 0.75-fold and 0.25-fold gain modes and to achieve a wide output voltage range, a full-bridge T-type three-level

LLC resonant converter is constructed, as shown in Figure 18. The key parameters are as follows: Lr = 8.2 μH (including 0.4 μH transformer leakage inductance), Cr = 0.88 μF, Lm = 31.2 μH, fr = 58 kHz, and the transformer turns ratio n = 2. The MOSFET switching device is the STW70N60DM2 from STMicroelectronics. The rectifier diode is the VS-E5PX3006L-N3, and the controller used is the DSP F28377D.

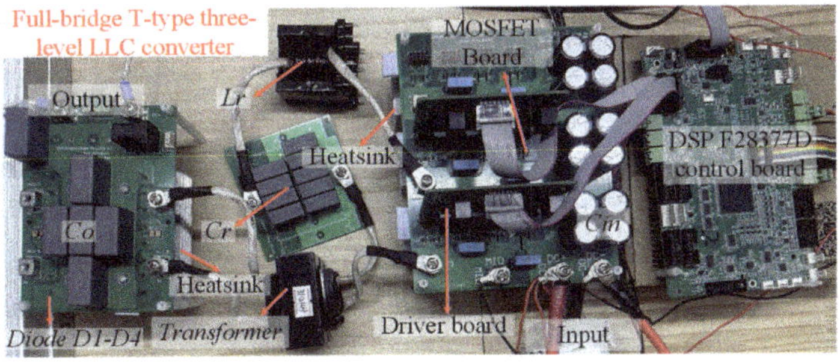

Figure 18. Full-bridge T-type three-level LLC resonant converter prototype.

As shown in Figures 19 and 20, the experimental waveforms are obtained under the traditional 1-times gain and 0.5-fold gain modes. The input voltage is a DC 200 V, and the switching frequency is 58 kHz. The output voltages are 400 V and 200 V, respectively, corresponding to the unity and 0.5-fold normalized voltage gain. The peak resonant currents are 26 A and 13 A, respectively, with the output load set to R_{load} = 120 Ω. The experimental waveforms show that the output voltages are stable at 400 V and 200 V, with very small ripple. Additionally, the resonant inductor current and excitation inductor current align with the theoretical analysis and simulation results, confirming the validity and feasibility of the modulation scheme.

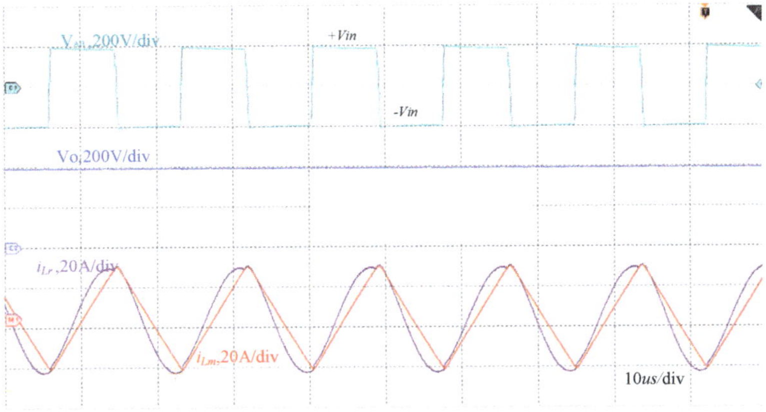

Figure 19. Experimental waveforms in conventional unity gain mode.

The dynamic experimental waveforms of the proposed 0.75-fold gain modulation mode, switching from sub-mode 2 to sub-mode 1, are shown in Figure 21. In sub-mode 2, the voltages of V_{AB} are 0.5 times and −1 times the input voltage. After switching to sub-mode 1, the voltages of V_{AB} become 1 and −0.5 times the input voltage. Both the theoretical analysis and simulation experiments validate the experimental waveforms, confirming their feasibility and correctness. During the switching process, there is no overcurrent in the resonant or excitation currents, which ensures the safety and reliability of the converter's operation. The input voltage remains at DC 200 V, and the output voltage is 300 V with minimal ripple. The output load is set to R_{load} = 120 Ω, and the switching frequency is 58 kHz, with a peak resonant current of 20.7 A.

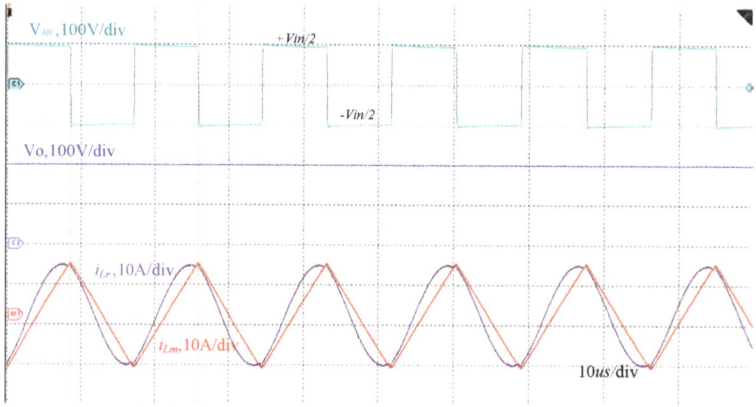

Figure 20. Experimental waveforms in conventional 0.5 gain mode.

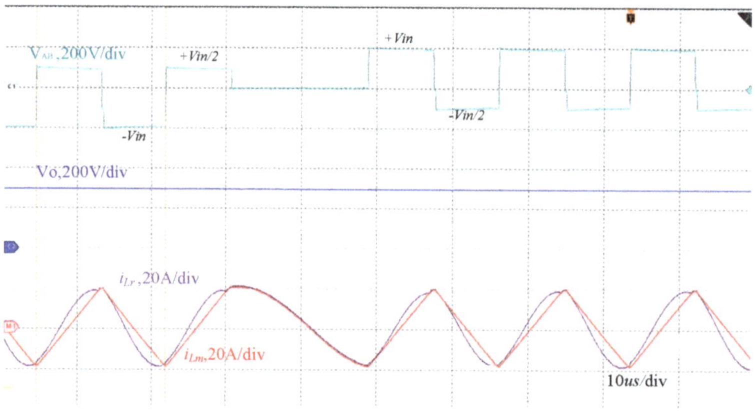

Figure 21. Dynamic experimental waveforms of the proposed 0.75-times gain mode switching from sub-mode 2 to sub-mode 1.

Figure 22 shows the steady-state experimental waveform after switching to sub-mode 2. In this sub-mode, the output voltage is 0.5 times and −1 times the input voltage, achieving 0.75 times the normalized gain of 300 V. The experimental waveforms align with the theoretical analysis and simulation results, proving that the modulation method is correct and feasible. This approach is conducive to achieving a wide voltage range output.

The dynamic experimental waveforms for the proposed 0.25-fold gain modulation mode, switching from sub-mode 2 to sub-mode 1, are shown in Figure 23. In sub-mode 2, the voltage of V_{AB} is −0.5 times the input voltage. After switching to sub-mode 1, the voltage of V_{AB} becomes 0.5 times the input voltage. Both theoretical analysis and simulation experiments validate the experimental waveforms, confirming their feasibility and correctness. During the switching process, neither the resonant current nor the excitation current exceeds the current limit, ensuring the safety and reliability of the converter's operation. The input voltage is maintained at 200 V DC, and the output voltage is 100 V with minimal ripple. The output load is set to R_{load} = 120 Ω, the switching frequency is 58 kHz, and the peak resonant current is 9.2 A.

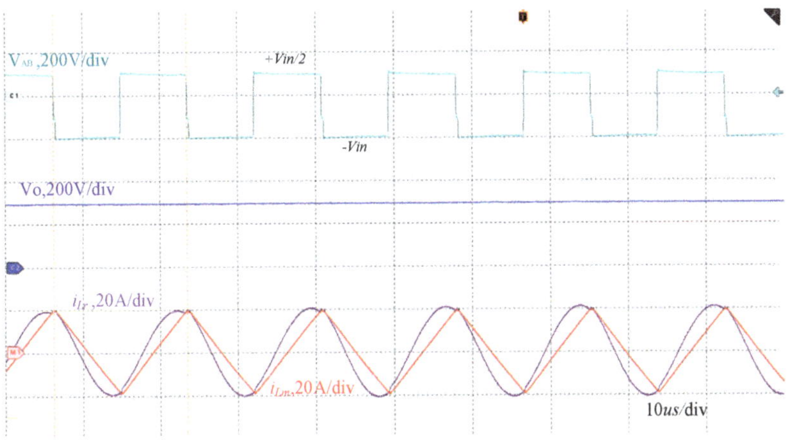

Figure 22. Proposed 0.75-times gain sub-mode 2 steady state experimental waveforms.

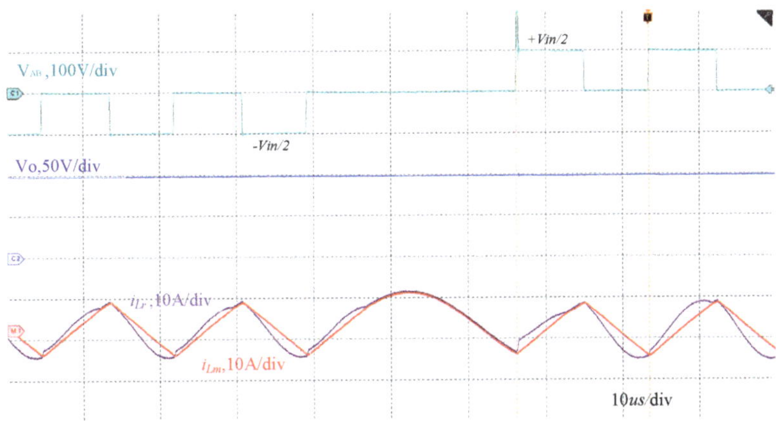

Figure 23. Dynamic experimental waveforms of the proposed 0.25-times gain mode switching from sub-mode 2 to sub-mode 1.

Figure 24 shows the steady-state experimental waveforms after switching to sub-mode 1. In this sub-mode, the output voltage is 0.5 times the input voltage, resulting in a normalized gain of 0.25 times 100 V. The experimental waveforms align with the theoretical analysis and simulation results, demonstrating the correctness and feasibility of the modulation method. This approach facilitates the realization of a wide voltage range output.

As shown in Figure 25, the heavy load waveform operates in 0.75-times gain mode with an input voltage of 100 V. Figure 26 presents the dynamic waveform of the output voltage, with the input voltage fixed at 100 V. Figure 26a shows the dynamic change of the voltage from 240 V to 180 V in 1-times gain mode, Figure 26b shows the dynamic change of the voltage from 180 V to 140 V in 0.75-times gain mode, Figure 26c shows the dynamic change of the voltage from 130 V to 90 V in 0.5-times gain mode, and Figure 26d shows the dynamic change of the voltage from 60 V to 50 V in 0.25-times gain mode. From the waveforms in these four sub-figures, it is evident that there is no over-voltage or over-current in the input voltage and resonant current during the voltage transition. This demonstrates that the converter can achieve a wide voltage range output by selecting different operating modes for voltage variation.

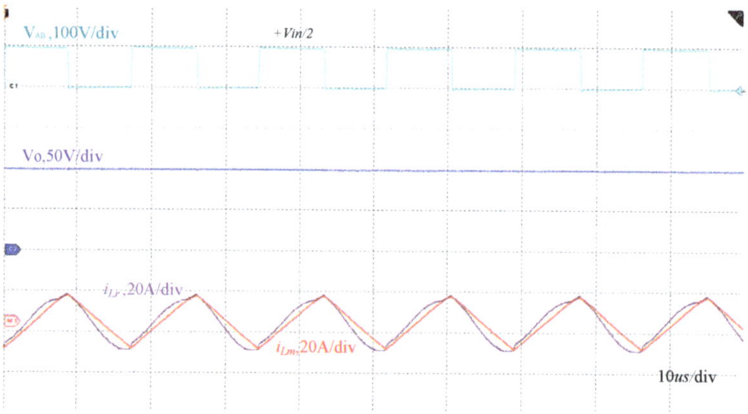

Figure 24. Proposed 0.25-times gain sub-mode 1 steady state experimental waveforms.

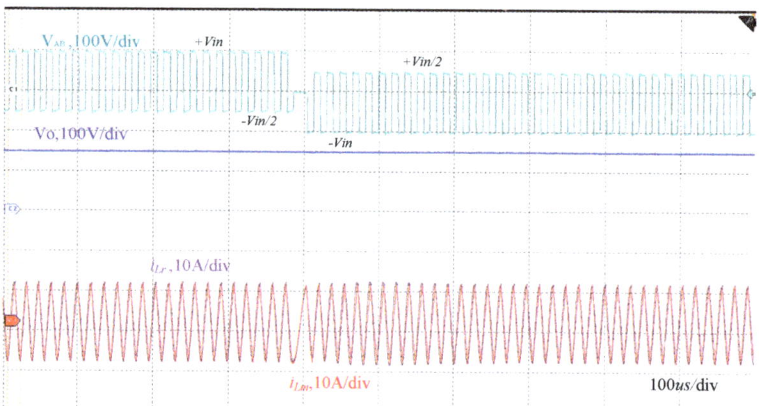

Figure 25. Heavy load waveforms in 0.75-times gain mode with 100 V input voltage.

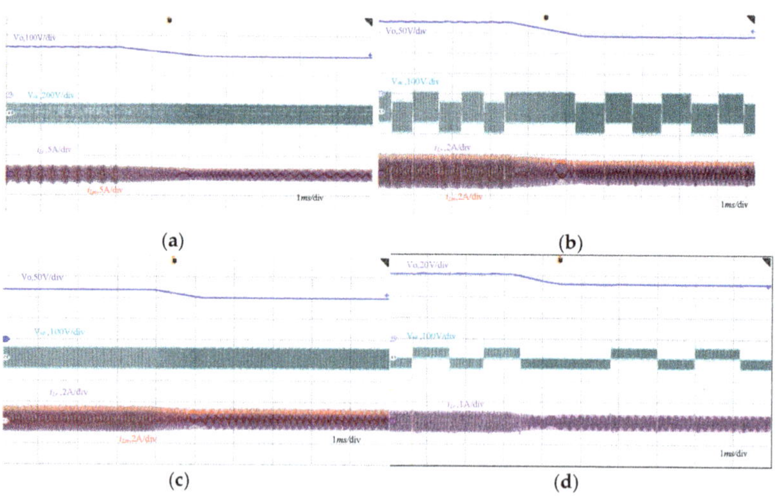

Figure 26. Output voltage dynamic response waveforms in different modes. (**a**) Conventional 1 gain mode; (**b**) 0.75 gain mode; (**c**) conventional 0.5 gain mode; (**d**) 0.25 gain mode.

The efficiency test results of the converter operating in the 0.75× gain mode are shown in Figure 27 with an input voltage of 200 V. At an output power of 1800 W and a switching frequency of 60 kHz, the converter achieves a conversion efficiency of 93.02%. These efficiency measurements for the proposed 0.75× gain modes at different output powers and frequencies verify the feasibility and effectiveness of the proposed method.

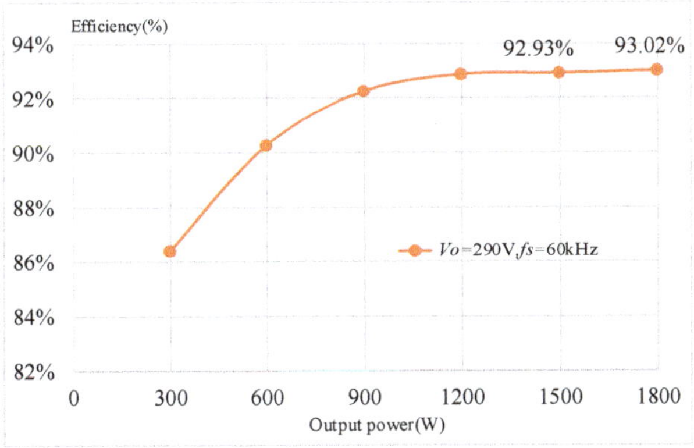

Figure 27. Efficiency at different output powers in 0.75× gain mode.

5. Conclusions

In this paper, a wide-output-voltage-range control method based on a full-bridge T-type three-level LLC resonant converter is proposed. The proposed 0.75-fold and 0.25-fold voltage gain modulation techniques effectively extend the output voltage range, addressing the challenges encountered by conventional LLC resonant converters in wide voltage applications. By combining the T-type three-level bridge arm with PWM modulation and employing a variable topology and frequency control strategy, the method successfully broadens the voltage gain range. Additionally, the PWM modulation method enhances the output voltage range by dynamically adjusting both the topology and frequency, effectively overcoming the limitations of traditional LLC converters. Simulation models and experimental prototypes developed in this study validate the effectiveness and feasibility of the proposed method. Experimental results demonstrate that the method achieves an input voltage of 100 V and an output voltage with a gain exceeding six times, reaching 40 V–240 V. This provides a reliable solution for applications requiring a wide output voltage range, such as new energy power generation systems and electric vehicle charging equipment. In the future, this converter can be applied to practical systems, such as electric vehicle charging or renewable energy interfaces.

Author Contributions: Conceptualization, K.Z. (Kangjia Zhang) and Z.Y.; Methodology, K.Z. (Kangjia Zhang) and Z.Y.; Software, M.L.; Validation, X.Y.; Formal analysis, K.Z. (Kun Zhao); Investigation, X.Y.; Resources, K.Z. (Kun Zhao); Data curation, K.Z. (Kangjia Zhang) and M.L.; Writing—original draft, K.Z. (Kangjia Zhang); Writing—review & editing, Z.Y. All authors have read and agreed to the published version of the manuscript.

Funding: This work was supported in part by Chengdu Guojia Electrical Engineering Company Ltd., under Grant NEEC-2022-B15.

Data Availability Statement: The original contributions presented in this study are included in the article. Further inquiries can be directed to the corresponding author.

Conflicts of Interest: The authors declare that this study received funding from Chengdu Guojia Electrical Engineering Company Ltd. The funder was not involved in the study design, collection, analysis, interpretation of data, the writing of this article, or the decision to submit it for publication.

References

1. Fei, C.; Lee, F.C.; Li, Q. High-Efficiency High-Power-Density LLC Converter with an Integrated Planar Matrix Transformer for High-Output Current Applications. *IEEE Trans. Ind. Electron.* **2017**, *64*, 9072–9082. [CrossRef]
2. Kundu, U.; Yenduri, K.; Sensarma, P. Accurate ZVS Analysis for Magnetic Design and Efficiency Improvement of Full-Bridge LLC Resonant Converter. *IEEE Trans. Power Electron.* **2017**, *32*, 1703–1706. [CrossRef]
3. Yao, Z.; He, X.; Liu, M.; Liu, J.; Xiao, Z.; Tang, Y. Fixed Switching Frequency Control Using Trapezoidal Current Mode to Achieve ZVS in Three-Level DC-DC Converters. *IEEE Trans. Ind. Electron.* **2025**, *72*, 4175–4185. [CrossRef]
4. Wang, H.; Dusmez, S.; Khaligh, A. Design and Analysis of a Full-Bridge LLC-Based PEV Charger Optimized for Wide Battery Voltage Range. *IEEE Trans. Veh. Technol.* **2014**, *63*, 1603–1613. [CrossRef]
5. Deng, J.; Li, S.; Hu, S.; Mi, C.C.; Ma, R. Design Methodology of LLC Resonant Converters for Electric Vehicle Battery Chargers. *IEEE Trans. Veh. Technol.* **2014**, *63*, 1581–1592. [CrossRef]
6. Mustafa, A.; Mekhilef, S. Dual Phase LLC Resonant Converter with Variable Frequency Zero Circulating Current Phase-Shift Modulation for Wide Input Voltage Range Applications. *IEEE Trans. Power Electron.* **2021**, *36*, 2793–2807. [CrossRef]
7. Lin, J.; Yueh, H.; Lin, Y.; Liu, P. Variable-Frequency and Phase-Shift with Synchronous Rectification Advance On-Time Hybrid Control of LLC Resonant Converter for Electric Vehicles Charger. *IEEE J. Emerg. Sel. Top. Ind. Electron.* **2023**, *4*, 348–356. [CrossRef]
8. Wei, Y.; Luo, Q.; Mantooth, A. Overview of Modulation Strategies for LLC Resonant Converter. *IEEE Trans. Power Electron.* **2020**, *35*, 10423–10443. [CrossRef]
9. Zhang, J.; Cai, Z. Hybrid Control Switching Technology for LLC Resonant Converter. *Energies* **2024**, *17*, 6250. [CrossRef]
10. Jiang, H.; Yue, X. Phase-Shifting Adaptive LLC Resonant Converter with Reduced Turn-Off Loss in Wide Voltage Application. *Energies* **2024**, *17*, 1555. [CrossRef]
11. Dou, Y.; Ai, S.; Cheng, S.; Fang, S. Hybrid Control and Optimal Parameters Design for Wide Input Range Series Half Bridge Three Level LLC Resonant Converter in UUV Power Supply System. *IEEE Trans. Transp. Electrif.* **2025**, *1*, 1. [CrossRef]
12. Yue, Y.; Liu, Y.; Zhang, J.; Zhao, H.; Yang, J. Hybrid Control Method of Full-Bridge LLC Resonant Converter Based on Electric Vehicle. *IEEE Access* **2024**, *12*, 51859–51868. [CrossRef]
13. Luo, G.; Yao, Z.; Jiang, L.; Xiao, Z.; Chen, W.; Tang, Y. Current Balancing Control Method with Reduced Number of Current Sensors for Interleaved LLC Converters. *IEEE J. Emerg. Sel. Top. Power Electron.* **2024**, *12*, 2604–2613. [CrossRef]
14. Gao, L.; Liu, K.; Chen, Y.; Luan, H.; Zhou, S.; Cai, H. Variable-Gain LLC Converter with Series–Parallel Hybrid Switch-Controlled Capacitor Network. *IEEE Trans. Power Electron.* **2025**, *40*, 3885–3898. [CrossRef]
15. Hou, Z.; Jiao, D.; Lai, J.-S. An Ultrawide Range Pulse Width Modulated LLC Converter with Voltage Multiplier Rectifiers. *IEEE Trans. Power Electron.* **2025**, *40*, 6216–6229. [CrossRef]
16. Ning, G.; Du, L.; Xu, G.; Sun, Y.; Liu, Y.; Han, H. Dual-Mode LLC Converter with Wide Voltage Gain in Narrow Switching Frequency Range. *IEEE Trans. Power Electron.* **2024**, *39*, 12058–12064. [CrossRef]
17. Wei, Y.; Luo, Q.; Mantooth, A. Comprehensive Analysis and Design of LLC Resonant Converter with Magnetic Control. *CPSS Trans. Power Electron. Appl.* **2019**, *4*, 265–275. [CrossRef]
18. Zhu, L.; Sheng, Z.; Peng, F.; Yang, L. Control Strategy of Half-Bridge Three-Level LLC Resonant Converters with Wide Output Voltage Range. *IEEE Trans. Plasma Sci.* **2022**, *50*, 4381–4386. [CrossRef]
19. Yao, Z.; Liu, M.; Li, Y.; Xiao, Z.; Jiang, Y.; Zhang, L. Generalized Trapezoidal Current Mode-Based Zero-Voltage Switching for Multilevel DC-DC Converters. *IEEE Trans. Power Electron.* **2025**, *40*, 8956–8961. [CrossRef]
20. Qin, J.; He, Z.; Weng, Z.; Guan, R.; Li, Z.; Li, Z. Serial Hybrid-Clamped Three-Level Half-Bridge LLC Resonant Converter with Wide Input Voltage Range. *IEEE Trans. Ind. Electron.* **2024**, *71*, 13991–14001. [CrossRef]
21. Chen, J.; Wang, Z.; Lan, Y.; Zeng, L.; Xiong, X.; Wang, W. A Novel Five-level LLC Resonant Converter and Its Modulation Strategy. In Proceedings of the 2024 IEEE 8th Conference on Energy Internet and Energy System Integration (EI2), Shenyang, China, 29 November–2 December 2024; pp. 3714–3719. [CrossRef]
22. Jiang, T.; Zhang, J.; Wu, X.; Sheng, K.; Wang, Y. A Bidirectional Three-Level LLC Resonant Converter with PWAM Control. *IEEE Trans. Power Electron.* **2016**, *31*, 2213–2225. [CrossRef]
23. Yao, Z.; He, X.; Lan, H.; Yang, H.; Xiao, Z.; Deng, F. A Trapezoidal Current Mode to Reduce Peak Current in Near-CRM for Three-Level DC–DC Converter. *IEEE Trans. Power Electron.* **2024**, *39*, 9446–9456. [CrossRef]
24. Jin, K.; Ruan, X. Hybrid Full-Bridge Three-Level LLC Resonant Converter—A Novel DC–DC Converter Suitable for Fuel-Cell Power System. *IEEE Trans. Ind. Electron.* **2006**, *53*, 1492–1503. [CrossRef]

25. Rashwan, A.; Sayed, M.A.; Mobarak, Y.A.; Shabib, G.; Senjyu, T. Predictive Controller Based on Switching State Grouping for a Modular Multilevel Converter with Reduced Computational Time. *IEEE Trans. Power Deliv.* **2017**, *32*, 2189–2198. [CrossRef]
26. Jia, P.; Liu, M. A Wide Range LLC Resonant Converter Realized by an Adjustable Turns Ratio Transformer. *IEEE Trans. Power Electron.* **2025**, *40*, 10396–10401. [CrossRef]
27. Li, G.; Yang, D.; Zhou, B.; Zhang, H. A Topology-Reconfigurable LLC Resonant Converter for Wide Output Range Applications. *IEEE Trans. Veh. Technol.* **2022**, *71*, 10389–10399. [CrossRef]
28. Zuo, Y.; Shen, X.; Du, B.; Sui, Q.; Geboers, T.; Martinez, W. Reconfigurable LLC Resonant Converter for Wide Voltage Range and Reduced Voltage Stress in DC-Connected EV Charging Stations. In Proceedings of the 2025 IEEE Applied Power Electronics Conference and Exposition (APEC), Atlanta, GA, USA, 16–20 March 2025; pp. 436–441. [CrossRef]
29. Xue, Y.; Yang, G. A Wide-Output-Range DC-DC Converter and Minimum Loss Collaborative Control Strategy. *Energies* **2024**, *17*, 499. [CrossRef]
30. Alaql, F.; Alfraidi, W.; Alhatlani, A.; Al-Shamma'a, A.A.; Hussein Farh, H.M.; Allehyani, A. A Detailed Analysis and Gain Derivation of Reconfigurable Voltage Rectifier-Based LLC Converter. *Electronics* **2024**, *13*, 3788. [CrossRef]
31. Hao, Z.; Yang, Y.; Li, H. Design of "十" Low Turn Ratio Planar Transformer in High Transformation Ratio LLC Resonant Converter. In Proceedings of the 2024 IEEE 10th International Power Electronics and Motion Control Conference (IPEMC2024-ECCE Asia), Chengdu, China, 17–20 May 2024; pp. 3781–3786. [CrossRef]
32. Khodabakhsh, J.; Moschopoulos, G. A T-Type DC-DC Converter with Reduced Middle-Leg Switch Current Stress and Improved Light-Load Efficiency. *IEEE Open J. Power Electron.* **2020**, *1*, 238–249. [CrossRef]
33. Sun, X.; Li, X.; Shen, Y.; Wang, B.; Guo, X. Dual-Bridge LLC Resonant Converter with Fixed-Frequency PWM Control for Wide Input Applications. *IEEE Trans. Power Electron.* **2017**, *32*, 69–80. [CrossRef]
34. Yao, Z.; Jiang, L.; Sun, T.; Xiao, Z.; Chen, W.; Tang, Y. Smooth Optimal Trajectory and Voltage Balancing Control for T-Type Three-Level LLC Resonant Converter in Alternating Mode. *IEEE Trans. Power Electron.* **2025**, *40*, 9278–9288. [CrossRef]
35. Jiang, L.; Yao, Z.; Luo, G.; Yang, B.; Chen, W. Design and Control Method of T-Type Three-Level LLC Converter for Wide Input Voltage Application. In Proceedings of the 2024 IEEE 10th International Power Electronics and Motion Control Conference (IPEMC2024-ECCE Asia), Chengdu, China, 17–20 May 2024; pp. 2122–2127. [CrossRef]
36. Ma, W.; Li, H.; Yin, S.; Pang, X.; Chu, D. Design of Hybrid SiC/Si Based T-Type Three-Level LLC Resonant Converter with Wide-Input Range and Low Conduction Loss for Automotive Auxiliary Power Module. *IET Power Electron.* **2023**, *16*, 209–226. [CrossRef]
37. Queiroz, S.S.; Costa, L.F. Optimal Design of a LLC Resonant DC-DC Converter Based on the Stacked Half-Bridge Concept for Solid-State Transformer. *IEEE Trans. Power Electron.* **2025**. [CrossRef]
38. Qin, J.; Li, Z.; He, Z.; Weng, Z.; Guan, R.; Li, Z. Thermal Balance Modulation of Switching Devices for Hybrid-Clamped Three-Level Full-Bridge LLC Resonant Converter. *IEEE Trans. Power Electron.* **2024**, *39*, 5487–5497. [CrossRef]
39. Yao, Z.; He, X.; Xiao, Z.; He, Z.; Jiang, Y.; Tang, Y. Full-Range ZVS Control Method Based on Near-CRM for T-Type Three-Level Inverter. In Proceedings of the 2023 IEEE Energy Conversion Congress and Exposition (ECCE), Nashville, TN, USA, 29 October–2 November 2023; pp. 2987–2992. [CrossRef]
40. Shi, Y.; Wang, L.; Xie, R.; Shi, Y.; Li, H. A 60-kW 3-kW/kg Five-Level T-Type SiC PV Inverter with 99.2% Peak Efficiency. *IEEE Trans. Ind. Electron.* **2017**, *64*, 9144–9154. [CrossRef]
41. Wei, Y.; Zhang, S.; Liu, J.; Mantooth, H.A. Hybrid PFM and PWM Modulation Strategy for Stacked Structure LLC Resonant Converter with Wide Input Voltage Range Application. *IEEE Trans. Ind. Appl.* **2024**, *60*, 4038–4053. [CrossRef]
42. Kirshenboim, O.; Peretz, M.M. Combined Multilevel and Two-Phase Interleaved LLC Converter with Enhanced Power Processing Characteristics and Natural Current Sharing. *IEEE Trans. Power Electron.* **2018**, *33*, 5613–5620. [CrossRef]

energies

MDPI

Article

State-Set-Optimized Finite Control Set Model Predictive Control for Three-Level Non-Inverting Buck–Boost Converters

Mingxia Xu [1], Hongqi Ding [2,*], Rong Han [3], Xinyang Wang [3], Jialiang Tian [1], Yue Li [1] and Zhenjiang Liu [1]

[1] School of Electrical Engineering, Dalian Jiaotong University, Dalian 116024, China; xumingxia@djtu.edu.cn (M.X.); 13546032451@163.com (J.T.); liyueoin@163.com (Y.L.); 19979978688@139.com (Z.L.)

[2] School of Electrical and Automation Engineering, Nanjing Normal University, Nanjing 210023, China

[3] School of Optoelectronic Engineering and Instrumentation Science, Dalian University of Technology, Dalian 116024, China; hanrong@dlut.edu.cn (R.H.); pjwxy990506@mail.dlut.edu.cn (X.W.)

* Correspondence: dinghq@nnu.edu.cn

Abstract

Three-level non-inverting buck–boost converters are promising for electric vehicle charging stations due to their wide voltage regulation capability and bidirectional power flow. However, the number of three-level operating states is four times that of two-level operating states, and the lack of a unified switching state selection mechanism leads to serious challenges in its application. To address these issues, a finite control set model predictive control (FCS-MPC) strategy is proposed, which can determine the optimal set and select the best switching state from the excessive number of states. Not only does the proposed method achieve fast regulation over a wide voltage range, but it also maintains the input- and output-side capacitor voltage balance simultaneously. A further key advantage is that the number of switching actions in adjacent cycles is minimized. Finally, a hardware-in-the-loop experimental platform is built, and the proposed control method can realize smooth transitions between multiple operation modes without the need for detecting modes. In addition, the state polling range and the number of switching actions are superior to conventional predictive control, which provides an effective solution for high-performance multilevel converter control in energy systems.

Keywords: buck–boost converters; finite control set model predictive control (FCS-MPC); switching actions; three-level converters; wide voltage range

1. Introduction

With the advantages of cleanliness, energy saving and high efficiency [1,2], new energy vehicles have gradually become an indispensable mode of transport for people who travel. However, due to the long charging time of conventional electric vehicles and the difference in charging voltage levels of different manufacturers, the voltage fluctuation ranges from 200 V to 920 V [3], and some of them even reach 1000 V [4]. Therefore, the solution of DC/DC fast-charging technology with adaptable high voltage, high power, and wide voltage range has received extensive attention from researchers.

The buck–boost converter has been widely adopted in electric vehicle charging stations, energy storage systems, and DC micro grids due to its non-inverting output and flexible voltage regulation capabilities [5,6]. However, as fast-charging power levels increase from tens of kilowatts to hundreds of kilowatts, there exists an urgent demand for power conversion solutions that support ultra-wide voltage ranges. A novel buck–boost

https://doi.org/10.3390/en18174481

flying capacitor converter for electric vehicle (EV) fast-charging station DC–DC stages has been proposed in [7]. On this basis, an improved DC–DC buck–boost partial power converter with better efficiency compared to [7] is subsequently developed in [8]. However, conventional two-level buck–boost converters with single-leg topology have inherent limitations—switching devices experience increased voltage stress at high output voltages, which raises the susceptibility to overvoltage damage. This operational constraint necessitates the selection of devices with higher voltage class, resulting in increased cost and challenges in increasing the switching frequency. Consequently, existing two-level buck–boost converter architectures are proving inadequate to meet the requirements of high-power, fast-charging systems.

To address the trade-offs between voltage stress and switching frequency, three-level buck–boost topologies have been investigated. These configurations achieve multi-level output characteristics through flying capacitors or cascaded structures [9,10], effectively reducing device voltage stress to 50% of conventional topologies. The three-level framework offers dual advantages: (1) filter size can be reduced, and power density can be improved by multi-level waveform generation, and (2) lower-voltage devices facilitate higher switching frequency operation, thereby enhancing the dynamic response. However, the three-level topology presents significant challenges in implementing control strategies such as multi-mode operation, input/output capacitor voltage balance, and switching state optimization.

In terms of control strategies, conventional two-level buck–boost converters have inherent defects during mode transitions due to minimum pulse width limitation [11]. Specifically, when the input and output voltages approach equilibrium, the voltage gain dead zone in the buck–boost mode transition region occurs in conventional control strategies, resulting in regulation failure [12]. To address this problem, a variety of strategies have been investigated to improve performance. For non-inverting buck–boost topologies, a multi-mode operation strategy has been proposed in [13], where an additional buck–boost mode is introduced alongside the conventional buck and boost modes to mitigate the dead-zone effects. However, this approach can lead to switching oscillations in the region of critical transition thresholds. To suppress such oscillations, a hysteresis control has been implemented in [14].

To further enhance the dynamic characteristics of the converter, the MPC strategy is widely adopted. A modulated MPC is proposed in [15], where the operating state is divided into four modes and the optimal duty cycle is selected via a voltage-based cost function. A key drawback, however, is that this algorithm requires explicit monitoring of the operating mode. To overcome this problem, three operating modes are classified in [16], and the cost function is selected as the minimum fluctuation of inductor current in each control cycle, so that the optimized duty cycle for the subsequent control interval is preferred. For FCS-MPC, a state-space model including switching states is established in [17], and four switching states are polled to optimize the output voltage tracking effect. Since the cost function does not include the inductor current, it is prone to cause current inrush. Building on [17], the work in [18] refines this approach by incorporating both the inductor current and its limit into the cost function. This dual-objective optimization ensures smooth output voltage transitions while effectively suppressing current inrush. Unlike that with two levels, the three-level buck–boost converter confronts the following problems: Firstly, the capacitor voltage balancing control on the input and output sides needs to be regulated. Secondly, the four half-bridge topologies can generate 16 switching states [19], and the global polling computational burden increases dramatically. Finally, the voltage oscillation phenomenon still exists when the input and output voltages are close to each other. To address these issues, Ref. [20] proposes a modulated MPC that introduces two extended operating modes and employs duty cycle hysteresis control to prevent mode oscillations. Notably, the monitor of the converter operating mode is also required, and the optimal number of switching actions cannot be guaranteed. For FCS-MPC, the duty cycle is solved in [21], and the optimal duty cycle is polled in the range of 0 to 1 as a way to achieve optimal control. However, the duty cycle has continuous characteristics, and its accuracy under discrete polling does not meet the demand for control accuracy.

Aiming to address limitations in existing FCS-MPC strategies, a switch-state-optimized FCS-MPC method is proposed in this paper. The contributions can be listed as follows:

(1) The total number of switching states is reduced from 16 to 11 while retaining full-range buck–boost operation and capacitor voltage balancing capabilities, thereby reducing control complexity. Additionally, the proposed adjacent-state polling strategy further limits candidate states per control cycle to a maximum of five, substantially decreasing computational load and switching losses.

(2) A multi-objective cost function integrating mode transitions, capacitor voltage balancing, and current tracking is established to enable systematic trade-offs among objectives.

(3) Seamless mode switching through shared bridging state (M1) ensures smooth transition for different load scenarios.

Ultimately, the proposed control dynamically optimizes switching actions and mode transitions, offering a novel solution for efficient control of three-level topologies. This paper is organized as follows: The principle and discrete state-space model of the three-level buck–boost converter are analyzed in Section 2. The switching state optimization method is proposed in Section 3. Next, the FCS-MPC algorithm is presented in Section 4. Section 5 validates the effectiveness of the proposed method through hardware-in-the-loop. Finally, the conclusions are presented in Section 6.

2. Three-Level Non-Inverting Buck–Boost Converter Working Principle

2.1. Subsection Working Principle of TLNIBBC

The topology of the three-level non-inverting buck–boost converter is shown in Figure 1. It consists of eight switches named S_1–S_8 and their anti-parallel diodes. Among them, R represents the load resistance and L is the inductor connecting the input and output sides. R_{eq} represents the equivalent resistance of the inductor and the loop. C_{i1} and C_{i2} are the input-side support capacitors with voltages V_{i1} and V_{i2}, respectively, and the sum of which is the total input-side voltage V_{in}. Similarly, C_{o1} and C_{o2} are the output-side capacitors of the converter, with voltages V_{o1} and V_{o2}, respectively, and the sum of which is the total output voltage V_o. In addition, i_i and i_o are the input-side and output-side currents, respectively. Points o_1 and o_2 are the midpoints of the input and output capacitors. Points a, b, c, and d represent the output terminals of the four half-bridge legs, and the potentials are controlled by switching-state pairs (S_1, S_2), (S_3, S_4), (S_5, S_6), and (S_7, S_8).

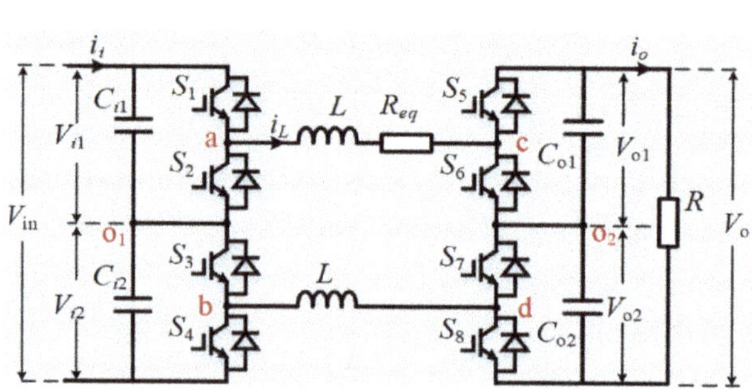

Figure 1. The topology of three-level buck–boost converter.

The three-level non-inverting buck–boost converter can be divided into a buck arm on the input side and a boost arm on the output side, of which both sides are symmetrical with respect to the inductance. Both buck and boost arms are cascaded by two half-bridge two-level converters. To avoid short-circuiting of capacitor C_{i1}, switching states S_1 and S_2 are complementary. Likewise, the switching states S_3 and S_4, S_5 and S_6, and S_7 and S_8 are each complementary. The three-level non-inverting buck–boost converter can be operated in either buck mode or boost mode.

As seen in Figure 2, the first row (modes ① to ④) forms the boost operating mode, while the first column (modes ①, ⑤, ⑨, and ⑬) constitutes the buck operating mode. Notably, mode ① is a shared state for both buck and boost operations. Therefore, theoretically, only these seven operating states are needed to achieve any step-up or step-down voltage conversion. However, these two operating modes alone cannot guarantee voltage balance across the input-side and output-side capacitors. Hence, additional modes must be incorporated. For instance, controlling the duty cycles of modes ⑦ and ⑪ enables voltage balancing for the input-side capacitors. Similarly, voltage balancing for the output-side capacitors can be achieved by controlling the duty cycles of modes ⑥ and ⑩. In summary, only 11 operating modes are required to implement the buck–boost conversion function.

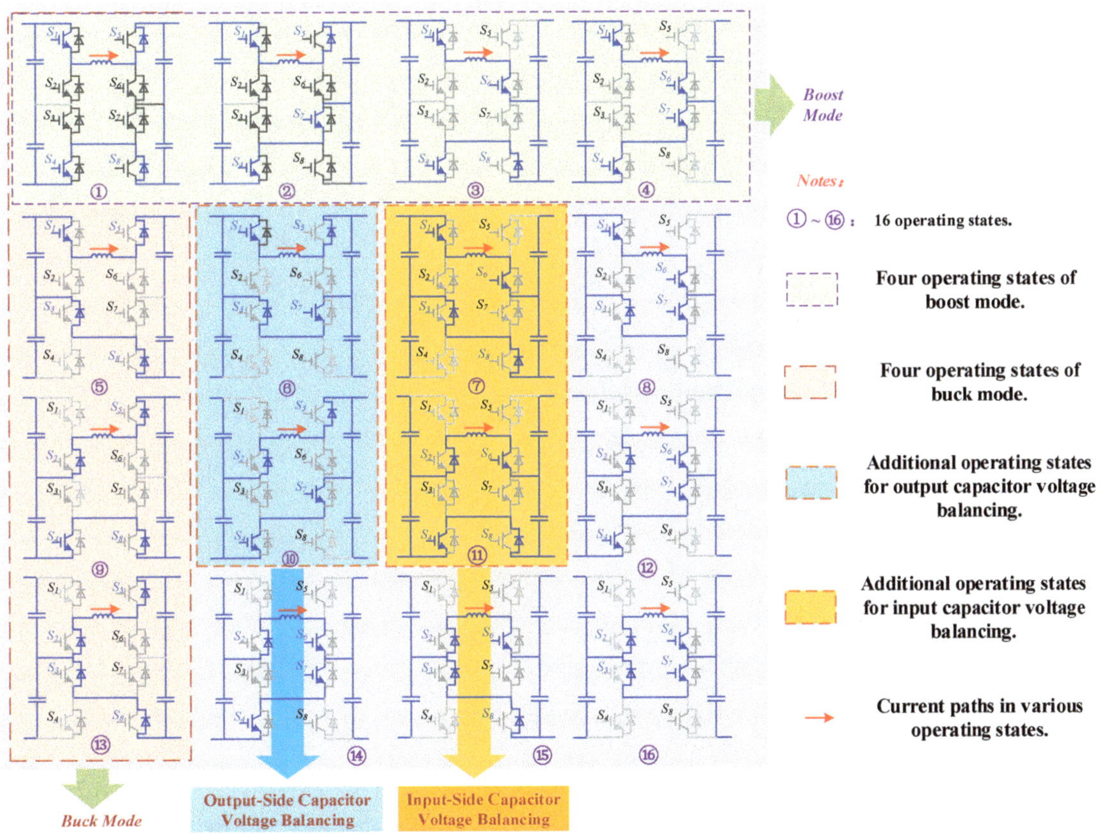

Figure 2. Summary of operating modes for the three-level buck–boost converter.

2.2. Time-Domain State-Space Model

Defining the switching states of the four half-bridges of the three-level non-inverting buck–boost as Q_1–Q_4, respectively, the following can be obtained:

$$Q_1 = \begin{cases} 1, & S_1\ on\ and\ S_2\ off \\ 0, & S_1\ off\ and\ S_2\ on \end{cases}, Q_2 = \begin{cases} 1, & S_3\ on\ and\ S_4\ off \\ 0, & S_3\ off\ and\ S_4\ on \end{cases}$$
$$Q_3 = \begin{cases} 1, & S_5\ on\ and\ S_6\ off \\ 0, & S_5\ off\ and\ S_6\ on \end{cases}, Q_4 = \begin{cases} 1, & S_7\ on\ and\ S_8\ off \\ 0, & S_7\ off\ and\ S_8\ on \end{cases} \quad (1)$$

According to the KVL equation, the input side midpoint voltage of the arm can be obtained as

$$v_{ab} = 2L\frac{di_L}{dt} + R_{eq} \cdot i_L + v_{cd} \quad (2)$$

Further simplification can be obtained:

$$v_{ab} = Q_1 \cdot V_{i1} + (1 - Q_2) \cdot V_{i2} \quad (3)$$

Similarly, the output side voltage can be obtained as follows:

$$v_{cd} = Q_3 \cdot V_{o1} + (1 - Q_4) \cdot V_{o2} \quad (4)$$

For the input-side capacitance, applying the KCL equation gives

$$\begin{cases} C_{i1}\frac{dV_{i1}}{dt} = i_i - Q_1 \cdot i_L \\ C_{i2}\frac{dV_{i2}}{dt} = i_i - (1 - Q_2) \cdot i_L \end{cases} \quad (5)$$

Likewise, the output side can be obtained:

$$\begin{cases} C_{o1}\frac{dV_{o1}}{dt} = Q_3 \cdot i_L - \frac{V_{o1}+V_{o2}}{R} \\ C_{o2}\frac{dV_{o2}}{dt} = (1 - Q_4) \cdot i_L - \frac{V_{o1}+V_{o2}}{R} \end{cases} \quad (6)$$

Combined with (2) to (6), the time-domain state-space model of a three-level non-inverting buck–boost DC–DC converter can be written in a unified matrix form:

$$\dot{\boldsymbol{x}} = \boldsymbol{A}\boldsymbol{x} + \boldsymbol{B}\boldsymbol{u} \quad (7)$$

where $\boldsymbol{x} = \begin{bmatrix} i_L & V_{i1} & V_{i2} & V_{o1} & V_{o2} \end{bmatrix}^T$, $\boldsymbol{u} = \boldsymbol{i_i}$.

$$\boldsymbol{A} = \begin{bmatrix} -\frac{R_{eq}}{2L} & \frac{Q_1}{2L} & \frac{(1-Q_2)}{2L} & -\frac{Q_3}{2L} & -\frac{(1-Q_4)}{2L} \\ -\frac{Q_1}{C_{i1}} & 0 & 0 & 0 & 0 \\ -\frac{(1-Q_2)}{C_{i2}} & 0 & 0 & 0 & 0 \\ \frac{Q_3}{C_{o1}} & 0 & 0 & -\frac{1}{RC_{o1}} & -\frac{1}{RC_{o1}} \\ \frac{(1-Q_4)}{C_{o2}} & 0 & 0 & -\frac{1}{RC_{o2}} & -\frac{1}{RC_{o2}} \end{bmatrix},$$
$$\boldsymbol{B} = \begin{bmatrix} 0 & \frac{1}{C_{i1}} & \frac{1}{C_{i2}} & 0 & 0 \end{bmatrix}^T$$

i_L represents the inductor current, and Q_1–Q_4 are the switching state variables, which reflect the on–off situation of the switches S_1–S_8. In summary, (7) is applicable whether the converter is in buck mode or boost mode.

2.3. Discrete State-Space Model

Based on the differential equations in the previous section, the continuous time model (7) is discretized by the forward Euler formula, and then the discrete time model of the converter can be obtained:

$$x^p(k+1) = G(T_s)x(k) + H(T_s)u(k) \tag{8}$$

where $G(T_s) = e^{AT_s}$, $H(T_s) = \int_0^{T_s} e^{At}dt \cdot B$. T_s is the sampling time. $x^p(k+1)$ is the predicted value of state variable x at time k + 1, where $x^p = [i_L{}^P, V_{i1}{}^P, V_{i2}{}^P, V_{o1}{}^P, V_{o2}{}^P]^T$. Substituting (7) into (8), the inductor current discrete-time model can be written as

$$\begin{aligned} i_L{}^P(k+1) = \left(1 - \tfrac{R_{eq}T_s}{2L}\right) \cdot i_L(k) + \tfrac{Q_1T_s}{2L} \cdot V_{i1}(k) - \tfrac{Q_3T_s}{2L} \cdot V_{o1}(k) \\ + \tfrac{(1-Q_2)T_s}{2L} \cdot V_{i2}(k) - \tfrac{(1-Q_4)T_s}{2L} \cdot V_{o2}(k) \end{aligned} \tag{9}$$

The discretized equations of the input- and output-side capacitor voltages are written as follows (10)–(13):

$$V_{i1}{}^P(k+1) = -\frac{Q_1T_s}{C_{i1}} \cdot i_L(k) + V_{i1}(k) + \frac{T_s}{C_{i1}} \cdot i_i(k) \tag{10}$$

$$V_{i2}{}^P(k+1) = -\frac{(1-Q_2)T_s}{C_{i2}} \cdot i_L(k) + \frac{T_s}{C_{i2}} \cdot i_i(k) + V_{i2}(k) \tag{11}$$

$$V_{o1}{}^P(k+1) = \frac{Q_3T_s}{C_{o1}} \cdot i_L(k) + \left(1 - \frac{T_s}{RC_{o1}}\right)V_{o1}(k) - \frac{T_s}{RC_{o1}} \cdot V_{o2}(k) \tag{12}$$

$$V_{o2}{}^P(k+1) = \frac{(1-Q_4)T_s}{C_{o2}} \cdot i_L(k) - \frac{T_s}{RC_{o2}} \cdot V_{o1}(k) + \left(1 - \frac{T_s}{RC_{o2}}\right)V_{o2}(k) \tag{13}$$

Equations (9)–(13) are discrete-time state variables, and according to this model, the MPC can be easily implemented.

3. Optimization of Multi-Mode Switching States

The discrete-time model is defined by switching state variables Q_1–Q_4, where each binary variable Q_i ($i = 1, 2, 3, 4$) is constrained to two states (0 or 1). Consequently, 11 distinct mathematical expressions are derived for the discrete-time model. When a conventional MPC algorithm is applied, the discrete state variables must be evaluated 16 times per control cycle, resulting in an excessive computational burden on the controller. More critically, if each switch state is polled exactly according to the inner-loop current prediction, the number of switching actions in the neighboring control cycle is severely increased, and the performance of the converter deteriorates. For instance, a direct transition from the switching state 1010 to 0101 would require all eight switches to be toggled simultaneously, causing significant switching losses that are highly detrimental to converter efficiency. To mitigate these issues, systematic classification and optimization of converter modes are necessitated. The optimization strategies under three operational scenarios—buck mode, boost mode, and transient alternating states—are elaborated in the following sections.

3.1. Switching State Optimization in Buck Mode

The current paths corresponding to the four modes (M1–M4) of the three-level non-inverting converter operating in buck mode are illustrated in Figure 3. In Figure 3a, the switching states of Q_1–Q_4 are configured as 1010, where the load power is supplied jointly by the input voltage source and the inductor. In M2, with switching state 1110, the input capacitor C_{i1} and the inductor power the load, causing C_{i1} to discharge. Similarly, for the switching state 0110 shown in Figure 3c, the load is supplied by the inductor and supporting capacitors, with both C_{i1} and C_{i2} being bypassed. Finally, in Figure 3d, state 0010 is activated, enabling the load power to be provided by the input-side capacitor C_{i2} and inductor, accompanied by the discharge of C_{i2}.

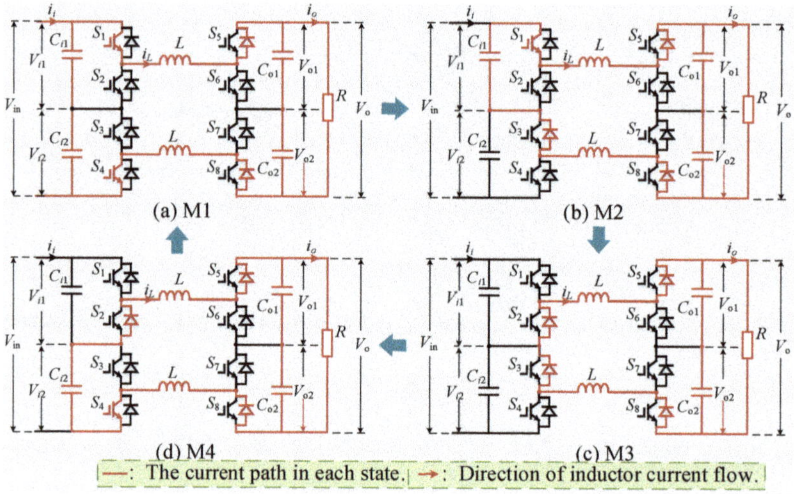

Figure 3. Four operation states of buck mode.

The four operational states of the buck mode are systematically concluded in Table 1. It is observed that the switching state transitions between adjacent operational states are characterized by a single-bit variation in the states of Q_1–Q_4. Therefore, in order to suppress the drastic increase in the number of switching actions in the adjacent control cycle of the converter, three alternative states in the next control cycle are polled: one is to keep the state of the previous cycle unchanged, and the other two are to go into the adjacent states. Thus, the schematic of the state transfer in buck mode of the three-level converter is obtained, as shown in Figure 4.

Table 1. Switch states and port voltage in buck mode.

Operation States	Q_1	Q_2	Q_3	Q_4	v_{ab}	v_{cd}
M1	1	0	1	0	V_{in}	V_o
M2	1	1	1	0	V_{i1}	V_o
M3	0	1	1	0	0	V_o
M4	0	0	1	0	V_{i2}	V_o

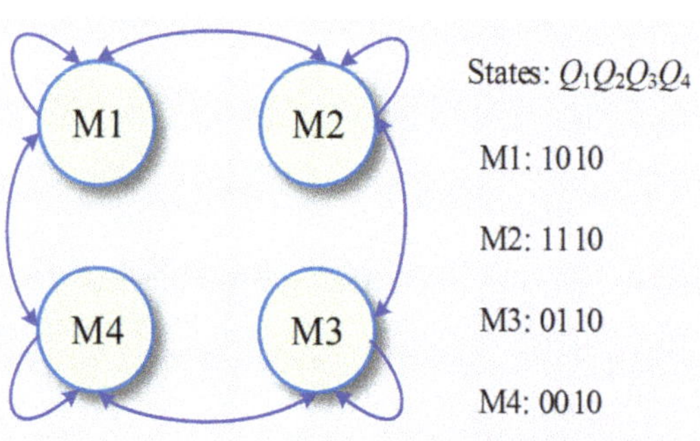

Figure 4. Schematic diagram of states transfers in buck mode.

If the current control cycle is in state M1, the range for state optimization in the next control cycle is M1, M2, and M4. As the number of switching actions between M1 and M3 is two, state M3 is excluded in the next cycle. In addition, the decision as to which of the modes, M1, M2, or M4, to select is determined by the cost function described later.

3.2. Switching State Optimization in Boost Mode

The current paths corresponding to the four operational states (M1, M5–M7) of the three-level non-inverting converter in boost mode are illustrated in Figure 5. In Figure 5a, the switching states of Q_1–Q_4 are configured as 1010, where the load power is supplied by both the input source and the inductor. In Figure 5b, the switching states of Q_1–Q_4 are set to 1011. Here, the load power is provided by the output capacitors, while capacitor C_{o1} is charged and C_{o2} discharged. As shown in Figure 5c, when the switching states of Q_1–Q_4 are 1001, the load is supplied by supporting capacitors, and energy is stored in both inductors during the state. Finally, in Figure 5d, the switching states of Q_1–Q_4 are 1000. The load power is provided by the supporting capacitors, with capacitor C_{o2} being charged and C_{o1} discharged.

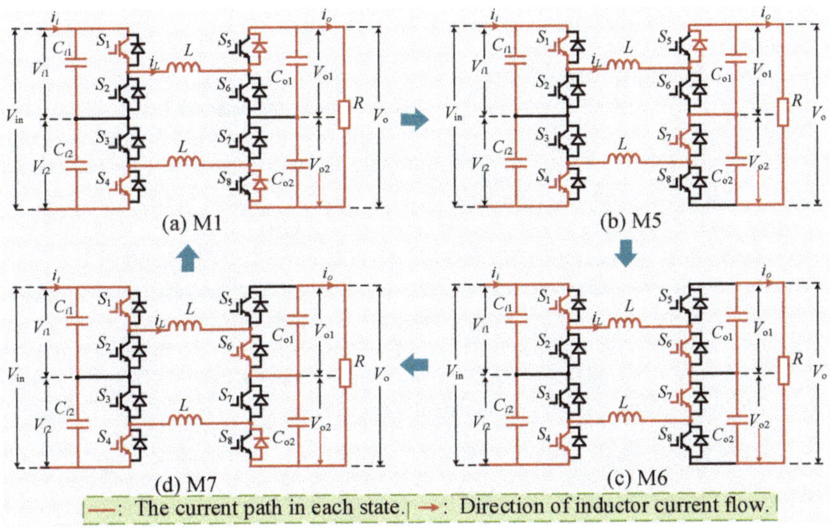

Figure 5. Four operation states of boost mode.

Four operational states of the three-level non-inverting converter in boost mode are summarized in Table 2. Similar to the buck mode, only one switching state variation is permitted between adjacent operational states for switching devices Q_1–Q_4. To mitigate excessive switching actions across consecutive control cycles, the candidate states for the next control cycle are restricted to three possibilities: one is to keep the state of the previous cycle unchanged, and the other two are to go into the adjacent states. A schematic representation of the state transition constraints for boost mode is illustrated in Figure 6.

Table 2. Switch states and port voltage of boost mode.

Operation States	Q_1	Q_2	Q_3	Q_4	v_{ab}	v_{cd}
M1	1	0	1	0	V_{in}	V_o
M5	1	0	1	1	V_{in}	V_{o1}
M6	1	0	0	1	V_{in}	0
M7	1	0	0	0	V_{in}	V_{o2}

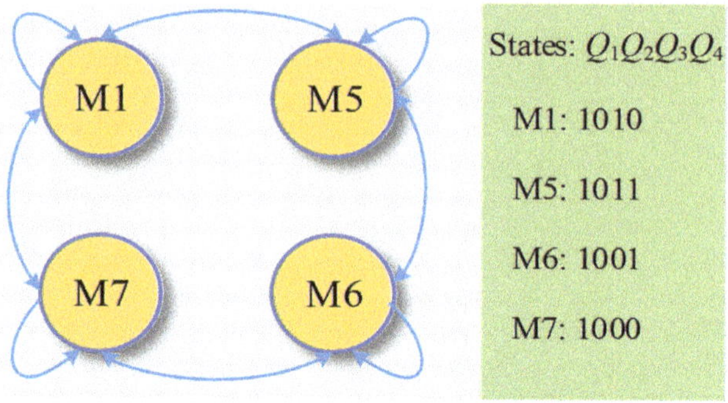

Figure 6. Schematic diagram of state transfers in boost mode.

If the current control cycle is in state M1, the range for state optimization in the next control cycle is M1, M5, and M7. As the number of switching actions between M1 and M6 is two, state M6 is excluded in the next cycle. In addition, the decision as to which of the modes, M1, M5, or M7, to select is determined by the cost function described later.

3.3. Transient Switching Process Within Two Operation Modes

The state transition processes of the converter in both buck and boost modes are analyzed separately in the preceding sections. However, in practical applications where the input or output voltage varies over a wide range, transitions between buck and boost modes must be considered. Under such conditions, the transitional process between the two modes termed the bridging mode is required to be addressed.

It is worth noting that state M1 is identified as a common mode shared by both buck and boost operations. Therefore, M1 serves as a critical intermediary bridge for mode transitions. When the converter operates in state M1, five potential operational states—M1, M2, M4, M5, and M7—are permitted in the subsequent control cycle. By leveraging M1 as the bridging state, it is ensured that only one switch variation occurs during transients, thereby effectively suppressing excessive switching actions.

4. The Proposed FCS-MPC

MPC has been widely adopted in power electronics and motor drive systems due to its fast dynamic response, simplified control structure, and straightforward implementation of switching states. For these reasons, the FCS-MPC architecture is adopted in this paper. The block diagram of the proposed FCS-MPC is shown in Figure 7.

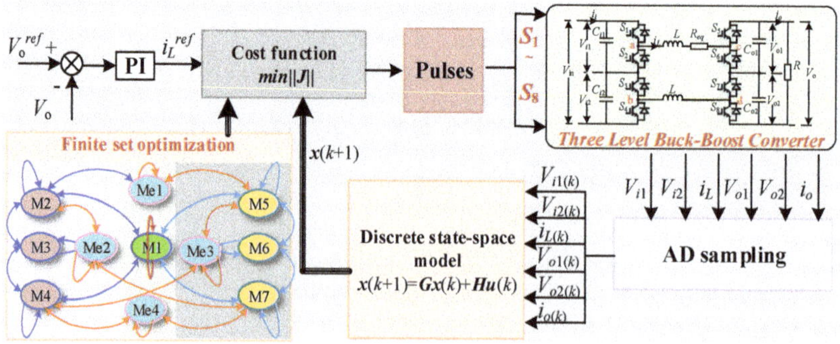

Figure 7. Block diagram of the proposed FCS-MPC.

The proposed FCS-MPC method is composed of two control loops: a voltage outer loop and a predictive current inner loop. First, the output voltage V_o is sampled and compared with its reference value V_{ref}. The resulting error is processed by a PI controller, thereby generating the reference current $i_L{}^{ref}$ for the inner loop. Furthermore, the input capacitor voltage V_{in}, output capacitor voltage V_o, and inductor current i_L are sampled and substituted into the discrete state-space prediction model (8) to derive the predicted state variables. These predicted values, along with their reference values, are incorporated into the cost function. Subsequently, the cost function is evaluated for all candidate switching states using the proposed operational state optimization strategy. The switching state that minimizes the cost function is selected as the optimal solution across the next control cycle. Finally, the corresponding pulse signals are distributed to the switching devices.

4.1. Selection of the Cost Function

As illustrated in Figure 7, the reference current for the inner loop is obtained via the outer-loop voltage PI controller. To ensure accurate tracking of the inductor current, the inductor current is incorporated into the cost function. However, a critical concern arises from the three-level topology: both the input and output voltages are the sum of two capacitor voltages. If the upper and lower capacitor voltages are not actively regulated, voltage unbalance may be induced, leading to capacitor voltage deviation. More critically, such deviation can result in uneven voltage stress across the switching devices, which may ultimately lead to device failure due to overvoltage under severe conditions. Therefore, to effectively balance the capacitor voltages on both the input and output sides, the capacitor voltages should also be included as control objectives in the cost function. Therefore, the cost function can be formulated as follows:

$$\begin{aligned} J(k) = {} & |i_L{}^{ref} - i_L{}^{\mathrm{P}}(k+1)| + \lambda_1(|V_{o1}{}^{\mathrm{P}}(\mathrm{k}+1) - V_{o2}{}^{\mathrm{P}}(\mathrm{k}+1)|) \\ & + \lambda_2(|V_{i1}{}^{\mathrm{P}}(\mathrm{k}+1) - V_{i2}{}^{\mathrm{P}}(\mathrm{k}+1)|) + \lambda_{iL\_limit} \end{aligned} \tag{14}$$

where λ_1 and λ_2 are designated as the weighting coefficients for output-side voltage balancing control and input-side voltage balancing control, respectively. The cost function is formulated with four terms: inductor current tracking, output-side voltage balancing, input-side voltage balancing, and inductor current limitation. To address overcurrent issues, λ_{iL_limit} is introduced as the weighting coefficient for inductor current protection. Specifically, the inductor current is constrained to prevent excessive values, and thus, λ_{iL_limit} is defined as follows:

$$\lambda_{iL\_limit} = \begin{cases} 0, & i_L(k+1) < i_{L\_\max} \\ \infty, & i_L(k+1) \geq i_{L\_\max} \end{cases} \tag{15}$$

where i_{L_max} is defined as the inductor current protection threshold. Once the predicted inductor current exceeds i_{L_max}, the weighting coefficient is set to infinity, thereby ensuring that the corresponding switching states are excluded from consideration. Conversely, when the current below i_{L_max}, the weighting coefficient is assigned as zero, eliminating its impact on the remaining terms of the cost function.

4.2. Switching State Optimization Based on Capacitor Voltage Balance Control

After the cost function is finalized, the switching states for each control cycle are optimized such that the following objectives are achieved:

$$\min J(k), subject\ to\ M_i, i \in (1, 2, \cdots, 8) \tag{16}$$

When the converter operates in buck mode, the switching states are polled and optimized among M1–M4. It is noteworthy that, in buck mode, voltage balancing of the input-side capacitors can be achieved by adjusting the duration of states M2 and M4. However, output-side capacitor voltage unbalance cannot be resolved through modes M1–M4. Therefore, to ensure simultaneous voltage balancing on both input and output sides, four additional modes (Me1–Me4) are introduced. The rules for the additional state introductions should be satisfied:

(1) Voltage balancing on both input and output sides must be achievable in any selected mode.
(2) The number of switching actions during transitions between adjacent modes must be minimized.

By adhering to these rules, the four supplementary states are incorporated into the control set, as illustrated in Figure 8.

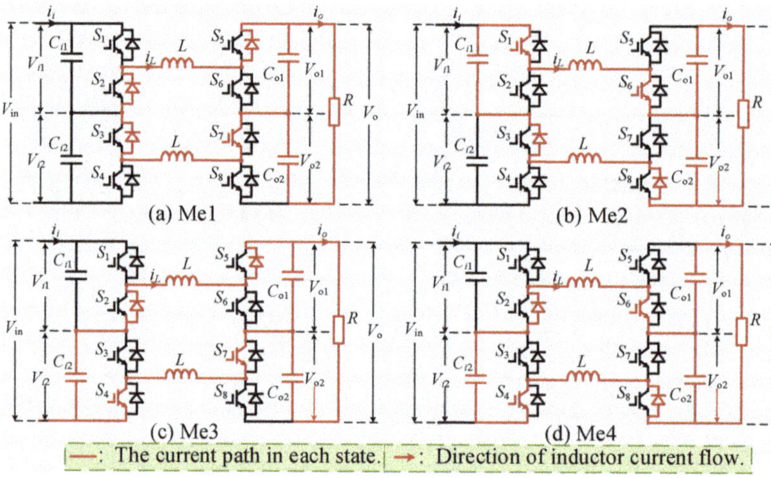

Figure 8. Voltage balance states in buck mode and boost mode.

Similarly, to ensure input-side capacitor voltage balancing in boost mode, the aforementioned four supplementary states (Me1–Me4) are incorporated into it. Consequently, the optimized mode selection diagram in Figure 7 is revised and reoptimized as shown in Figure 9. By evaluating all candidate states through (16), the state that satisfies the predefined conditions is selected as the optimal solution for the subsequent control cycle. Notably, out of the total 16 possible switching states, only 11 states are utilized in the proposed control strategy, with the remaining 5 states excluded from consideration.

Since only one switching state variation is permitted between consecutive control cycles, the subsequent candidate states are restricted to M1, M2, M3, Me1, and Me2 when the current state is M2 and input voltage unbalance occurs. If voltage deviation dominates the cost function, the converter may oscillate among M2, M3, and Me2, failing to reach M4 and causing optimization failure. To ensure effective optimization, mode M4 is forcibly included in the candidate set when the converter operates in M2 at moment k. If mode M4 is determined as the optimal state for moment $k + 1$ via cost function evaluation, mode M1 is temporarily applied at moment $k + 1$, enabling a feasible transition to M4 at moment $k + 2$ while maintaining the single switch variation constraint. This strategy prevents optimization failure while ensuring minimal switching actions. A similar approach is adopted for modes M5 and M7.

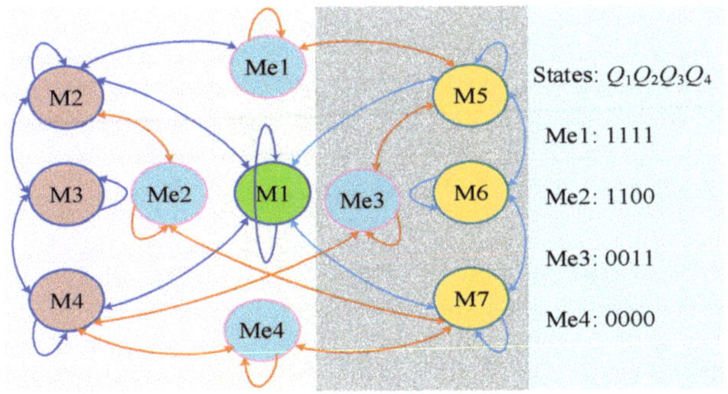

Figure 9. Schematic diagram of state transfer considering capacitor voltage balance on both sides.

Therefore, switching transitions are strictly confined to adjacent modes, significantly reducing the computational overhead of state polling and suppressing unnecessary switching state changes.

4.3. Optimal Control Algorithm for Three-Level Non-Inverting Buck–Boost Converter

To elaborate on the proposed MPC algorithm, the procedural flowchart of the control strategy is illustrated in Figure 10. The steps of the FCS-MPC are detailed as follows:

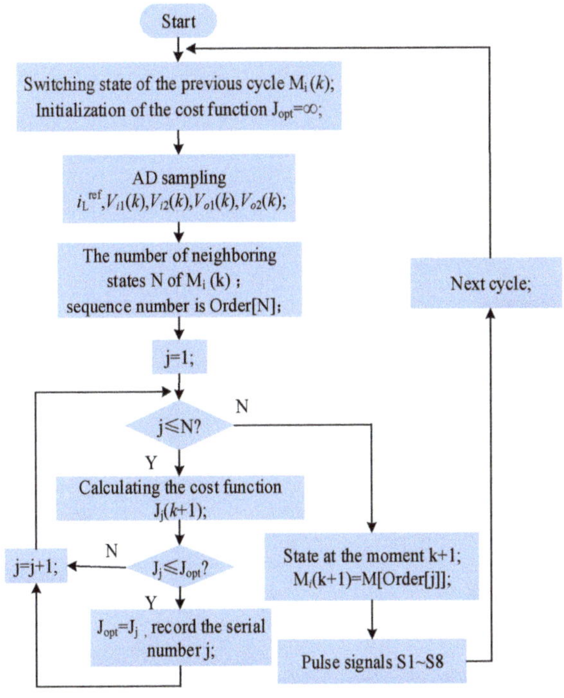

Figure 10. Flow chart of the proposed FCS-MPC.

First, the previous switching state $M_i(k)$ is analyzed, and the cost function is initialized with $J_{opt} = \infty$. Subsequently, the input voltage, output voltage, and inductor current of the buck–boost converter are sampled. Based on the state transition diagram in Figure 9, the candidate switching states for the current control cycle are identified, and the total number of candidate modes is denoted as N. The indices of all candidate states are stored in an array order [N]. Following this, the cost function values for all N candidate states are computed using (14). If the candidate cost function value J_j is found to be less than J_{opt}, J_j is assigned to J_{opt}, and the index j is recorded. After all N states are polled, the j-th state is selected as the optimal state for the next control cycle, and the corresponding eight switching pulse signals are generated. Finally, the switching devices are driven via the gate driver circuit.

The aforementioned process is repeated for each control cycle, iteratively determining the optimal state while minimizing computational overhead and switching transients.

5. Experimental Validation

To validate the effectiveness of the FCS-MPC algorithm for three-level non-inverting buck–boost converters, an experimental platform based on the Starsim HIL (Hardware-in-the-Loop) system was constructed in the laboratory. As shown in Figure 11, the digital controller configuration employed a dual-core architecture comprising DSP and FPGA. The Texas Instruments (TI) C2000-series TMS320C28346 DSP is implemented for voltage/current sampling at converter terminals and execution of the FCS-MPC algorithm. The FPGA is utilized for multi-channel pulse processing, including pulse coordination and dead-time configuration. The principles for parameter design of the main circuit can be obtained in [10]. The main parameters of the experiment are summarized in Table 3.

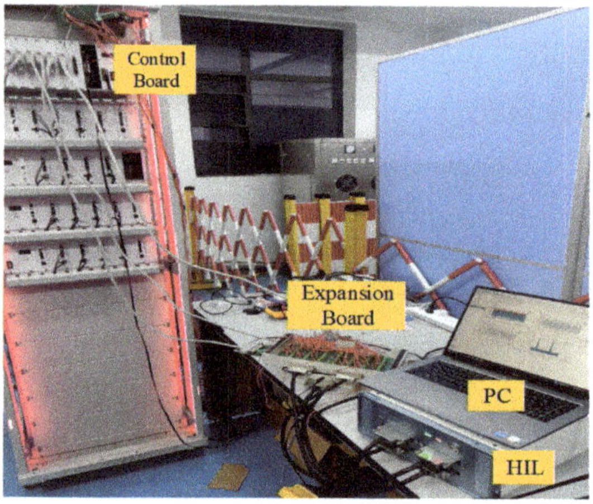

Figure 11. Photograph of experimental platform.

Table 3. Parameters of three level buck–boost converter.

Parameters	Values
Input Voltage/V	400
Output Voltage/V	200~800
Inductor/mH	0.5
Input-Side Capacitor C_i/mF	1.0
Output-Side Capacitor C_o/mF	1.0
Equivalent resistance R_{eq}/Ω	0.001
Load R/Ω	50
Sampling Frequency f/kHz	100.0
Control Period Ts/us	25.0

The effectiveness of the proposed control strategy is validated through a series of experimental verifications under the following operating scenarios: buck mode, near-unity voltage ratio conversion, boost mode, input-side voltage balancing, output-side voltage balancing, and wide-range voltage dynamic performance.

5.1. Buck Mode Verification

The waveforms of the three-level non-inverting buck–boost converter in buck mode are shown in Figure 12a–d, which corresponding to the input capacitor voltages, output capacitor voltages, total output voltage and inductor current, respectively. As shown in Figure 12a, the voltage fluctuations of the upper and lower input capacitors are maintained between 199.6 V and 200.4 V, demonstrating the effectiveness of the input-side voltage balancing control. The output capacitor voltages shown in Figure 12b are maintained within 99.6 V to 100.4 V, confirming effective voltage balancing at the output side. In addition, the total output voltage shown in Figure 12c is stabilized at 200 V with little fluctuation. In particular, the bridge inductor current shown in Figure 12d is smooth with no apparent current spikes. Therefore, the buck mode experimental results demonstrate that the proposed FCS-MPC strategy achieves effective step-down operation while ensuring balanced voltage control on both input and output capacitors.

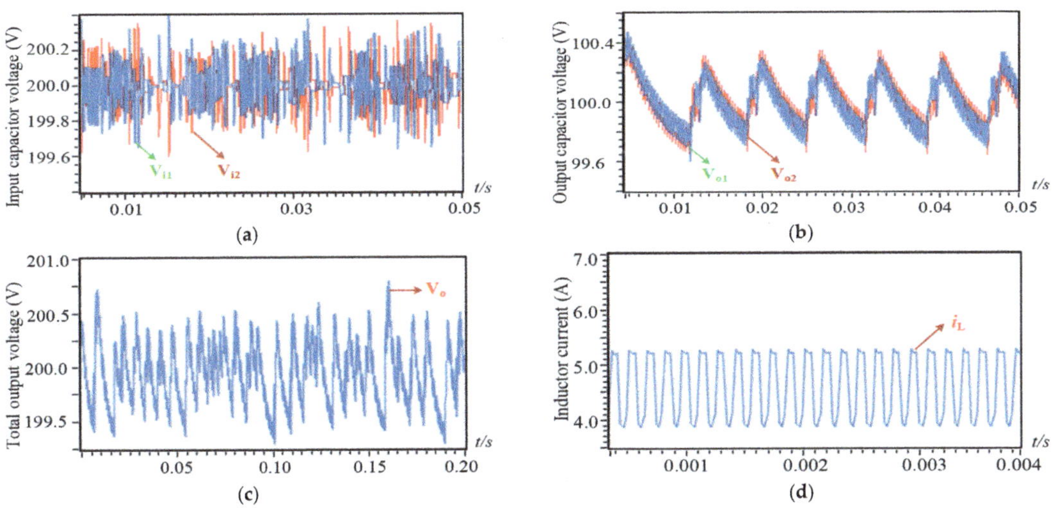

Figure 12. Three-level converter operating in buck mode: (**a**) voltage of the upper and lower capacitors on the input side; (**b**) voltage of the upper and lower capacitors on the output side; (**c**) total output voltage; (**d**) inductor current.

5.2. Close Input and Output Voltages

The voltage regulation characteristics of the three-level non-inverting buck–boost converter under near unity voltage ratio operation are shown in Figure 13a–d. As seen in Figure 13a, the upper and lower input capacitor voltages are maintained between 199.8 V and 200.2 V, indicating the effectiveness of voltage balancing at the input-side capacitors. The output capacitor voltages presented in Figure 13b are stabilized around 200 V, with the fluctuations maintained between 198 V and 203 V, confirming effective voltage balancing at the output side. In addition, the total output voltage, shown in Figure 13c, is regulated at 400 V with accurate steady-state operation. In particular, the inductor current shown in Figure 13d demonstrates smooth operation with no apparent current spikes. These experimental results confirm that the proposed FCS-MPC strategy effectively achieves output voltage regulation and bilateral capacitor voltage balancing when input and output voltages approach approximate magnitudes.

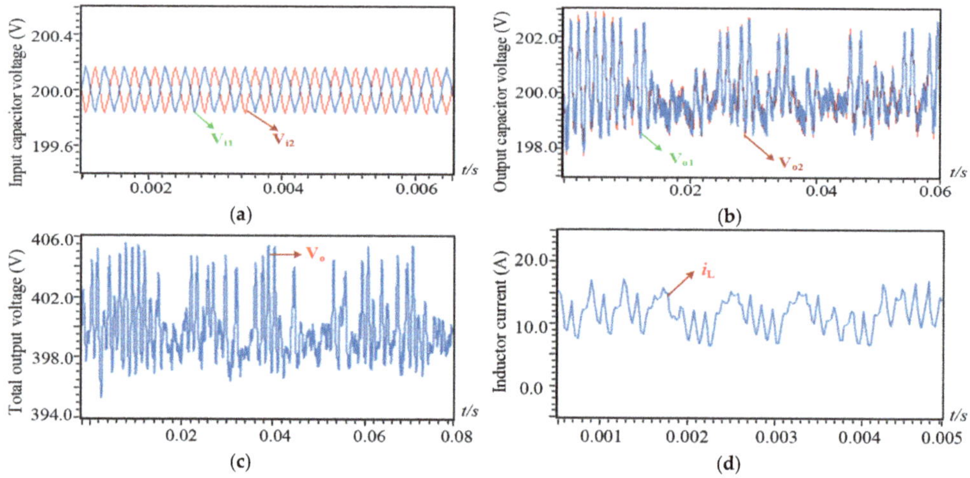

Figure 13. The waveforms of the three-level converter when the input and output voltages are equal: (**a**) voltage of the upper and lower capacitors on the input side; (**b**) voltage of the upper and lower capacitors on the output side; (**c**) total output voltage; (**d**) inductor current.

5.3. Boost Mode Verification

The waveforms of the three-level buck–boost converter in boost mode are shown in Figure 14a–d. As shown in Figure 14a, the input capacitor voltages are maintained between 199.4 V and 200.6 V, indicating effective voltage balancing between the upper and lower capacitors. The output capacitor voltages shown in Figure 14b are stabilized near 400 V, with fluctuations regulated between 398.0 V and 402.0 V, confirming voltage balancing at the output stage. In addition, the total output voltage shown in Figure 14c is maintained at 800 V with little voltage ripple. Notably, the inductor current shown in Figure 14d exhibits a smooth dynamic behavior without abrupt spikes. These results confirm that the proposed FCS-MPC strategy is capable of achieving boost mode operation, while ensuring balanced voltage regulation across both input and output capacitors.

5.4. Dynamic Processes

(1) Wide Voltage Range Variation

To validate the smooth control functionality of the three-level buck–boost converter for wide-range output applications, a dynamic experimental verification is conducted by linearly increasing the output reference voltage from 200 V to 800 V, with corresponding experimental waveforms provided in Figure 15a. As revealed in the results, the output voltage achieves smooth transition during the transient process from 200 V to 800 V. Furthermore, the dynamic capacitor voltage balance is maintained between input and output sides throughout the entire operation, demonstrating superior voltage equalization performance. In addition, no significant current surges are detected in the inductor current during the transient process, confirming the effectiveness of the proposed control strategy.

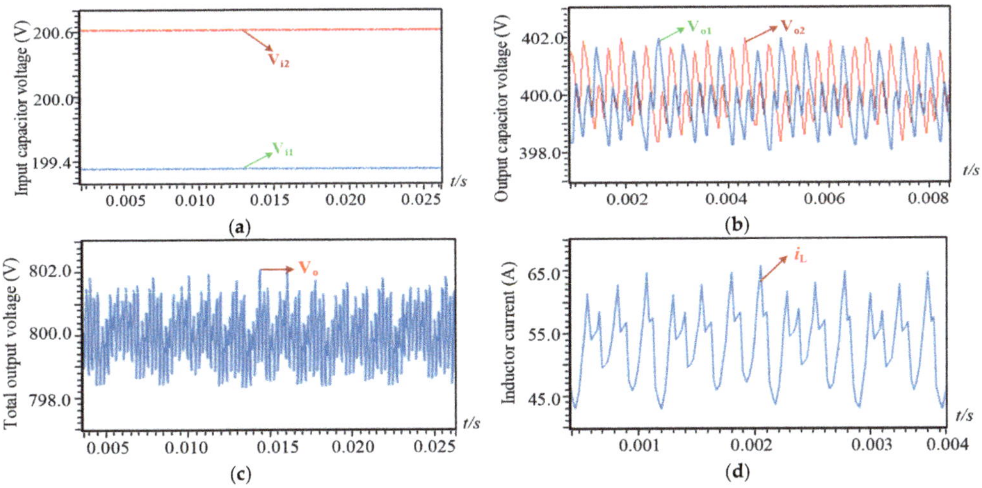

Figure 14. Three-level converter operating in boost mode: (**a**) voltage of the upper and lower capacitors on the input side; (**b**) voltage of the upper and lower capacitors on the output side; (**c**) total output voltage; (**d**) inductor current.

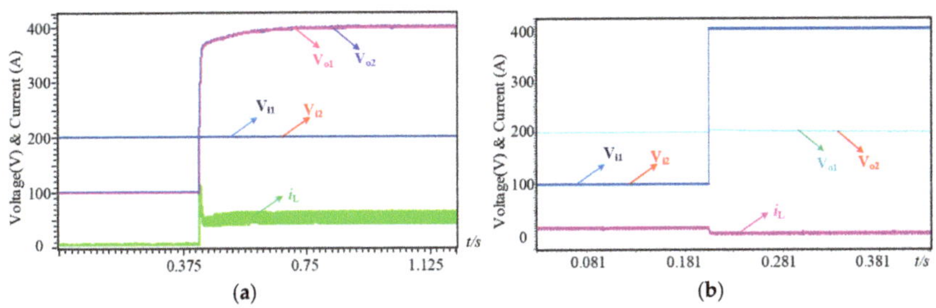

Figure 15. Dynamic process: (**a**) wide-range output voltage; (**b**) wide-range input voltage.

In Figure 15b, the output voltage remains stable at 400 V under a wide range of input voltages from 200 V to 800 V. During the input voltage increase process, the upper and lower capacitor voltages on the input side remain balanced. As the input voltage increases and the load power remains unchanged, the inductor current decreases. In addition, the output voltage remains stable at 400 V, and the upper and lower capacitor voltages of the output-side capacitor are balanced.

(2) Load Variation

The voltage and current waveforms on the input and output sides under load changes are shown in Figure 16a–d. At t = 0.399 s, the output side load resistance changes from 100 Ω to 50 Ω. Obviously, the inductor current increases instantaneously after the resistance changes. In addition, even though there are transient fluctuations in the input and output voltages, the balance of the capacitor voltages on the input and output sides can still be achieved. The waveform of the total output voltage is shown in Figure 16c. At the moment of load change, the output voltage drops shortly and then quickly stabilizes at 400 V. The inductor current is shown in Figure 16d. At the moment of transient switching, the inductor current increases and gradually goes stable. Therefore, the adaptability to dynamic load fluctuations is validated under the proposed control.

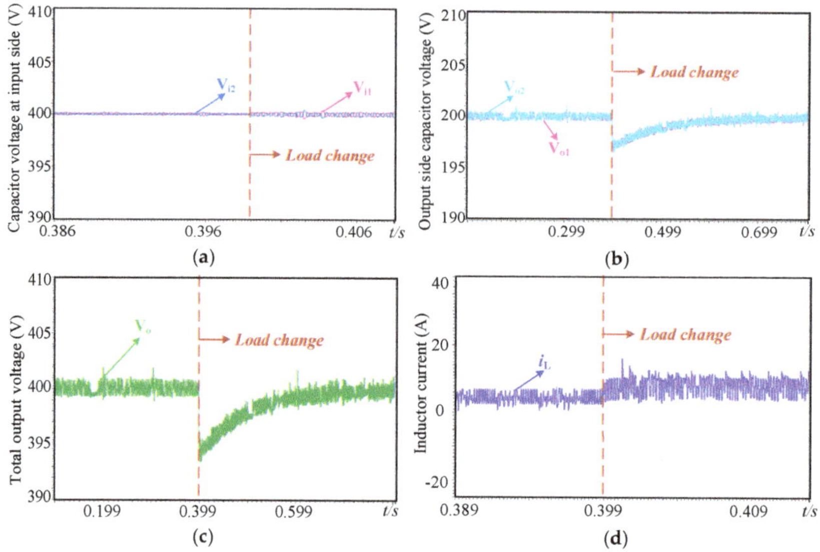

Figure 16. Voltage and current under load changes: (**a**) upper and lower capacitor voltages on the input side; (**b**) upper and lower capacitor voltages on the output side; (**c**) total output voltage; (**d**) inductance current.

5.5. Statistics on the Number of Switching Actions

(1) Pulse Signals of Each Device in the Conventional Global Polling Strategy

As shown in Figure 17a, the pulse signals of switches S_1–S_7 are presented under the conventional global polling strategy in buck mode. Notice that the switching action count ΔN between adjacent control intervals is either 2 or 3. Similarly, the corresponding pulse signals in boost mode are shown in Figure 17b, where ΔN likewise falls within {2, 3}. Therefore, under the global polling strategy, multiple devices switching simultaneously are observed during individual control cycles.

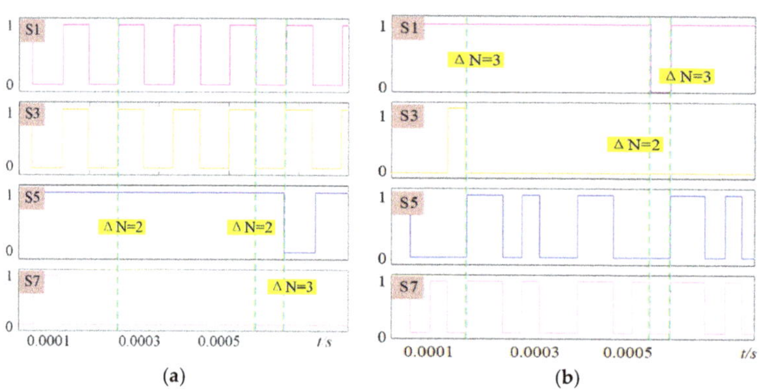

Figure 17. Pulse signals under the global polling strategy: (**a**) buck mode; (**b**) boost mode.

(2) Pulse Signals of Each Device in the Proposed Control Strategy

In the buck mode, the pulse signals S_1–S_7 of the proposed control strategies are shown in Figure 18a. The number of switching actions between two adjacent cycles is ΔN = 1. Therefore, there is only one switching action in each control cycle. Similarly, the pulse signals in the boost mode are shown in Figure 18b, and the characteristic of ΔN = 1 can also be obtained. Therefore, the feasibility of the proposed control strategy to reduce the number of switching actions is verified.

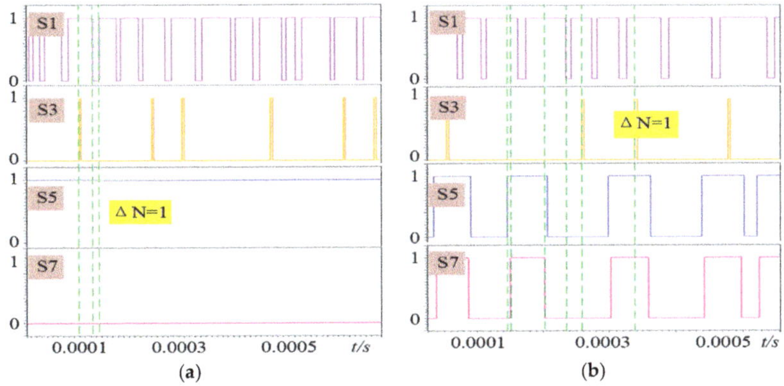

Figure 18. Pulse signals under the proposed control: (**a**) buck mode; (**b**) boost mode.

The efficiency under the proposed FCS-MPC and global polling control is shown in Figure 19. It is clear that the efficiency of the system under the proposed FCS-MPC is always higher than that under the global polling strategy, which demonstrates that the proposed control has remarkable advantages in terms of efficiency improvement.

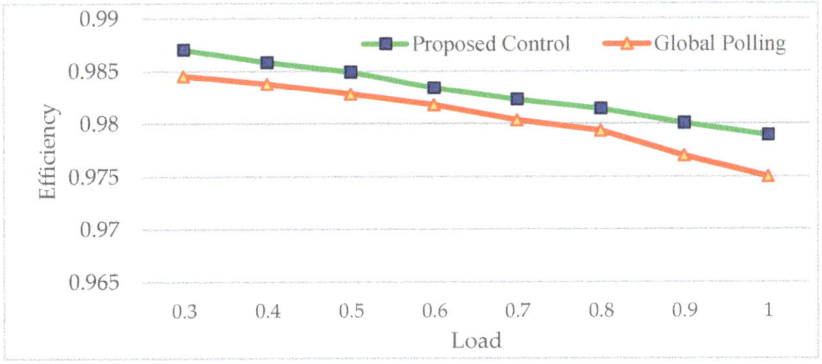

Figure 19. The efficiency comparison of converters under different control strategies.

6. Conclusions

An FCS-MPC strategy was proposed for three-level non-inverting buck–boost converters, achieving seamless wide-range voltage regulation through optimized switching action reduction and strategic integration of voltage balancing. By incorporating capacitor voltage equilibrium constraints into the cost function, consistent voltage balancing on both sides was ensured without additional operational mode detection. Experimental results confirm that the proposed strategy provides superior dynamic performance under wide input/output voltage variations and load changes, while maintaining a balanced voltage across the input and output capacitors. Furthermore, switching losses can be significantly reduced and efficiency of the converter improved by minimizing the number of device actions between adjacent control intervals.

Author Contributions: Conceptualization, M.X. and H.D.; methodology, M.X.; validation, R.H., M.X., and H.D.; formal analysis, M.X. and J.T.; investigation, Y.L. and Z.L.; resources, M.X. and X.W.; writing—original draft preparation, M.X.; writing—review and editing, R.H., J.T., and Y.L.; super vision, M.X. and H.D.; project administration, M.X., R.H., and X.W. All authors have read and agreed to the published version of the manuscript.

Funding: This research was funded by the Fundamental Research Funds for the Provincial Universities of Liaoning, grant number LJ212410150040, and the Natural Science Foundation of Liaoning Province, grant number 2025-BS-0448, and the Liaoning Province Transportation Science and Technology Project, grant number 2024066.

Data Availability Statement: The original contributions presented in the study are included in the article, further inquiries can be directed to the corresponding author.

Conflicts of Interest: The authors declare no conflicts of interest.

References

1. Yuan, J.; Dorn-Gomba, L.; Callegaro, A.D.; Reimers, J.; Emadi, A. A Review of Bidirectional On-Board Chargers for Electric Vehicles. *IEEE Access* **2021**, *9*, 51501–51518. [CrossRef]
2. Wouters, H.; Martinez, W. Bidirectional Onboard Chargers for Electric Vehicles: State-of-the-Art and Future Trends. *IEEE Trans. Power Electron.* **2024**, *39*, 693–716. [CrossRef]
3. Adnan, A.; Qin, Z.; Wijekoon, T.; Bauer, P. An Overview on Medium Voltage Grid Integration of Ultra-Fast Charging Stations: Current Status and Future Trends. *IEEE Open J. Ind. Electron. Soc.* **2022**, *3*, 420–447. [CrossRef]
4. Available online: https://insidechinaauto.com/2025/03/24/byd-announces-first-megawatt-fast-charging-ev-platform/ (accessed on 24 March 2025).
5. Sharida, A.; Bayhan, S.; Abu-Rub, H. Novel Multi-Mode DC-DC Converter for Battery Storage Applications. In Proceedings of the IECON 2023-49th Annual Conference of the IEEE Industrial Electronics Society, Singapore, 16–19 October 2023. [CrossRef]
6. Kitazoe, K.; Natori, K.; Sato, Y. Bidirectional Power Flow Control Using Four-Switch Buck-Boost Converter in DC Power Network. *IEEE J. Ind. Appl.* **2024**, *13*, 618–624. [CrossRef]
7. Pesantez, D.; Rodriguez, F.; Renaudineau, H.; Rivera, S.; Kouro, S. Buck-Boost Flying Capacitor DC-DC Converter for Electric Vehicle Charging Stations. In Proceedings of the 2023 IEEE 17th International Conference on Compatibility, Power Electronics and Power Engineering, Tallinn, Estonia, 14–16 June 2023. [CrossRef]
8. Muñoz, R.V.; Renaudineau, H.; Rivera, S.; Kouro, S. Evaluation of DC-DC buck-boost partial power converters for EV fast charging application. In Proceedings of the IECON 2021–47th Annual Conference of the IEEE Industrial Electronics Society, Toronto, ON, Canada, 13–16 October 2021. [CrossRef]
9. Eull, M.; Preindl, M. Bidirectional three-level DC-DC converters: Sum-difference modeling and control. In Proceedings of the 2017 IEEE Transportation Electrification Conference and Expo, Chicago, IL, USA, 22–24 June 2017. [CrossRef]
10. Bi, K.; Xu, C.; Ai, J.; Li, J.; Fan, Q. Decoupling Model Predictive Controlled Three-Level Noninverting Buck–Boost Converter Applied in DC Energy Storage System. *IEEE Trans. Power Electron.* **2024**, *39*, 6677–6687. [CrossRef]
11. Sun, B.; Trintis, I.; Munk-Nielsen, S.; Guerrero, J.M. Effects and analysis of minimum pulse width limitation on adaptive DC voltage control of grid converters. In Proceedings of the 2016 IEEE Applied Power Electronics Conference and Exposition, Long Beach, CA, USA, 20–24 March 2016. [CrossRef]
12. Jia, P.; Hao, Y. Research on the Dead-Time Control Scheme for the Four-Switch Buck-Boost Converter. In Proceedings of the 2020 IEEE 9th International Power Electronics and Motion Control Conference, Nanjing, China, 29 November–2 December 2020. [CrossRef]
13. Jia, L.; Sun, X.; Zheng, Z.; Ma, X.; Dai, L. Multimode Smooth Switching Strategy for Eliminating the Operational Dead Zone in Noninverting Buck–Boost Converter. *IEEE Trans. Power Electron.* **2020**, *35*, 3106–3113. [CrossRef]
14. Cao, E.; Ding, W.; Duan, B.; Song, J.; Zhan, C.; Wan, D. Predictive Control Strategy for Four-switch Buck-Boost Converter with Mode Smooth Transition. In Proceedings of the 2022 6th CAA International Conference on Vehicular Control and Intelligence, Nanjing, China, 28–30 October 2022. [CrossRef]
15. Wu, Y.; Wang, W.; Zeng, G.; Wu, X.; Tang, F. Multi-Mode Model Predictive Control Strategy for the Four-Switch Buck-Boost Converter. *Trans. China Electrotech. Soc.* **2022**, *37*, 2572–2583. [CrossRef]
16. Bai, Y.; Hu, S.; Yang, Z.; Zhu, Z.; Zhang, Y. Model Predictive Control for Four-Switch Buck–Boost Converter Based on Tuning-Free Cost Function with Smooth Mode Transition. *IEEE J. Emerg. Sel. Top. Power Electron.* **2022**, *10*, 6607–6618. [CrossRef]

17. Ullah, B.; Ullah, H. Finite Control Set Model Predictive Control of Noninverting Buck-Boost DC-DC Converter. In Proceedings of the 2022 International Conference on Technology and Policy in Energy and Electric Power, Jakarta, Indonesia, 18–20 October 2022. [CrossRef]
18. Li, X.; Liu, Y.; Xue, Y. Four-Switch Buck–Boost Converter Based on Model Predictive Control with Smooth Mode Transition Capability. *IEEE Trans. Ind. Electron.* **2021**, *68*, 9058–9069. [CrossRef]
19. Zhu, Z.; Ma, F.; Lin, L.; Wang, L.; Kuang, D.; Xu, Q. Multimode Smooth Transition Technique for Three-Level Cascaded Noninverting Buck–Boost DC–DC Converter. *IEEE Trans. Power Electron.* **2021**, *36*, 14409–14419. [CrossRef]
20. Peng, Q.; Zhou, S.; Ma, F.; Huang, G.; Fan, R. A Model Predictive Control of Three-Level Cascaded Noninverting Buck–Boost Converter for Energy Storage System. *IEEE Trans. Ind. Electron.* **2025**, *72*, 4933–4942. [CrossRef]
21. Jayan, V.; Ghias, A.; Merabet, A. Fixed Frequency Model Predictive Control of Three-level Bi-directional Flying Capacitor DC-DC converter in DC microgrid. In Proceedings of the IECON 2019-45th Annual Conference of the IEEE Industrial Electronics Society, Lisbon, Portugal, 14–17 October 2019. [CrossRef]

energies

MDPI

Article

Carrier Reconfiguration for Improving Output Voltage Quality and Balancing Capacitor Voltages in MMDTC-Based STATCOM

Fengxiang Xie [1], Yuantang Qi [1], Yongdong Ji [1], Xiaofan Ji [2], Xiangzheng Cui [3], Shuo Liu [3] and Decun Niu [3,*]

[1] Guoneng (Gonghe) New Energy Development Co., Ltd., Hainan Tibetan Autonomous Prefecture 813000, China; 12068234@ceic.com (F.X.); 20109584@ceic.com (Y.Q.); 20084510@ceic.com (Y.J.)
[2] National Institute of Clean-and-Low-Carbon Energy, Beijing 102209, China; 20024057@ceic.com
[3] School of Automation and Electronic Engineering, Qingdao University of Science and Technology, Qingdao 266061, China; 2023041004@mails.qust.edu.cn (X.C.); 4022040077@mails.qust.edu.cn (S.L.)
* Correspondence: niudecun@qust.edu.cn; Tel.: +86-1506-912-8650

Abstract

For Modular Multilevel DC-Link T-Type Converter (MMDTC)-based STATCOMs, under identical operating conditions, the submodule (SM) capacitor voltage ripple is inversely proportional to its capacitance value. A configuration with a lower capacitance will inevitably result in significant capacitor voltage ripples. During the PWM modulation process, these ripples can lead to distortions in the output voltage waveform. To address this issue, this paper proposes an innovative carrier reconfiguration method that not only compensates for the output voltage pulse deviation caused by SM capacitor voltage ripples but also achieves effective balancing of the SM capacitor voltages. Finally, the validity and performance of the proposed carrier reconfiguration method are verified through both simulations and experimental results.

Keywords: STATCOM; MMDTC; carrier reconfiguration; capacitor voltage ripple

1. Introduction

Modular Multilevel DC-Link T-Type Converters (MMDTCs) present significant advantages for high-voltage, high-capacity Static Synchronous Compensators (STATCOMs), including reduced component count, low power loss, and robust tolerance to grid imbalances [1]. A critical challenge in conventional MMDTC-STATCOM design, however, is the reliance on large electrolytic capacitors within submodules (SMs) to suppress capacitor voltage ripples and ensure high-quality output waveforms. While effective, electrolytic capacitors suffer from inherent limitations such as shorter lifespan, an ability to withstand lower voltage/current, and reduced reliability compared to film capacitors. Film capacitors offer superior performance in these aspects but are characterized by lower capacitance density and higher cost [2–5]. To facilitate the adoption of more reliable film capacitors, there is a compelling need to develop operational strategies that enable MMDTC-STATCOMs to function effectively with significantly reduced SM capacitance values.

Addressing this need, the concept of a low-capacitance operation mode for MMDTC-STATCOMs has been introduced [6]. This mode exploits the phase relationship between the SM capacitor voltage and the arm voltage during reactive power compensation, allowing for substantial SM capacitor voltage ripples while maintaining the fundamental output performance. Crucially, this permits the use of low-capacitance, film-based capacitors. However, operating with low capacitance inherently introduces significant SM capacitor voltage ripples. These ripples propagate through the Pulse Width Modulation (PWM) process, directly distorting the output voltage waveform and degrading its quality Furthermore, ensuring stable operation demands effective balancing of the SM capacitor voltages under these high-ripple conditions.

Existing SM capacitor voltage balancing strategies primarily fall into two categories: the sorting method and Carrier Phase-Shifted PWM (CPS-PWM). Sorting methods dynamically select which SMs to insert based on capacitor voltage measurements and the arm current direction within each control cycle [7–16]. While various enhancements exist (e.g., incorporating preservation factors [6], equalization coefficients [13], or fundamental frequency operation [16]), sorting methods inherently suffer from drawbacks: they often lead to uneven switching distribution and power loss among SMs, compromise long-term reliability, and do not guarantee a fixed switching frequency. Conversely, CPS-PWM techniques inherently maintain a fixed switching frequency and promote balanced SM utilization. Numerous CPS-PWM variants have been developed for modular converters, achieving voltage balancing through modulation wave adjustments [17–23], as shown in Figure 1. Techniques include controlling high-frequency current components [17], decoupled sampling schemes [18], linearized pulse sorting [19], hybrid control strategies [20], generalized balancing algorithms [21], methods addressing asynchronous sampling [22], and flexible voltage control via reference signal manipulation [23].

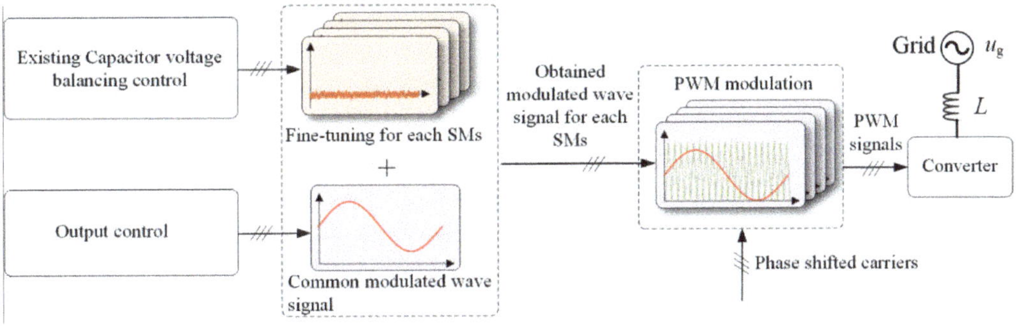

Figure 1. Example of the existing PWM modulation methods by fine-tuning the modulated wave signal (References [17–23]).

However, a critical limitation persists across both sorting methods and CPS-PWM variants: none comprehensively address the detrimental impact of large SM capacitor voltage ripples—inherent in low-capacitance operation—on the quality of the output voltage waveform. While focused on achieving voltage balance, these methods neglect the fact that the fluctuating capacitor voltage itself becomes the source voltage during PWM switching. This ripple directly modulates the amplitude of the output voltage pulses, introducing significant harmonic distortion that existing balancing controls do not compensate for. Therefore, operating MMDTC-STATCOMs in a low-capacitance mode necessitates a control strategy that simultaneously achieves two tightly coupled objectives: (1) effective balancing of SM capacitor voltages, and (2) compensation for the output voltage distortion caused by the permitted capacitor voltage ripples.

To bridge this gap, this paper proposes an innovative carrier reconfiguration method specifically designed for low-capacitance MMDTC-STATCOMs. The core insight is that dynamically adjusting the carrier signal, rather than the modulation wave, offers a pathway to independently control both objectives. The key contributions are as follows:

(1) Problem Analysis and Coupling Mechanism: A detailed analysis of how the carrier signal influences both output voltage quality and capacitor charging/discharging

durations, revealing the inherent coupling and potential conflict between these two control objectives under carrier adjustment.

(2) Dual-Objective Carrier Reconfiguration: A novel strategy involving two sequential carrier adjustments per control cycle:

The first reconfiguration targets the AC component of the capacitor voltage ripple to compensate for output voltage pulse deviations, preserving waveform quality.

The second reconfiguration targets the DC component deviation to achieve precise SM capacitor voltage balancing.

This paper is organized to provide a comprehensive understanding of the proposed carrier reconfiguration method. Section 2 introduces the MMDTC-STATCOM topology and its low-capacitance operation—a relatively new concept that creates unique control challenges. Section 3 presents novel analysis revealing the coupling between output quality and voltage balancing objectives under high-ripple conditions. Building on these insights, Section 4 develops the carrier reconfiguration method. Readers familiar with MMDTC fundamentals may proceed directly to Section 3, while those seeking only the method implementation can start from Section 4, though the full context enhances understanding.

2. The Operating Principle of an MMDTC-STATCOM with Low Capacitance

As depicted in Figure 2, the MMDTC-STATCOM consists of two arms and a T-type switching structure. The upper and lower arms are constructed using Half-Bridge Submodules (HBSMs). Each phase of the T-type structure comprises three high-voltage switches (T_{ku}, T_{kn}, T_{kl}; k = a, b, c). These high-voltage switches can be implemented using series-connected IGBTs, series-connected IGCTs, or series-connected thyristors [1,6].

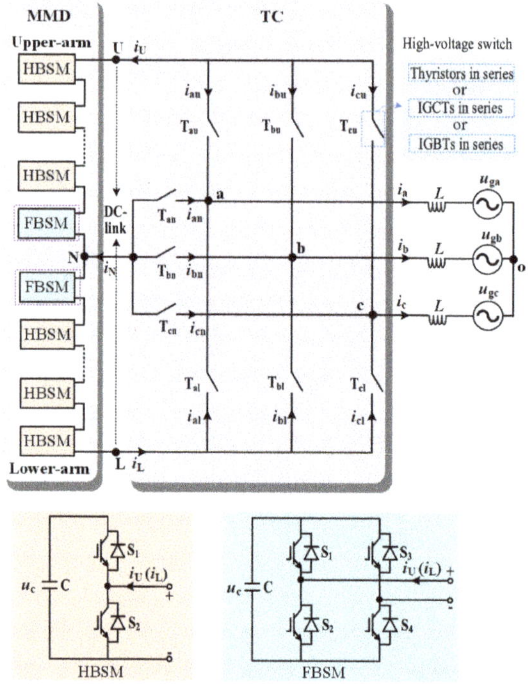

Figure 2. Configuration of MMDTC-based STATCOM.

As depicted in Figure 3, the switching functions for each phase of the T-type structure are defined as S_x (x = a, b, c). Specifically, S_x = 1, 0, and −1 correspond to the conduction states of the upper, middle, and lower switches of phase x, respectively, and then the relationship between the arm voltages and the line-to-line voltages can be derived [1,2].

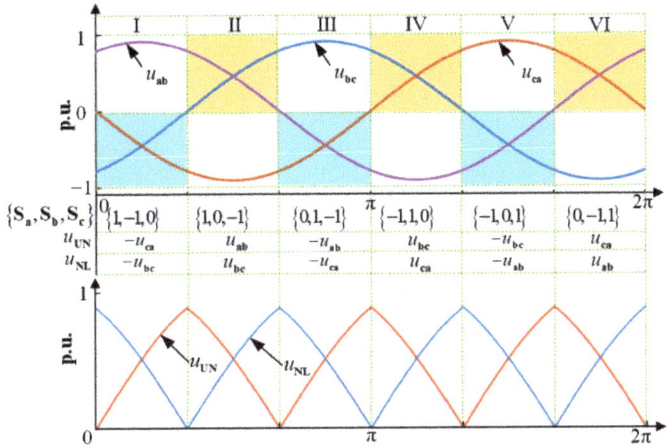

Figure 3. Switching sequences of T-structure and arm voltages along with line-to-line voltages.

The output phase voltages of the MMDTC can be defined as follows:

$$u_{\text{ao, bo, co}} = V_o \sin(\omega t + \pi/6 + 2k\pi/3) \quad (1)$$

where k takes the values of 0, −1, and 1 accordingly; V_0 represents the amplitude of the phase voltage; and $\omega = 2\pi f$.

Then, the output line voltage can be obtained:

$$u_{\text{ab, bc, ca}} = \sqrt{3}V_o \sin(\omega t + \pi/3 + 2k\pi/3) \quad (2)$$

The instantaneous expressions of the arm voltages are as follows:

$$u_{\text{UN}} = \begin{cases} \sqrt{3}V_o \sin\theta, & \theta \in [0, \frac{\pi}{3}) \\ \sqrt{3}V_o \sin(\theta + \frac{\pi}{3}), & \theta \in [\frac{\pi}{3}, \frac{2\pi}{3}] \end{cases} \quad (3)$$

$$u_{\text{NL}} = \begin{cases} -\sqrt{3}V_o \sin(\theta - \frac{\pi}{3}), & \theta \in [0, \frac{\pi}{3}) \\ \sqrt{3}V_o \sin(\theta - \frac{\pi}{3}), & \theta \in [\frac{\pi}{3}, \frac{2\pi}{3}] \end{cases} \quad (4)$$

where $\theta = \omega t - floor(\omega t/(\frac{2\pi}{3}))$, $floor$ (α) refers to rounding down the elements of α to the nearest integers.

As depicted in Figure 4, the MMDTC-STATCOM features two reactive power compensation modes: the capacitive compensation mode and the inductive compensation mode. In the capacitive compensation mode, the output voltage of the MMDTC is marginally higher than the grid voltage. Conversely, in the inductive compensation mode, the output voltage of the MMDTC is slightly lower than the grid voltage. The inductive voltage drop associated with AC filtering inductor influences this difference, as follows:

$$V_o = \begin{cases} (1+\lambda)V_g, \text{ In capacitive mode} \\ (1-\lambda)V_g, \text{ In inductive mode} \end{cases} \quad (5)$$

where V_g refers to the amplitude of the grid phase voltage, and $\lambda = \omega L I_o / V_g$ refers to the ratio of the voltage drop across AC filtering inductor.

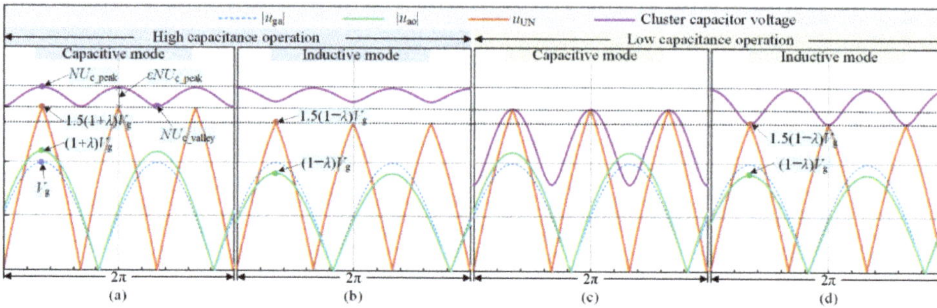

Figure 4. Schematic diagrams of MMDTC-STATCOM operation (**a**,**b**) in high-capacitance operation and (**c**,**d**) in low-capacitance operation.

As depicted in Figure 4, a comparison is presented between the high-capacitance operation and the low-capacitance operation of the MMDTC-STATCOM [6]. In the high-capacitance operation, capacitors with larger capacitance values are employed to suppress voltage ripples, ensuring that the cluster capacitor voltage remains consistently higher than the arm voltage, regardless of whether in capacitive mode (Figure 4a) or in inductive mode (Figure 4b). Specifically, the valley value (NU_{c_valley}) of the cluster capacitor voltage must always exceed the peak value of the arm voltage of $1.5(1 + \lambda)V_g$ in the capacitive compensation mode. Conversely, the low-capacitance operation leverages the characteristic wherein the SM capacitor voltage and the arm voltage are in phase during capacitive compensation, allowing for a relatively large ripple in the capacitor voltage, as shown in Figure 4c. In the inductive compensation mode, as shown in Figure 4d, as the SM capacitor voltage is inverted relative to the arm voltage, it is essential to ensure that the valley value (NU_{c_valley}) of the cluster capacitor voltage is not less than the peak value of $1.5(1 - \lambda)V_g$ of the arm voltage. This represents the condition that must be satisfied in low-capacitance operation.

From the above analysis, it can be concluded that by utilizing the phase characteristics of the capacitor voltage and the arm voltage, a considerable ripple in the capacitor voltage is permissible. This mode of operation is referred to as low-capacitance operation. By modulating the width of a series of pulses in the PWM modulation process, the desired waveform shape and amplitude can be effectively achieved. Taking Figure 5a as an example, when the ripple of the capacitor voltage is minimal (typically within 10%), the ripple component can be neglected. In other words, the capacitor voltage only contains the DC component ($u_c = U_{dc}$). Assuming the modulated wave is V_{ref}, when the carrier frequency is sufficiently high, the switching harmonics can be ignored. Consequently, the output voltage after modulation can be expressed as:

$$u_o(t) = u_c V_{ref}(t) = U_{dc} V_{ref}(t) \tag{6}$$

However, when the capacitor voltage exhibits the larger ripples, the ripple component becomes non-negligible. Owing to the triple-frequency ripple of the SM capacitors in the MMDTC-STATCOM, expressed a: $u_c = U_{dc} + U_3\sin(3\omega t)$, the same PWM modulation method is applied. As illustrated in Figure 5b, the resulting output voltage after modulation can be represented as:

$$u_o(t) = u_c V_{ref}(t) = \underbrace{U_{dc} V_{ref}(t)}_{\text{Designed voltage}} + \underbrace{U_3 \cos(3\omega t) V_{ref}(t)}_{\text{Undesigned voltage}} \quad (7)$$

Equation (7) reveals a critical phenomenon: the fluctuating component of the capacitor voltage is propagated to the output side through the PWM modulation process, thereby increasing the harmonic distortion in the output voltage. This indicates that under the low-capacitance operation mode, the influence of capacitor voltage ripples on the output voltage quality must be carefully assessed. To address this issue, this paper proposes an innovative SM capacitor-voltage-balancing control technique based on carrier reconfiguration.

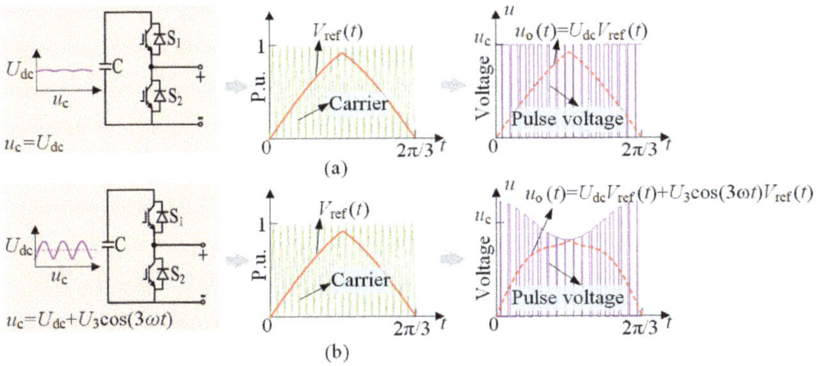

Figure 5. PWM modulation process under conditions where (**a**) the capacitor voltage ripple is minimal, and (**b**) the capacitor voltage ripple is significant.

3. Impact of Carrier Signal on Output Voltage Quality and Capacitor Voltage Balance

In the PWM modulation process, maintaining a constant modulated wave signal while dynamically adjusting the carrier signal can achieve capacitor voltage balancing and compensation for the influence of the SM capacitor voltage ripple on the output voltage quality. However, when pursuing the dual control objectives of capacitor voltage balancing and output voltage quality simultaneously, complex scenarios involving mutual constraints or mutual enhancements often arise. The following section will present a detailed analysis.

3.1. Impact of Carrier Signal on Output Voltage Quality

In fact, ripples in the capacitor voltage modify the amplitude of the output voltage pulse signal during the PWM modulation process, leading to deviations from the desired value and consequently affecting the total harmonic distortion (THD) of the output voltage. To effectively compensate for this deviation, the pulse width can be adjusted by fine-tuning the carrier signal. As illustrated in Figure 6, within any carrier period T_c, when the instantaneous value u_c of the capacitor voltage is lower than the DC component U_{dc}^*, it results in a reduced amplitude of the output voltage pulse signal. In this case, the pulse width can be increased by appropriately decreasing the carrier amplitude, thereby compensating for the pulse deviation. Conversely, when u_c exceeds the DC component U_{dc}^*, the amplitude of the output voltage pulse signal increases. At this point, the carrier amplitude should be increased accordingly to decrease the pulse width, thereby reducing the pulse deviation.

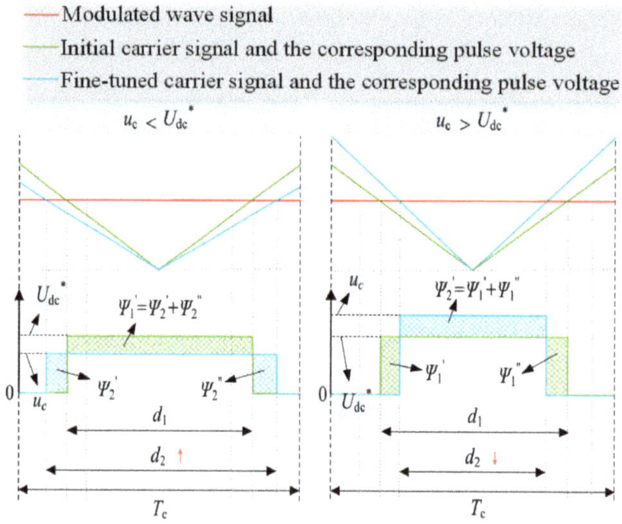

Figure 6. Impact of carrier signal on SM output voltage pulse.

3.2. Impact of Carrier Signal on Capacitor Voltage Balancing

In essence, capacitor voltage balancing can be achieved by precisely controlling the charging and discharging durations of each SM. For analytical purposes, the direction of the arm current is defined as illustrated in Figure 7. Since the MMDTC's arm comprises multiple SMs connected in series, the current through each SM corresponds to the arm current i_{arm}. When the arm current direction is defined as positive, the capacitor enters a charging state; conversely, it transitions to a discharging state.

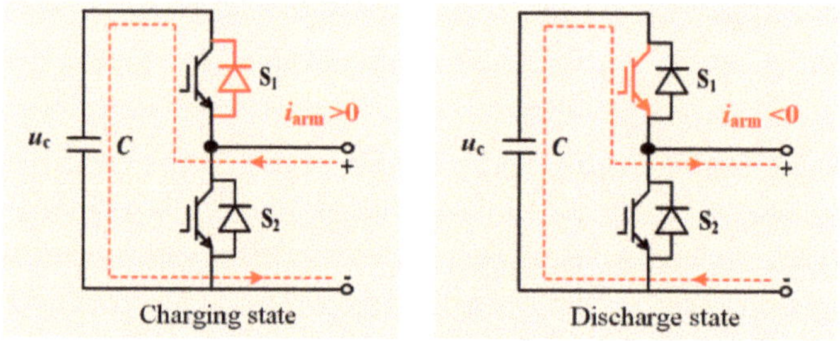

Figure 7. Schematic diagram of the current direction of the SM.

As shown in Figure 8, the detailed implementation for achieving capacitor voltage balancing is as follows:

(1) When the capacitor voltage is lower than its reference value u_c*:

- (a) If the arm current i_{arm} is positive (indicating a charging state), the carrier amplitude should be decreased to extend the capacitor's charging duration, as shown in Figure 8a.
- (b) If the arm current i_{arm} is negative (indicating a discharging state), the carrier amplitude should be increased to decrease the amount of charge released from the capacitor, as shown in Figure 8b.

(2) Conversely, when the capacitor voltage is higher than its reference value u_c*:

(a) If the arm current i_{arm} is positive, the carrier amplitude should be increased to restrict further charging of the capacitor, as shown in Figure 8c.

(b) If the arm current i_{arm} is negative, the carrier amplitude should be reduced to enhance the capacitor's discharge process, as shown in Figure 8d.

This control precisely regulates the conduction time of the SMs by dynamically adjusting the carrier amplitude, thereby ensuring accurate and balanced control of the capacitor voltage.

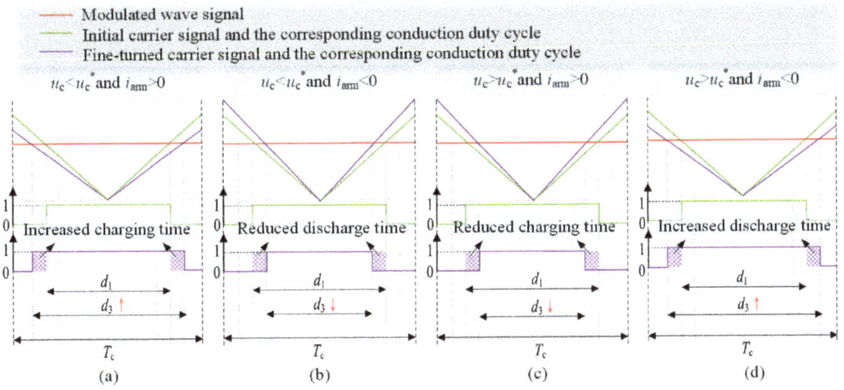

Figure 8. Impact of carrier signal on capacitor voltage balancing under the conditions of **(a)** $u_c < u_c^*$ and $i_{\text{arm}} > 0$, **(b)** $u_c < u_c^*$ and $i_{\text{arm}} < 0$, **(c)** $u_c > u_c^*$ and $i_{\text{arm}} > 0$, and **(d)** $u_c > u_c^*$ and $i_{\text{arm}} < 0$, respectively.

3.3. Mutual Influence and Coupling Between Control Objectives

From a fundamental perspective, regardless of whether it is pulse deviation compensation or capacitor voltage balancing control, both are achieved by modulating the carrier amplitude to adjust the pulse width. Since both control objectives are regulated via the carrier signal, a coupling relationship inherently exists between them, as illustrated in Figure 9.

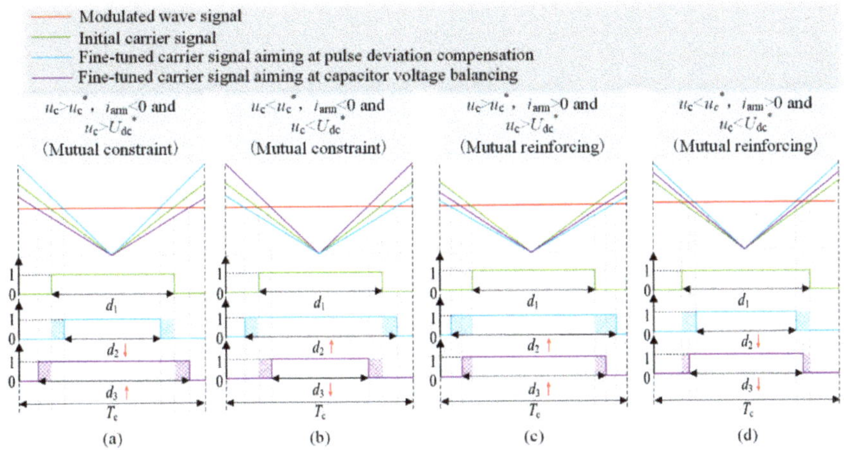

Figure 9. The mutual influence between capacitor voltage balancing and voltage quality improvement using carrier signal regulation under the conditions of **(a)** $u_c < u_c^*$, $i_{\text{arm}} < 0$ and $u_c > U_{\text{dc}}^*$, **(b)** $u_c < u_c^*$, $i_{\text{arm}} < 0$ and $u_c < U_{\text{dc}}^*$, **(c)** $u_c > u_c^*$, $i_{\text{arm}} > 0$ and $u_c > U_{\text{dc}}^*$, and **(d)** $u_c < u_c^*$, $i_{\text{arm}} > 0$ and $u_c < U_{\text{dc}}^*$.

Figure 9a,b illustrate the relationship between carrier fine-tuning and pulse width variation under different conditions. It is evident that there is a mutual constraint between voltage pulse deviation compensation and capacitor voltage balancing control. Specifically, in the case shown in Figure 9a, pulse deviation compensation requires increasing the carrier amplitude to reduce the pulse width, whereas capacitor voltage balancing control necessitates decreasing the carrier amplitude to increase the pulse width. These two control objectives impose entirely opposing requirements on the carrier signal. Similarly, in the scenario depicted in Figure 9b, pulse deviation compensation demands reducing the carrier amplitude, while capacitor voltage balancing control requires increasing the carrier amplitude to narrow the pulse width. This further underscores the mutually restrictive nature of the two control objectives.

Conversely, when the arm current $i_{\text{arm}} > 0$, as shown in Figure 9c,d, the requirements of the two control targets for carrier fine-tuning exhibit consistency, establishing a mutually reinforcing relationship. However, this mutual enhancement could potentially result in an overestimation of the required carrier adjustment amount, thereby introducing additional modulation errors. Therefore, the coupling effect between these two control objectives must be carefully analyzed and fully considered.

4. Proposed Carrier Reconfiguration Method

To address the aforementioned issues, this paper proposes a reconfiguration of the carriers for each SM. As illustrated in Figure 10, when the capacitor voltage becomes unbalanced, it essentially indicates that the DC component U_{dc} deviates from the reference U_{dc}^*. In other words, as long as the measured value U_{dc} accurately tracks the reference value U_{dc}^*, the balance of the capacitor voltage can be effectively achieved.

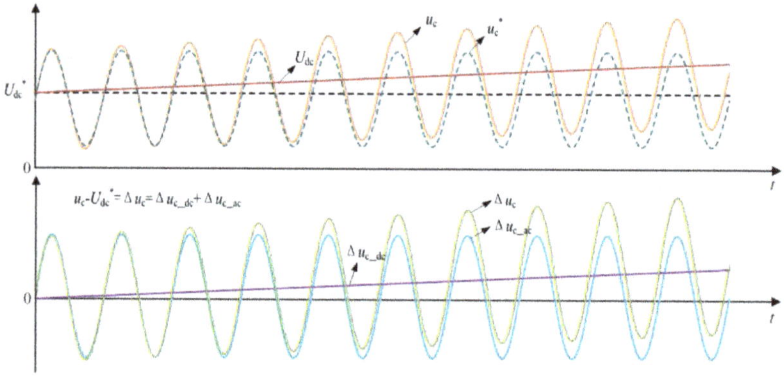

Figure 10. Schematic diagram of the AC and DC component decomposition of the SM capacitor voltage.

The voltage difference Δu_{c} can be expressed as:

$$\Delta u_{\text{c}} = u_{\text{c}} - U_{dc}^* \tag{8}$$

And then, Δu_{c} can be further divided into two parts:

$$\Delta u_{\text{c}} = \Delta u_{\text{c\_}dc} + \Delta u_{\text{c\_}ac} \tag{9}$$

Based on the preceding analysis, it is clear that the pulse deviation in the output voltage of the SM predominantly stems from the AC component of the capacitor voltage. Therefore, to compensate for this pulse deviation, only $\Delta u_{\text{c\_ac}}$ needs to be considered.

Based on the analytical conclusions presented in the previous section, the carrier can be reconfigured twice to achieve independent control. The first reconfiguration focuses on the AC component Δu_{c_ac}, aiming to compensate for the output voltage pulse. The second reconfiguration targets the DC component Δu_{c_dc}, aiming to implement capacitor voltage balancing control.

The specific implementation of carrier reconfiguration is illustrated in Figure 11. The carrier reconfiguration process will dynamically reconstruct the carrier for the next period within the current sampling period. The predicted value of the capacitor voltage is as follows:

$$u_c(k+1) = u_c(k) + \frac{T_s}{C} i_{arm}(k) S(k) \tag{10}$$

where $u_c(k)$ denotes the measured capacitor voltage at time k; $u_c(k+1)$ represents the predicted capacitor voltage for the next sampling period at time k; $i_{arm}(k)$ signifies the arm current at time k. The symbol T_s stands for the sampling period, and S is a switching function. Specifically, when the SM is active at time k, $S(k) = 1$; otherwise, $S(k) = 0$.

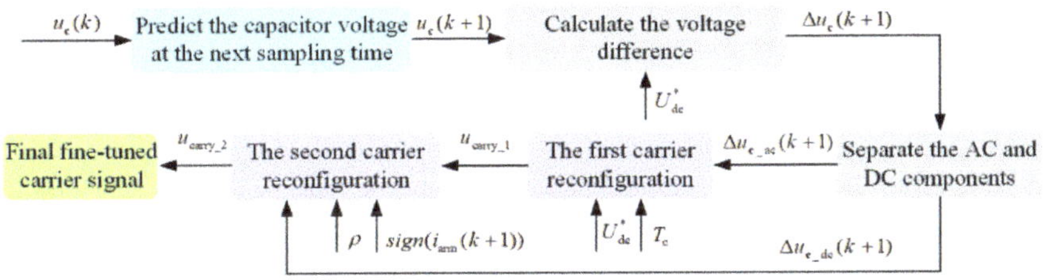

Figure 11. Schematic diagram of carrier signal reconfiguration steps.

After obtaining $u_c(k+1)$, $\Delta u_c(k+1)$ at the moment of $k+1$ can be calculated as follows:

$$\Delta u_c(k+1) = u_c(k+1) - U_{dc}^* \tag{11}$$

Further signal separation of $\Delta u_c(k+1)$ can yield $\Delta u_{c\_ac}(k+1)$ and $\Delta u_{c\_dc}(k+1)$ at the $(k+1)^{th}$ moment. These are then utilized for the first carrier reconfiguration and the second carrier reconfiguration, respectively.

4.1. The First Carrier Reconfiguration

The implementation of the first carrier reconfiguration is depicted in Figure 12. As can be observed, according to the voltage-second balance principle, introducing Δu_{c_ac}(k + 1) to adjust the carrier allows for the compensation of the SM's output voltage. By employing the similar triangle rule, the mathematical expression for the carrier u_{carry_1} after the first carrier reconfiguration can be derived from Figure 12:

$$u_{carry\_1} = \begin{cases} \dfrac{U_{dc}^* + \Delta u_{c\_ac}(k+1)}{U_{dc}^*} - \dfrac{2}{T_c U_{dc}^*} \displaystyle\int_{k_0 - \frac{T_c}{2}}^{k+1} [U_{dc}^* + \Delta u_{c\_ac}(k+1)] d(k+1), k_0 - \frac{T_c}{2} \le k+1 < k_0 \\ \dfrac{2}{T_c U_{dc}^*} \displaystyle\int_{k_0}^{k+1} [U_{dc}^* + \Delta u_{c\_ac}(k+1)] d(k+1), k_0 \le k+1 \le k_0 + \frac{T_c}{2} \end{cases} \tag{12}$$

Also by the similar triangle rule, it can be concluded in Figure 12 that:

$$\frac{d_1/2}{V_{ref}} = \frac{\Delta d}{\frac{2}{T_c} \int_{k_0}^{k_1} \Delta u_{c\_ac}(k+1) d(k+1)} \tag{13}$$

Transforming Equation (13) yields:

$$\gamma = \frac{2\Delta d}{d_1} = \frac{\frac{2}{T_c}\int_{k_0}^{k_1} \Delta u_{\text{c\_ac}}(k+1)d(k+1)}{V_{\text{ref}}} \tag{14}$$

It can be further concluded that:

$$d_2 = (1-\gamma)d_1 \tag{15}$$

where γ is the rate of change in d_2 with respect to d_1. From this analysis, it can be concluded that when $\Delta u_{\text{c\_ac}}(k + 1) > 0$, $\gamma > 0$ and $d_2 < d_1$; when $\Delta u_{\text{c\_ac}}(k + 1) = 0$, $\gamma = 0$ and $d_2 = d_1$; when $\Delta u_{\text{c\_ac}}(k + 1) < 0$, $\gamma < 0$ and $d_2 > d_1$.

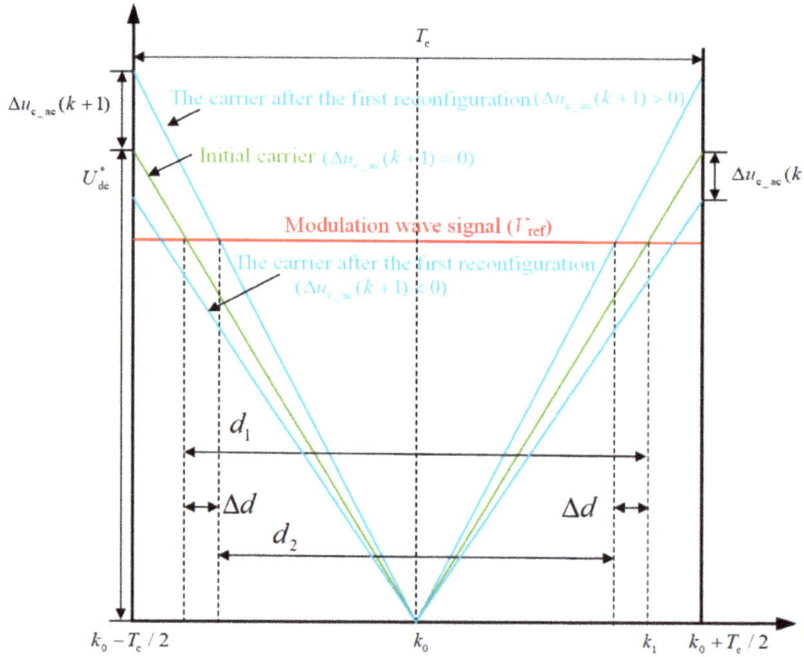

Figure 12. Schematic diagram of the first reconfiguration of carrier signal.

Through the first carrier reconfiguration, the expression of the arm voltage can be expressed as:

$$u_{\text{arm}} = \sum_{j=1}^{N} [S_j(1-\gamma_j)u_{cj}] \tag{16}$$

where S_j represents the switching function of the jth SM; γ_j represents the rate of change in the jth SM d_2 with respect to d_1, and u_{cj} represents the capacitance voltage of the jth SM.

Let $\Delta u_{\text{arm}} = -\sum\limits_{j=1}^{N} (S_j\gamma_j u_{cj})$. The aforementioned equation can be further reformulated as follows:

$$u_{\text{arm}} = \sum_{j=1}^{N} (S_j u_{cj}) + \Delta u_{\text{arm}} \tag{17}$$

That is to say, Δu_{arm} is a compensation variable added to u_{arm} to eliminate the influence of capacitor voltage ripples on the output voltage.

4.2. The Second Carrier Reconfiguration

The second carrier reconfiguration is specifically designed to eliminate Δu_{c_dc}. Given that Δu_{c_dc} is relatively small, it can be effectively mitigated by adjusting the error proportional coefficient ρ. Notably, the precise value of ρ can be dynamically optimized based on the voltage balancing effect of the capacitor.

Figure 13 illustrates the implementation of the second carrier reconfiguration, which is performed based on the outcome of the first carrier reconfiguration. The specific formula is as follows:

$$u_{\text{carry\_2}} = u_{\text{carry\_1}} + \rho \Delta u_{\text{c\_dc}}(k+1) * sign(i_{\text{arm}}(k+1)), k_0 - \frac{T_c}{2} \le k+1 \le k_0 + \frac{T_c}{2} \quad (18)$$

where $\text{sign}(i_{\text{arm}}(k+1))$ represents the direction of the arm current at the $(k+1)^{\text{th}}$ moment.

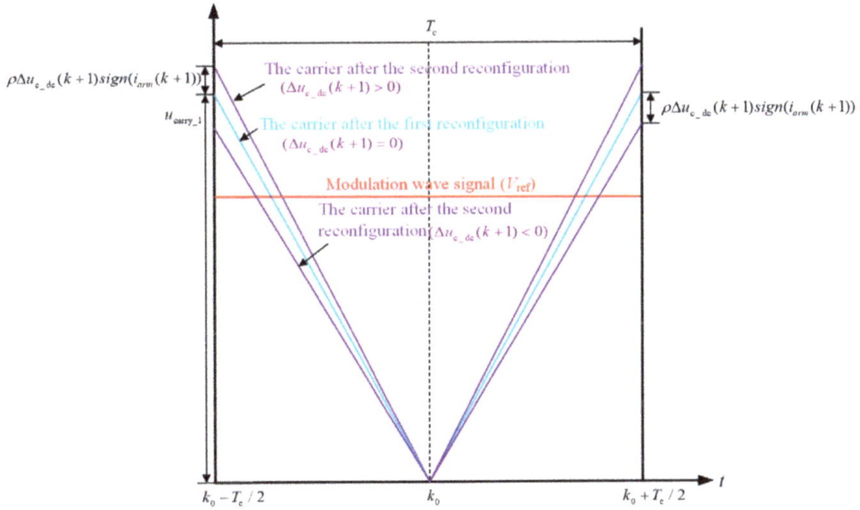

Figure 13. Schematic diagram of the second reconfiguration of carrier signal.

From Equation (18), it can be inferred that when $\Delta u_{cj\_dc}(k+1) = 0$, no secondary reconfiguration is required. If $\Delta u_{cj\_dc}(k+1) > 0$ or $\Delta u_{cj\_dc}(k+1) < 0$, the carrier is dynamically compensated by constructing the product of ρ, $\Delta u_{cj\_dc}(k+1)$, and $\text{sign}(i_{\text{arm}}(k+1))$. This method ensures precise dynamic compensation for the carrier signal, ultimately eliminating the DC offset in the capacitor voltage. After undergoing two reconfigurations, the carrier can be expressed as follows:

$$u_{\text{carry}} = \begin{cases} \dfrac{U_{\text{dc}}^* + \Delta u_{\text{c\_ac}}(k+1)}{U_{\text{dc}}^*} - \dfrac{2}{T_c U_{\text{dc}}^*} \displaystyle\int_{k_0 - \frac{T_c}{2}}^{k+1} [U_{\text{dc}}^* + \Delta u_{\text{c\_ac}}(k+1)] d(k+1) + \rho \Delta u_{\text{c\_dc}}(k+1) * sign(i_{\text{arm}}(k+1)), k_0 - \dfrac{T_c}{2} \le k+1 < k_0 \\ \dfrac{2}{T_c U_{\text{dc}}^*} \displaystyle\int_{k_0}^{k+1} [U_{\text{dc}}^* + \Delta u_{\text{c\_ac}}(k+1)] d(k+1) + \rho \Delta u_{\text{c\_dc}}(k+1) * sign(i_{\text{arm}}(k+1)), k_0 \le k+1 \le k_0 + \dfrac{T_c}{2} \end{cases} \quad (19)$$

The reconfigured carriers, as depicted in Figure 14, are utilized for PWM modulation. This approach significantly differs from the conventional PWM modulation method illustrated in Figure 1. It can be observed that in the proposed method, each SM employs a shared modulation wave signal, and capacitor voltage balancing is achieved through precise adjustment of the carrier wave signal. Simultaneously, this method mitigates output voltage distortion caused by capacitor voltage ripples.

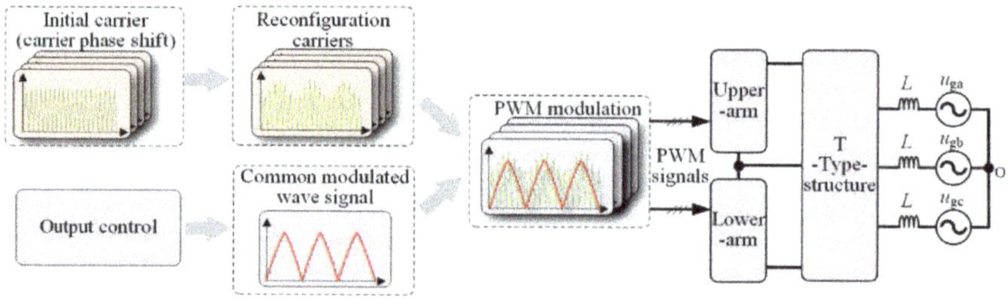

Figure 14. PWM modulation process using the reconfiguration carrier signals.

The proposed carrier reconfiguration method represents a paradigm shift from existing CPS-PWM strategies in several key aspects. Firstly, while conventional methods [17–23] achieve voltage balancing by adjusting the modulation wave signal, the proposed approach manipulates the carrier signal directly. This fundamental difference enables addressing not only voltage balancing but also the output voltage distortion caused by capacitor voltage ripples—a critical issue that existing methods overlook. Secondly, the proposed two-stage reconfiguration process (targeting AC and DC components separately) provides a systematic solution to the inherent coupling between voltage balancing and output quality objectives. This is particularly crucial for low-capacitance operation with a large ripple across the capacitor voltage. Table 1 summarizes the key differences between our method and existing approaches, demonstrating both conceptual innovations and quantitative improvements.

Table 1. Comparison between existing CPS-PWM methods and the proposed carrier reconfiguration method.

Aspect	Existing CPS-PWM Methods [17–23]	Proposed Carrier Reconfiguration Method
Control Variable	Modulation wave adjustment	Carrier signal adjustment
Primary Objective	Capacitor voltage balancing only	Dual objectives: voltage balancing and output quality improvement
Handling of Voltage Ripples	Neglects impact on output quality	Actively compensates for ripple-induced distortions
Low-Capacitance Operation	Not specifically designed for high ripple conditions	Optimized for low-capacitance operation with high ripples
Control Structure	Single-stage adjustment	Two-stage reconfiguration
Pulse Deviation Compensation	No	Yes, through first carrier reconfiguration
Switching Frequency	Fixed	Fixed

5. Simulation and Experimental Verification

5.1. Simulation Verification

To validate the correctness of the carrier reconfiguration proposed in this paper, a simulation model was developed in MATLAB/SIMULINK R2023b. The primary simulation parameters are presented in Table 2. Crucially, for both the proposed method and the existing CPS-PWM method, an identical carrier frequency (f_c = 500 Hz) and sampling period (T_s = 10 μs) were strictly maintained throughout all simulations.

Table 2. Main parameters of simulation.

Variable	Symbol	Values
Rated reactive power	Q_{rated}	10 MVar
Rated grid line voltage	V_{line}	10 kV
Rated grid phase voltage	V_{g}	$\sqrt{2}V_{line}/\sqrt{3}$
AC filtering inductor	L	2.5 mH
Number of SMs per arm	N	18
Rated phase current amplitude	I_{o}	817 A
SM capacitance	C_1	2.6 mF
Peak voltage of SM capacitor	$U_{\text{c\_peak}}$	735/800 V
Carrier frequency	f_{c}	500 Hz
Sampling period	T_s	10 µs

(a) Verification of Capacitor Voltage Balancing

To evaluate the control effectiveness of carrier reconfiguration on capacitor–voltage balancing, simulation analyses were conducted for scenarios both without and with carrier reconfiguration. The absence of carrier reconfiguration indicates that the carrier remains unmodified; specifically, the initial carrier is used without incorporating any control strategy for capacitor voltage balancing. Figure 15a–c present the simulation results of output phase voltages, output currents, and SM capacitor voltages, respectively. It can be observed from Figure 15c that without carrier reconfiguration, the performance of capacitor voltage balancing is inadequate, resulting in significant distortion in both the output voltage and current waveforms, as shown in Figures 15a and 15b, respectively. In contrast, the implementation of the proposed carrier reconfiguration technique markedly improves the capacitor voltage balancing across all SMs.

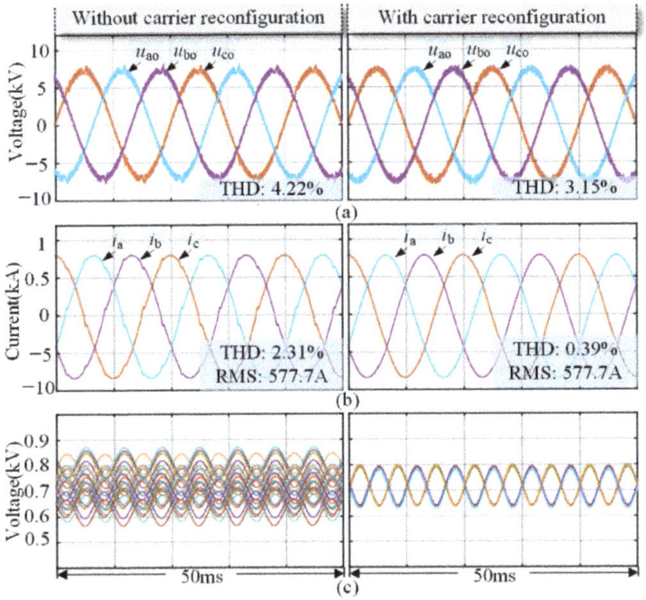

Figure 15. Simulation results of (**a**) output phase voltages, (**b**) output currents, and (**c**) SM capacitor voltages without and with carrier reconfiguration, respectively.

(b) Verification of Voltage Quality Improvement

To verify the compensation effect of the proposed carrier reconfiguration on the output voltage pulse, a comparison with the existing control method (shown in Figure 1, conventional CPS-PWM with modulation wave adjustment for voltage balancing [17–23]) is presented. The number of SMs N increases progressively from 3 to 18 in increments of 3. Notably, during this process, the amplitude of the rated phase current, the peak value of the capacitor voltage, and the ripple rate of the capacitor voltage remain unchanged. Figures 16 and 17 illustrate the simulation results for both the existing control method and the proposed method under varying numbers of SMs. It can be seen that the THD of both output voltages (Figure 16a,b) and output currents (Figure 16c,d) decrease with the increase in the number of SMs. However, using the proposed carrier reconstruction method, the THD value is lower than that using the existing method under the same conditions.

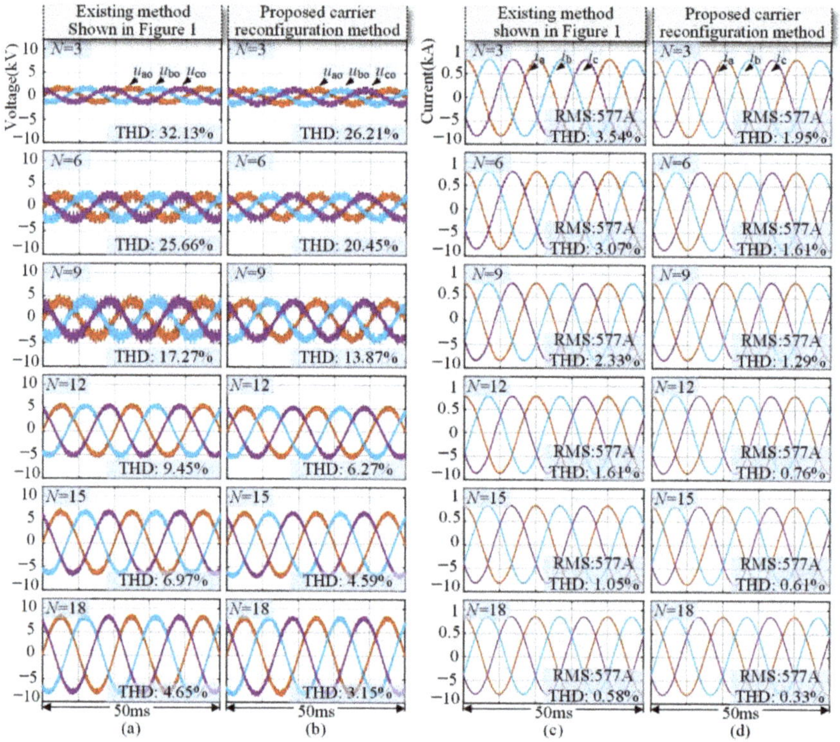

Figure 16. Simulation results comparison of (**a**,**b**) output phase voltages and (**c**,**d**) output currents following with different numbers of HBSMs each arm between the existing and the proposed method.

As can be observed from Figure 17a,b, the voltage pulse compensation does not compromise the capacitor voltage balancing effect. Furthermore, Figure 17c,d illustrate the arm current waveforms. The findings indicate that after the implementation of carrier reconfiguration method, the arm current remains largely unaffected.

A comprehensive statistical analysis was performed on the THD of both the output voltage and current, as shown in Figure 18a,b, respectively, for the two methods. The results indicate that the THD under the proposed method is significantly lower than that under the traditional method approach. Notably, when the number of SMs is constrained, this disparity becomes increasingly evident.

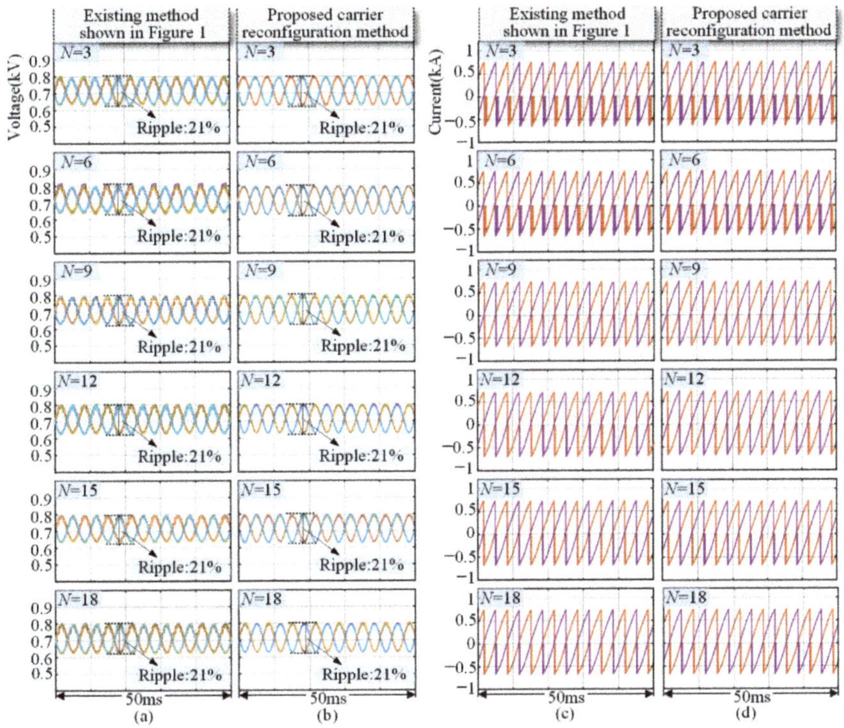

Figure 17. Simulation results comparison of (**a**,**b**) HBSM capacitor voltages and (**c**,**d**) arm currents following with different numbers of HBSMs each arm between the existing and the proposed method.

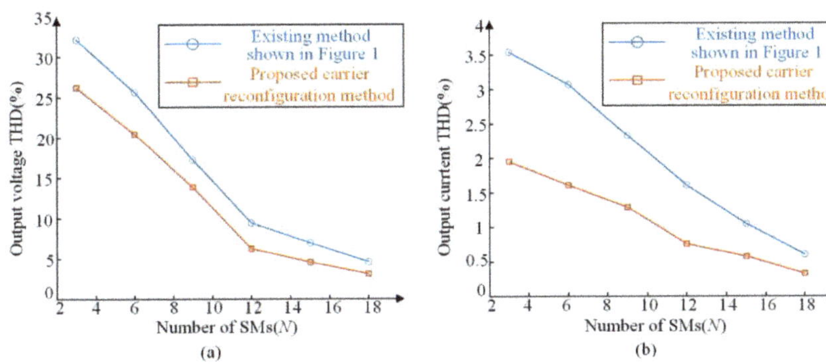

Figure 18. THD comparison of (**a**) output phase voltage and (**b**) output current between the existing and the proposed method.

In addition, the proposed method is comprehensively compared with existing methods in both the capacitive compensation mode and inductive compensation mode. The simulation results are illustrated in Figure 19. As depicted in Figure 19a,b,e, the proposed method achieves accurate compensation for voltage pulse deviations caused by capacitor voltage ripples, irrespective of whether the converter operates in the capacitive or inductive reactive power compensation mode. Compared to the existing method, the waveform quality of the output voltage and current under the proposed method has been significantly enhanced.

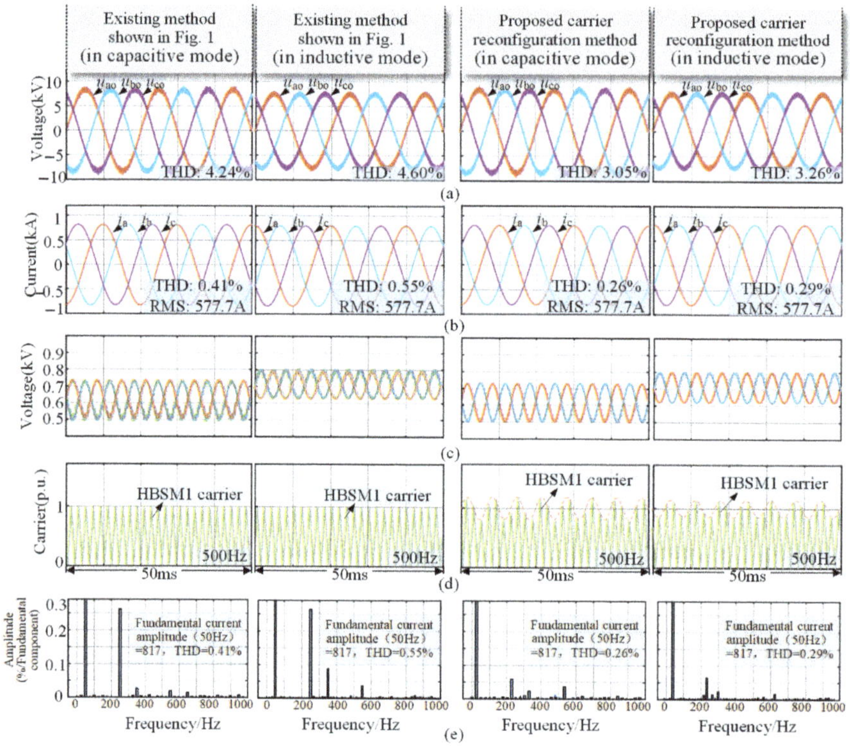

Figure 19. Comparison of simulation results between the existing and proposed methods under different compensation modes: (**a**) output phase voltage, (**b**) output current, (**c**) all capacitor voltages, (**d**) HBSM1 carrier signal, and (**e**) Fourier analysis of output current.

By analyzing Figure 19c,d, it can be observed that both methods achieve capacitor voltage balancing under the two compensation modes. However, the proposed method introduces real-time carrier reconfiguration, dynamically adjusting the carrier amplitude in response to variations in capacitor voltage ripples. This adjustment aligns closely with the trend of capacitor voltage ripples, enabling the SM to achieve an optimized conduction time and effectively compensating for voltage pulse deviations.

5.2. Experimental Verification

The experimental platform constructed in the laboratory is illustrated in Figure 20. In this setup, RT-LAB functions as the controller, and the specific experimental parameters are outlined in Table 3. Identical hardware and timing parameters, specifically the carrier frequency (f_c = 3000 Hz) and sampling period (T_s = 10 μs), were applied to both the proposed carrier reconfiguration method and the existing CPS-PWM method with modulation wave adjustment during comparative testing.

(1) Verification of Capacitor Voltage Balancing

To evaluate the effectiveness of the capacitor voltage balancing control, two operating conditions were considered: (1) inductive compensation mode without carrier reconfiguration, and (2) inductive compensation mode with carrier reconfiguration. As illustrated in Figure 21, the experimental results are presented for the output current (Figure 21a), output phase voltages (Figure 21b), and HBSM capacitor voltages (Figure 21c), respectively.

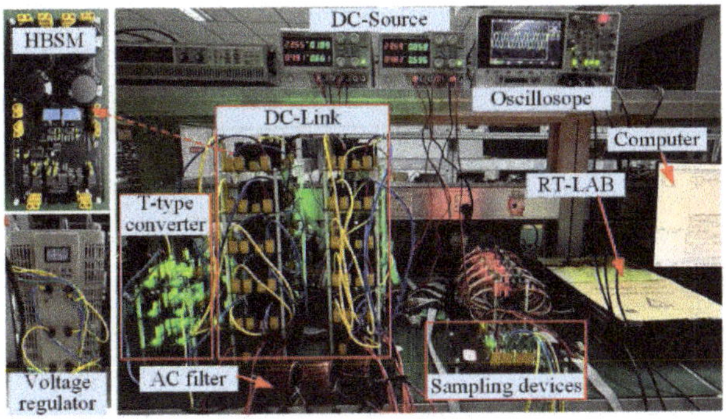

Figure 20. Photograph of MMDTC-STATCOM prototype.

Table 3. The main experimental parameters.

Variable	Symbol	Value
Grid phase voltage	V_g	$110\sqrt{2}$ V
Rated amplitude of phase current	I_o	5 A
Fundamental frequency	f	50 Hz
SM capacitance	C_1	200 µF
AC filtering inductor	L	10 mH
Number of SMs per arm	N	5
Peak voltage of SM capacitor	U_{c_peak}	55 V
Carrier frequency	f_c	3000 Hz
Sampling period	T_s	10 µs

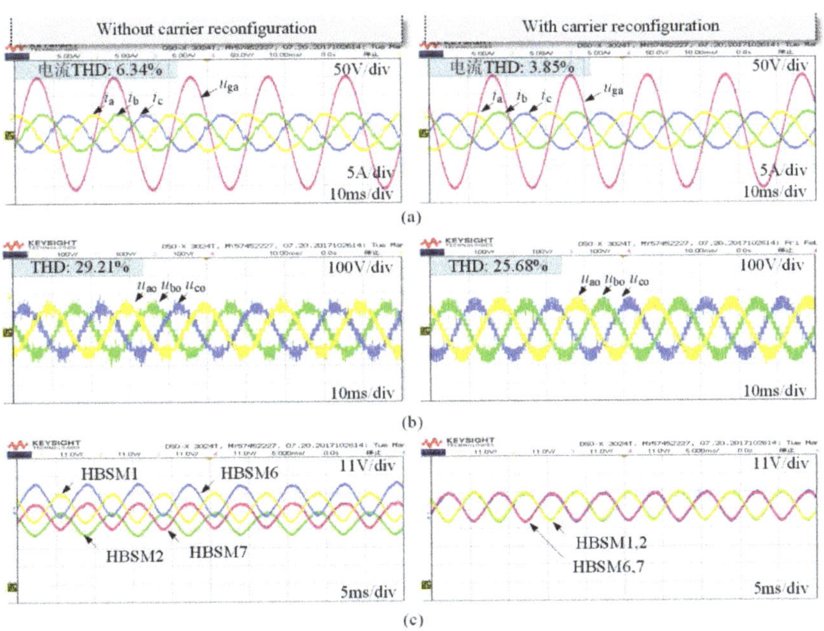

Figure 21. Experimental results of (**a**) grid voltage of phase-a and output currents, (**b**) output phase voltages, and (**c**) HBSM capacitor voltages without and with carrier reconfiguration, respectively.

As illustrated in the left half of Figure 21, without carrier reconfiguration, no balancing measures were implemented for the capacitor voltages, leading to a divergent state in their operation. This divergence subsequently induced distortions in both the output voltage and output current. In contrast, as shown in the right half of Figure 21, the proposed method effectively achieves superior voltage balancing for the SM capacitors. As a result, the total THD of the output voltage is reduced from 29.21% to 25.68%, and the THD of the output current decreases from 6.34% to 3.85%.

(2) Verification of Output Voltage Quality Improvement

Verification of pulse compensation for the output voltage of SMs was conducted under both the capacitive and inductive compensation modes. To demonstrate the improvement achieved, an experimental comparison was performed between the existing method and the proposed method. The experimental results are presented in Figure 22.

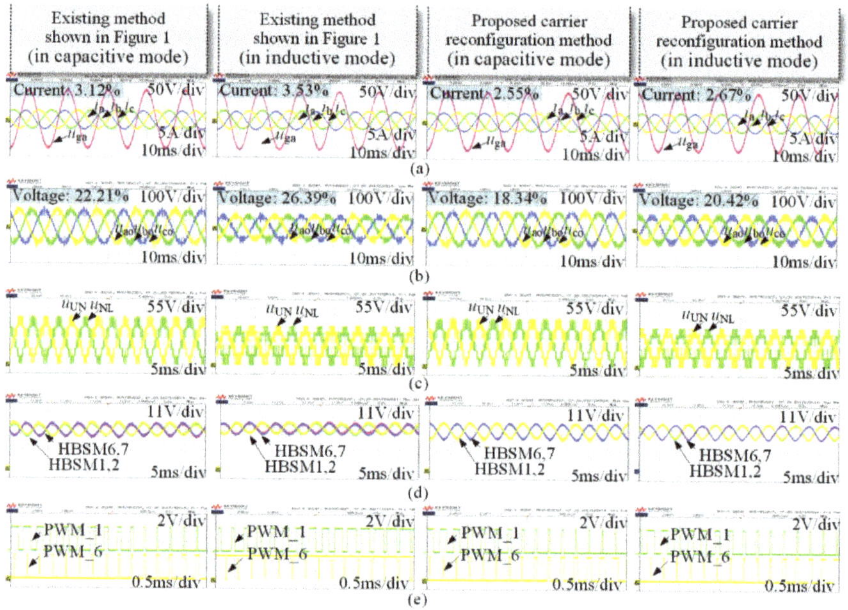

Figure 22. Comparative experimental results of proposed versus existing methods under different compensation modes: (**a**) grid voltage of phase-a and output currents, (**b**) output phase voltages, (**c**) arm voltages, (**d**) representative capacitor voltages from the upper arm (HBSM1, HBSM2) and the lower arm (HBSM6, HBSM7), and (**e**) PWM signals for upper-arm HBSM1 and lower-arm HBSM6.

From Figure 22a,b, it can be clearly observed that the proposed control method reduces the THD of the output voltage by approximately 15% and the THD of the output current by approximately 25% compared to the existing method. This improvement is mainly attributed to the compensation for pulse deviation caused by ripples in the capacitor voltage during carrier reconfiguration. By precisely adjusting the carrier amplitude, the pulse width of the output voltage is regulated to effectively mitigate this deviation.

Figure 22c,d depict the waveforms of the arm voltages and the SM capacitor voltages, respectively. The results demonstrate that when the capacitor voltage undergoes significant ripples, applying the proposed method to compensate for the SM voltage pulse not only markedly enhances the quality of the arm voltage waveform but also effectively improves the balancing of the capacitor voltage.

Figure 22e illustrates the PWM signals for the upper-arm HBSM1 and lower-arm HBSM6. It can be observed that regardless of the method employed, the switching frequency of the submodule remains unchanged. In other words, the proposed method modifies only the amplitude of the carrier signal, without affecting its frequency.

6. Conclusions

This paper proposes a carrier reconfiguration method designed to simultaneously achieve capacitor voltage balancing and improved output voltage quality in low-capacitance MMDTC-STATCOM systems. First, the study highlights how the carrier signal affects the output waveform quality when significant capacitor voltage ripples occur. Then, it investigates voltage pulse deviation compensation and capacitor voltage balancing through precise carrier fine-tuning while also analyzing the intrinsic relationship between these two control objectives. The key innovation involves decoupling these dual control targets by implementing two sequential carrier adjustments per cycle: (1) the first adjustment compensates for AC-component-induced pulse deviations (Equation (12)), thereby directly enhancing voltage waveform fidelity; (2) the second adjustment eliminates DC offset errors through dynamic carrier amplitude modulation (Equation (18)), enabling accurate capacitor voltage balancing. The simulation results (Figure 18) show that the proposed method reduces the output voltage THD by 18–32% compared to conventional CPS-PWM methods under the same number of submodules. The experimental validation further demonstrates a 17% reduction (from 22.21% to 18.34%) and a 23% reduction (from 26.39% to 20.42%) in the output voltage THD under the capacitive and inductive operating modes, respectively, as shown in Figure 22.

Author Contributions: Conceptualization, F.X.; Methodology, F.X. and Y.Q.; Software, Y.Q., X.C. and S.L.; Validation, Y.Q., X.C. and S.L.; Formal analysis, X.J.; Investigation, Y.J.; Resources, Y.J.; Data curation, X.J. and X.C.; Writing—original draft, S.L. and D.N.; Writing—review & editing, F.X. and D.N. All authors have read and agreed to the published version of the manuscript.

Funding: This research was funded by China Energy Investment Corporation Co., Ltd. grant number GJNY-22-109 And The APC was funded by China Energy Investment Corporation Co., Ltd.

Data Availability Statement: Data is contained within the article.

Conflicts of Interest: Author Fengxiang Xie, Yuantang Qi and Yongdong Ji was employed by the company Guoneng (Gonghe) New Energy Development Co., Ltd. The remaining authors declare that the research was conducted in the absence of any commercial or financial relationships that could be construed as a potential conflict of interest.

References

1. Niu, D.; Gao, F.; Xu, T.; Ma, Z. Thyristor-based T-type converter with modular multilevel DC-link. *IEEE Trans. Power Electron.* **2022**, *37*, 6792–6806. [CrossRef]
2. Wang, H.; Blaabjerg, F. Reliability of capacitors for DC-link applications in power electronic converters—An overview. *IEEE Trans. Ind. Appl.* **2014**, *50*, 3569–3578. [CrossRef]
3. Farivar, G.; Hredzak, B.; Agelidis, V.G. Reduced-capacitance thin-film H-bridge multilevel STATCOM control utilizing an analytic filtering scheme. *IEEE Trans. Ind. Electron.* **2015**, *62*, 6457–6468. [CrossRef]
4. Deng, R.; He, G.; Zeng, J.; Liu, X.; Wang, Q.; Huang, Y.Z.G.; Huang, Y.; Pu, Z. Voltage utilization and torque promotion for small film capacitor IPMSM drive considering the load characteristics. *IEEE Trans. Ind. Appl.* **2025**, *61*, 3931–3943. [CrossRef]
5. Zhao, J.; Liu, S.; Zhao, R.; An, J.; Xu, Y. Design and implementation of high-performance tantalum thin-film capacitors for embedded technologies. *IEEE Trans. Compon. Packag. Manuf. Technol.* **2025**, *15*, 868–873. [CrossRef]
6. Liu, S.; Niu, D.; Zhang, L.; Lin, J.; Fang, J.; Li, Z. Low-capacitance operation and its hierarchical control for STATCOM based on T-type converter with MMDTC. *IEEE Trans. Power Electron.* **2025**, *40*, 1153–1167. [CrossRef]
7. Hu, P.; Teodorescu, R.; Wang, S.; Li, S.; Guerrero, J.M. A currentless sorting and selection-based capacitor-voltage-balancing method for modular multilevel converters. *IEEE Trans. Power Electron.* **2019**, *34*, 1022–1025. [CrossRef]

8. Lyu, Y.; Zi, Y.; Li, X. Improved control of capacitor voltage balancing in modular multilevel converter submodules. *IEEE Access* **2024**, *12*, 19510–19519. [CrossRef]
9. Sukanya, V.; Bijukumar, B.; Fahad, N.A.; Mukundan, C.N.; Al-Durra, A. A sensor-optimized MPC strategy for voltage balancing in medium-voltage grid-tied MMCs using state matrix reduction and PS-PWM. *IEEE J. Emerg. Sel. Top. Power Electron.* **2025**, *13*, 3766–3780. [CrossRef]
10. Wu, W.; Qiu, L.; Rodriguez, J.; Liu, X.; Ma, J.; Fang, Y. Data-driven finite control-set model predictive control for modular multilevel converter. *IEEE J. Emerg. Sel. Top. Power Electron.* **2023**, *11*, 523–531. [CrossRef]
11. Rodriguez, E.; Leyva, R.; Townsend, C.D.; Farivar, G.G.; Tafti, H.D.; Pou, J. Constrained control of low-capacitance delta cascaded H-bridge STATCOMs: A model predictive control approach. *IEEE Trans. Power Electron.* **2021**, *36*, 14312–14328. [CrossRef]
12. Zhang, Y.; Wu, X.; Yuan, X.; Wang, Y.; Dai, P. Fast model predictive control for multilevel cascaded H-bridge STATCOM with polynomial computation time. *IEEE Trans. Ind. Electron.* **2016**, *63*, 5231–5243. [CrossRef]
13. Li, Z.; Gao, F.; Xu, F.; Ma, X.; Chu, Z.; Wang, P.; Gou, R.; Li, Y. Power module capacitor voltage balancing method for a ± 350-kV/1000-MW modular multilevel converter. *IEEE Trans. Power Electron.* **2016**, *31*, 3977–3984. [CrossRef]
14. Darus, R.; Pou, J.; Konstantinou, G.; Ceballos, S.; Picas, R.; Agelidis, V.G. A modified voltage balancing algorithm for the modular multilevel converter: Evaluation for staircase and phase-disposition PWM. *IEEE Trans. Power Electron.* **2015**, *30*, 4119–4127. [CrossRef]
15. Luo, Y.; Li, Z.; Xu, L.; Xiong, X.; Li, Y.; Zhao, C. An adaptive voltage-balancing method for high-power modular multilevel converters. *IEEE Trans. Power Electron.* **2018**, *33*, 2901–2912. [CrossRef]
16. Peng, H.; Xie, R.; Wang, K.; Deng, Y.; He, X.; Zhao, R. A capacitor voltage balancing method with fundamental sorting frequency for modular multilevel converters under staircase modulation. *IEEE Trans. Power Electron.* **2016**, *31*, 7809–7822. [CrossRef]
17. Deng, F.; Chen, Z. A control method for voltage balancing in modular multilevel converters. *IEEE Trans. Power Electron.* **2014**, *29*, 66–76. [CrossRef]
18. Dekka, A.; Wu, B.; Zargari, N.R. A novel modulation scheme and voltage balancing algorithm for modular multilevel converter. *IEEE Trans. Ind. Appl.* **2016**, *52*, 432–443. [CrossRef]
19. Deng, F.; Chen, Z. Voltage-balancing method for modular multilevel converters under phase-shifted carrier-based pulsewidth modulation. *IEEE Trans. Ind. Electron.* **2015**, *62*, 4158–4169. [CrossRef]
20. Nguyen, M.H.; Kwak, S.; Kim, T. Phase-shifted carrier pulse-width modulation algorithm with improved dynamic performance for modular multilevel converters. *IEEE Access* **2019**, *7*, 170949–170960. [CrossRef]
21. Dekka, A.; Wu, B.; Zargari, N.R.; Fuentes, R.L. Dynamic voltage balancing algorithm for modular multilevel converter: A unique solution. *IEEE Trans. Power Electron.* **2016**, *31*, 952–963. [CrossRef]
22. Wang, W.; Ma, K.; Cai, X. Efficient capacitor voltage balancing method for modular multilevel converter under carrier-phase-shift pulsewidth modulation. *IEEE Trans. Power Electron.* **2021**, *36*, 1553–1562. [CrossRef]
23. Tai, B.; Gao, C.; Liu, X.; Chen, Z. A novel flexible capacitor voltage control strategy for variable-speed drives with modular multilevel converters. *IEEE Trans. Power Electron.* **2017**, *32*, 128–141. [CrossRef]

energies

MDPI

Article

Low Voltage Ride Through Coordination Control Strategy of DFIG with Series Grid Side Converter

Xin Qi [1], Can Ding [2,*], Jun Zhang [1], Quan Wang [1] and Wenhui Chen [2]

1 China Electric Power Research Institute Wuhan Branch, Wuhan 430074, China; qixin@epri.sgcc.com.cn (X.Q.); zhangjunwhu@163.com (J.Z.); wqwqwlw@163.com (Q.W.)
2 College of Electrical Engineering and New Energy, China Three Gorges University, Yichang 443000, China; chenwenhui@ctgu.edu.cn
* Correspondence: dingcan@ctgu.edu.cn

Abstract: The present study investigates the control strategy of a novel doubled-fed induction generator (DFIG) with a series grid-side converter (SGSC) during grid faults. The rotor-side inverter is subject to a control strategy derived from the Model Predictive Current Control (MPCC) theory, which is implemented during periods of fault occurrence; for the series grid-side converter, the positive and negative sequence component control is implemented during both steady state and fault periods to enhance system stability and performance. The proposed coordinated control strategy is implemented on a doubly fed turbine with SGSC, while taking into account different degrees of symmetric and asymmetric faults to further evaluate the efficacy of the proposed method. The results of the simulations demonstrate the efficacy of the model-predictive current control scheme applied to the rotor-side converter under conditions of asymmetric faults. This enables the suppression of a range of phenomena, including rotor overcurrent, stator overcurrent, and overvoltage, electromagnetic torque ripple, and DC bus voltage during low-voltage ride-through (LVRT), among others. The present study confirms the viability of implementing positive and negative sequences of voltage separation control in the SGSC during both grid faults and steady state. This approach is expected to minimize the switching of SGSC control strategies and thereby reduce output power fluctuations. The Rotor Side Converter (RSC) and SGSC can perform coordinated control during faults, and the proposed method is able to improve low-voltage ride-through performance compared to existing methods, thereby preventing damage to the converter under multiple fault conditions.

Keywords: SGSC (series grid-side converter); positive and negative sequence voltage separation; MPCC (model predictive current control); LVRT (low voltage ride through); DFIG (doubly fed induction generator)

1. Introduction

The recent augmentation in wind power grid-connected capacity has been accompanied by the formulation of novel grid operation guidelines that explicitly delineate the requirements for the operation of grid-connected reactive power in large-capacity wind power systems. A further consideration is imperative to guarantee that grid-connected power quality requirements are satisfied. It is therefore essential that the negative sequence content of the grid-connected current remains within a specified range. To fulfill the voltage ride-through technical requirements of the power generation system under the recently enacted power grid regulations, scholars have proposed a new topology to enhance the performance of the DFIG system when connected to the grid. Most scholars firstly optimize the control strategy of SGSC under grid imbalance [1–6]. The LVRT operation of DFIG is considered to be the biggest challenge of wind power grid-connected technology [7]. To meet the technical requirements for low-voltage ride-through capability of power generation systems under the new grid regulations, some scholars have also investigated the DFIG performance with improved topologies. For instance, as posited by literature [8], a shift from a parallel to a series configuration is proposed for the connection of the grid-side converter to the grid. This novel configuration offers distinct advantages in terms of fault ride-through capability. However, related studies have identified challenges associated with the doubly fed converter system employing this scheme, including inadequate power handling capacity [9]. Consequently, the extant literature [10,11] displays the integration of a parallel grid-side converter with the series grid-side converter. This configuration is employed to enhance the system's power handling capacity, thereby facilitating ride-through under balanced three-phase short-circuit conditions. However, the existing literature has primarily focused on the investigation of the controls for the purpose of total output power control in doubly fed converter systems that utilize SGSC. A paucity of in-depth research exists with respect to the control of the system for the purpose of suppressing the output current in the event of grid faults.

Model Predictive Control (MPC) [12] is a technique that employs a system model to forecast future states and output variables. By continuously optimizing control inputs based on an objective function, it determines the most effective control strategy. As detailed in Reference [13], the utilization of MPC strategies within the domain of power electronic systems has been thoroughly investigated and presented. This paper provides a comprehensive explanation of how Finite Control Set Model Predictive Control (FCS-MPCC) can be applied to power converters. In addition, various studies have introduced anticipatory current regulation strategies designed specifically for doubly fed induction machines operating within wind energy generation frameworks. This model is based on the maximum wind energy tracking model of double-fed wind turbines. Reference [14] applied MPC theory to a solid-state transformer and effectively suppressed power quality harmonics. Reference [15] designed MPCC based on doubly fed wind turbine to solve the problem of DC bus voltage overrun.

To maintain generator voltage and ensure DFIG connectivity during voltage dips, coordinated control of the RSC and SGSC is employed for effective voltage dip mitigation. The new control strategy proposed here comes from the perspective of limiting the transient DC component and the negative sequence power frequency component. To manage the operation of the RSC and the SGSC in the DFIG configuration, a control strategy is formulated that incorporates model predictive current regulation in conjunction with a T/4 delay-based separation technique for decomposing the voltage into its symmetrical positive and negative sequence components. In the following, it is verified that the use of MPCC during the fault can effectively suppress the overcurrent and reduce the DC bus voltage amplitude, so that the effectiveness of the GSC can be guaranteed to achieve power support. Furthermore, it is demonstrated that separating the positive and negative sequences enables smooth SGSC control in steady-state conditions. The study indicates that this work enhances the LVRT capability of the DFIG system, improving the adaptability of grid-integrated doubly fed wind turbines to grid disturbances.

The innovations of this paper are as follows:

1. A voltage ride-through strategy is developed through coordinated control of the SGSC and RSC, where regulation of the SGSC output voltage suppresses the stator flux transient component, and the MPCC strategy enables the RSC to directly track the rotor current, effectively mitigating overcurrent issues. For example, in a single-phase ground fault, the stator current fluctuation range under traditional PI control is −7.14

to 7.14 kA, whereas the improved coordinated control reduces it to −4.3 to 4.3 kA, a reduction of about 40%;

2. This paper also introduces the T/4 delayed voltage separation control strategy applied to SGSC for positive and negative sequence component separation under both steady-state and fault conditions. Equation (12) shows how to achieve voltage positive and negative sequence separation, thereby avoiding the time delay problem of the traditional PI controller. Simulation results indicate that this strategy can limit the DC-link voltage fluctuation range between 1.17 and 1.22 kV during a single-phase short circuit fault, significantly improving system stability compared to traditional PI control.

2. The DFIG Model with SGSC

Compared with the mainstream DFIG system, the topology of the DFIG system using SGSC has changed. It is an SGSC and a series transformer added to the generator stator side in the original topology. The topology is shown in the Figure 1a.

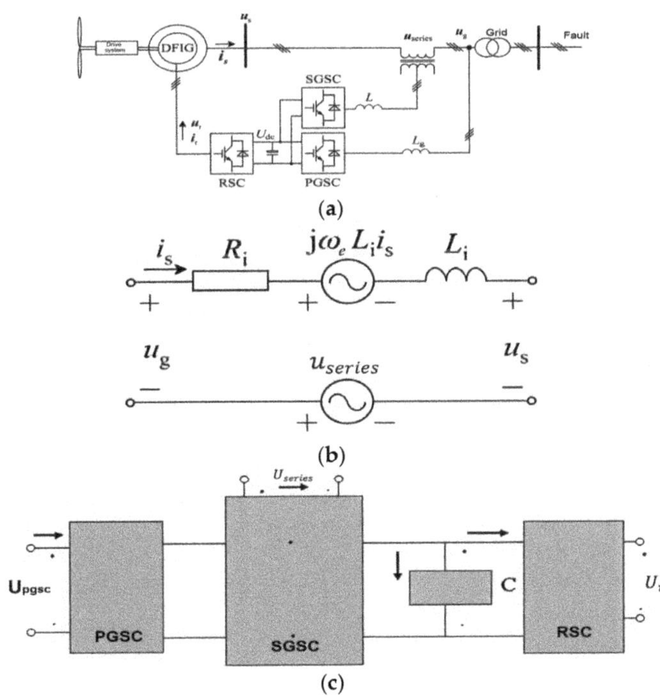

Figure 1. DFIG topology with SGSC and equivalent circuit. (**a**) DFIG topology with SGSC. (**b**) Equivalent circuit of the series grid-side converter in the synchronously rotating reference frame. (**c**) Equivalent circuit of the dc-link.

As shown in Figure 1a, the DFIG stator voltage vector results from the superposition of the grid voltage and the compensation voltage injected through the series transformer.

The stator voltage space vector of the DFIG can be manipulated by regulating the voltage generated by the SGSC, thereby optimizing the transition process of the DFIG when the power grid is faulty. But whether a good low-voltage transition operation can be achieved depends on the optimization result of the control strategy. In order to facilitate the analysis of the problem and simplify the design of the control scheme, a simplified equivalent circuit is used in the series transformer, that is, the excitation branch of the series transformer is ignored. As depicted in Figure 1b, the SGSC voltage within the synchronous rotating coordinate system can be mathematically described as follows:

$$u_g = R_i i_s + j\omega_e L_i i_s + L_i \frac{di_s}{dt} + u_s - u_{series} \tag{1}$$

$$P_{series} + jQ_{series} = u_{series}\hat{i}_s \tag{2}$$

where: u_s is the generator stator voltage space vector in the static coordinate axis system; u_{series} is the series voltage space vector injected by the series transformer; u_g is the grid voltage space vector; R_i is the series transformer resistance and input filter resistance; L_i is the series transformer leakage inductance and filter inductance; P_{series}, Q_{series} represent the real and reactive power transmitted through the series-connected grid converter, respectively.

The mathematical representations of both the stator and rotor sides of the DFIG are formulated as follows:

$$\begin{array}{l} u_s = R_s i_s + \frac{d\varphi_s}{dt} \\ u_r = R_r i_r + \frac{d\varphi_r}{dt} \end{array} \tag{3}$$

In the rotor coordinate system, the flux linkage of stator and rotor is expressed as:

$$\begin{array}{l} \varphi_s = L_s i_s + L_m i_r \\ \varphi_r = L_r i_r + L_m i_s \end{array} \tag{4}$$

In Equations (4) and (5): u_s, u_r and i_s, i_r denote the stator and rotor voltages and currents in the stationary reference frame, respectively; φ_s, φ_r are their flux linkages; R_s, R_r and L_s, L_r represent the stator and rotor resistances and self-inductances; L_m is the mutual inductance; ω_r is the rotor speed. From the state-space Equation (5), the stator and rotor flux linkage dynamics can be derived:

$$\varphi_r = \frac{L_m}{L_s}\varphi_s + \sigma L_r i_r \tag{5}$$

In this context, σ denotes the leakage inductance coefficient.

$$\sigma = \left(1 - L_m^2 / L_s L_r\right) \tag{6}$$

Substituting Equation (6) into Equation (4) enables the derivation of the connection between the rotor voltage and the stator flux.

$$u_r = \frac{L_m}{L_s}\frac{d\varphi_s}{dt} + (\sigma L_r + R_r) i_r \tag{7}$$

As shown in Figure 1c, when the DC bus voltage of the DFIG system is constant, the following relationship exists:

$$P_g - P_r - P_{series} = 0 \tag{8}$$

In the above equation: P_g is the power of the parallel grid-side converter, and P_r is the power of the rotor-side converter.

3. Coordinated Control of Voltage Ride Through Based on SGSC and RSC

In a typical DFIG setup, the generator stator is directly connected to the power grid. When a grid fault causes voltage sag, the stator flux will include a transient DC component, along with both positive and negative components (the latter occurring during unbalanced faults). The transient DC component is one of the main causes of rotor overcurrent. For the

DFIG system using SGSC, the introduction of SGSC can make the stator voltage flexible and controllable. If the output voltage is adjusted to suppress the stator flux's DC offset during faults, rotor over-current protection is achieved. However, the PI controller with sequence separation requires decomposing the current and voltage feedback into their respective components. It will cause a time delay in the feedback amount, thereby reducing the responsiveness and overall dynamic efficiency of the control system [16]. Therefore, it is necessary to introduce RSC control on the basis of SGSC control to suppress overvoltage and overcurrent.

3.1. Improved Control of SGSC During Grid Steady State and Fault

When the grid voltage is normal, RSC and PGSC employ vector control for generator power regulation and DC link voltage stabilization. The RSC uses traditional vector control. In the conventional vector control strategy, the three-phase stator current of the asynchronous machine is transformed into current components in the two-phase stationary reference frame (αβ axis). Subsequently, a coordinate transformation (Park transformation) is applied to equivalently convert these components into direct current components in the synchronous rotating reference frame (dq axis). The PGSC employs a traditional PI control strategy, while the SGSC adopts positive and negative sequence separation control during asymmetric faults and PI control during symmetric conditions. For SGSC, some scholars proposed that the control expression of positive and negative sequence separation can be used for the control of DFIG when the power grid is normal and unbalanced at the same time [17]. However, the author did not conduct valid verification. Therefore, it is inspired to apply the T/4 delay voltage separation control to the SGSC control under steady state and fault conditions in this paper. This is to reduce the switching of control strategies in case of failure, and its effectiveness is verified in the following simulations.

In order to keep the generator stator voltage consistent with the secondary voltage of the grid step-up transformer when the grid voltage is normal, and to eliminate the influence of the leakage impedance of the series transformer, that is to achieve:

$$u_s = u_g \tag{9}$$

Since most of the wind turbine grid-connected transformers are Yd-connected windings, during abnormal grid conditions, such as a fault occurrence, the voltage signal can be decomposed to extract its zero-sequence component independently, thereby enabling targeted analysis and control [6]. In addition, the grid connection method of DFIG is usually three-phase three-wire system, and the zero-sequence current channel cannot be formed. Therefore, the zero-sequence components of the system will not be considered in the following control, and only the positive and negative sequence components will be considered. For the convenience of distinction, in the following analysis, the superscript + and − represent the forward and reverse synchronous rotation coordinate systems; the subscript + and − represent the positive and negative sequence components. The grid voltage vector can be expressed as the following relationship:

$$u_g = u_{g+} + u_{g-} \tag{10}$$

In the event of a grid voltage failure, it is imperative to ensure that the terminal voltage vector and the grid positive sequence voltage vector are congruent. This is achieved by controlling the stator negative sequence voltage vector to zero, thereby eliminating the impact of the grid's negative sequence component on the DFIG during faults. The control target of SGSC can be expressed as follows:

$$\begin{aligned} u_{sdq+} &= u_{gdq+} \\ u_{sdq-} &= 0 \end{aligned} \tag{11}$$

In the equation, u_{sdq+}, u_{sdq-} and u_{gdq+} are the positive sequence and negative sequence components of the stator voltage vector and the positive sequence component of the grid voltage vector in the two-phase rotating coordinate system, respectively.

To fulfill the control objectives outlined in Equation (11), this study employs the T/4 delay technique to effectively decouple the positive and negative sequence components of the voltage:

$$\begin{bmatrix} u_{g\alpha+}(t) \\ u_{g\beta+}(t) \\ u_{g\alpha-}(t) \\ u_{g\beta-}(t) \end{bmatrix} = \frac{1}{2}\begin{bmatrix} 1 & 0 & 0 & -1 \\ 0 & 1 & 1 & 0 \\ 1 & 0 & 0 & 1 \\ 0 & 1 & -1 & 0 \end{bmatrix}\begin{bmatrix} u_{g\alpha}(t) \\ u_{g\beta}(t) \\ u_{g\alpha}(t-T/4) \\ u_{g\beta}(t-T/4) \end{bmatrix} \tag{12}$$

$u_{g\alpha+}, u_{g\beta+}, u_{g\alpha-}, u_{g\alpha-}$ are the positive and negative sequence components of the grid voltage in the $\alpha\beta$ frame.

Based on the derivation from Equation (12), the positive and negative sequence components of the grid voltage can be extracted within the two-phase stationary reference frame. Subsequently, through coordinate transformation involving rotational operations, these components are mapped into the synchronous reference frames—yielding the corresponding voltage components in both the forward and reverse rotating coordinate systems:

$$\begin{bmatrix} u_{gd+}^{+} \\ u_{gq+}^{+} \end{bmatrix} = \sqrt{\frac{2}{3}}\begin{bmatrix} \cos\theta_{g+} & \sin\theta_{g+} \\ -\sin\theta_{g+} & \cos\theta_{g+} \end{bmatrix}\begin{bmatrix} u_{g\alpha+} \\ u_{g\beta+} \end{bmatrix} \tag{13}$$

$$\begin{bmatrix} u_{gd-}^{-} \\ u_{gq-}^{-} \end{bmatrix} = \sqrt{\frac{2}{3}}\begin{bmatrix} \cos(-\theta_{g+}) & \sin(-\theta_{g+}) \\ -\sin(-\theta_{g+}) & \cos(-\theta_{g+}) \end{bmatrix}\begin{bmatrix} u_{g\alpha-} \\ u_{g\beta-} \end{bmatrix} \tag{14}$$

In the equation: T is the grid period. θ_{g+} is the positive sequence grid voltage vector angle, which can be obtained by the phase-locked loop [18]. However, under unbalanced grid conditions, the conventional basic PLL structure is easily affected by negative-sequence voltage components and harmonic disturbances, leading to significant errors in the extracted θ_{g+}. These errors negatively impact the accuracy of the dq-axis coordinate transformation and deteriorate the current regulation performance of the DFIG controller, potentially causing system oscillations.

To enhance the stability and robustness of the control system under asymmetric fault conditions, this study adopts an improved PLL structure to obtain θ_{g+} during the separation of positive and negative sequence components. Common enhanced PLL algorithms, Techniques like DSRF-PLL, DSC-PLL, and EPLL can reliably extract the positive-sequence phase angle, even with negative-sequence components and grid disturbances. For the purpose of simplifying implementation and simulation modeling, the DSRF-PLL structure is selected in this study. It effectively mitigates the impact of reverse-sequence components on the phase-locking outcomes, ensuring precise transformation between forward and reverse sequence coordinate systems, thereby enhancing the system's performance and stability under unbalanced grid conditions.

The positive and negative sequence components of the stator and grid voltages are derived from Equations (10) to (14) under the forward and reverse synchronous rotating axes.

The mathematical models for the forward and reverse sequence components of the series grid-side converter can be expressed in the direct and inverse synchronous reference frames under grid fault conditions.

$$\begin{cases} u^{+}_{seriesd+} = R_i i^{+}_{sd+} - \omega_e L_i i^{+}_{sq+} + L_i \frac{di^{+}_{sd+}}{dt} + u^{+}_{sd+} - u^{+}_{gd+} \\ u^{+}_{seriesq+} = R_i i^{+}_{sq+} + \omega_e L_i i^{+}_{sq+} + L_i \frac{di^{+}_{sq+}}{dt} + u^{+}_{sq+} - u^{+}_{gq+} \end{cases} \tag{15}$$

$$\begin{cases} u^{-}_{seriesd-} = R_i i^{-}_{sd-} + \omega_e L_i i^{-}_{sq-} + L_i \frac{di^{-}_{sd-}}{dt} + u^{-}_{sd-} - u^{-}_{gd-} \\ u^{-}_{seriesq-} = R_i i^{-}_{sq-} - \omega_e L_i i^{-}_{sq-} + L_i \frac{di^{-}_{sq-}}{dt} + u^{-}_{sq-} - u^{-}_{gq-} \end{cases} \tag{16}$$

Combining Equations (15) and (16), the voltage control is realized by using the PI control of the voltage closed-loop. The SGSC positive and negative sequence voltage control equations are:

$$\begin{aligned} u^{+}_{seriesdq+} &= \left[\frac{K_{p1}(\tau_{i1}s+1)}{\tau_{i1}s}\right]\left(u^{+}_{gdq+} - u^{+}_{sdq+}\right) \\ u^{-}_{seriesdq-} &= \left[\frac{K_{p2}(\tau_{i2}s+1)}{\tau_{i2}s}\right]\left(0 - u^{-}_{sdq-}\right) \end{aligned} \tag{17}$$

In Equation (17), K_{p1}, τ_{i1} and K_{p2}, τ_{i2} are the PI controller parameters for the positive- and negative-sequence voltages, respectively. Notably, both $K_{p1} < 0$ and $K_{p2} < 0$.

This paper applies the above sub-sequence control to the SGSC in both normal and fault modes, simplifying its control algorithm. The control flow is shown in Figure 2.

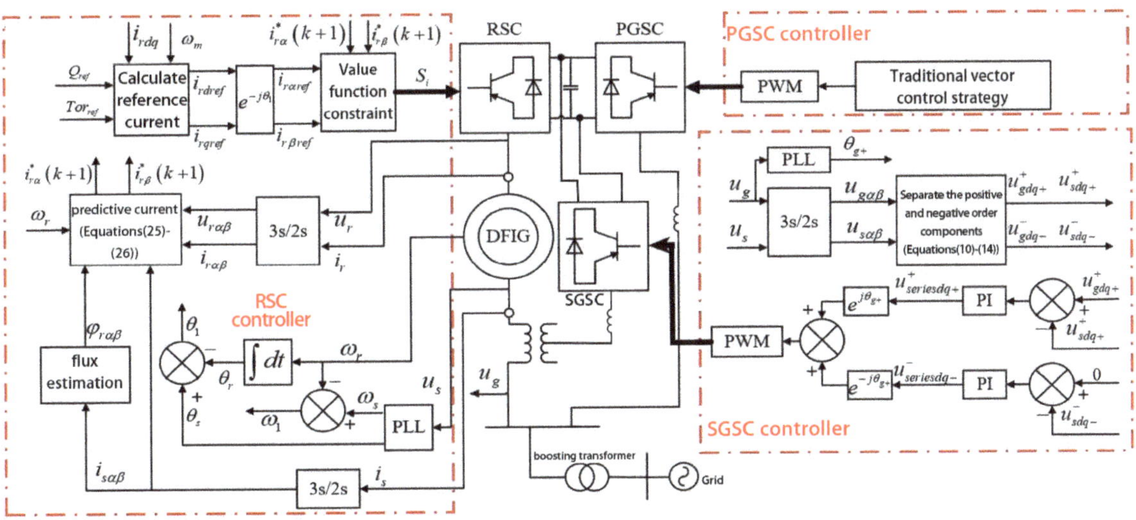

Figure 2. DFIG topology with SGSC.

In Figure 2, a comprehensive set of electrical and control variables is employed to achieve reliable operation under both low and high voltage ride-through (LVRT/HVRT) conditions. The system state variables include the rotor current in the dq reference frame i_{rdq}, the mechanical angular velocity of the wind turbine ω_m, and the reference values for reactive power and electromagnetic torque, denoted by Q_{ref} and Tor_{ref}, respectively. The control structure utilizes both dq-axis current references i_{rdref}, i_{rqref} and their corresponding $\alpha\beta$-axis references $i_{r\alpha ref}$, $i_{r\beta ref}$, while the angle θ is used for coordinate transformation. The optimal switching state determined by the predictive controller is represented by S_i, and the predicted rotor current at the next sampling instant is given by $i^{*}_{r\alpha}(k+1)$, $i^{*}_{r\beta}(k+1)$.

The measured electrical quantities include the rotor electrical angular velocity ω_m, rotor voltages and currents in the $\alpha\beta$ frame $u_{r\alpha\beta}$, $i_{r\alpha\beta}$ and the stator voltages and currents u_s, i_s along with their $\alpha\beta$ components $u_{s\alpha\beta}$, $i_{s\alpha\beta}$. The reference angular speed and angle used for transformations are denoted by ω_1 and θ_1, while the electrical angles of the rotor and stator are θ_r and θ_s, respectively. The rotor flux linkage is represented by $\varphi_{r\alpha\beta}$.

The angle corresponding to the fundamental balanced component of the grid voltage, obtained through phase-locked loop (PLL) synchronization, is indicated by θ_g^+. The grid voltage and its $\alpha\beta$ components are denoted as u_g, $u_{g\alpha\beta}$. To enable symmetrical component-based regulation, the dq-domain representations corresponding to the fundamental and unbalanced parts of the grid and stator voltages are introduced as u_{gdq+}^+, u_{gdq-}^-, u_{sdq+}^+, u_{sdq-}^-. In addition, the SGSC-injected voltage for voltage compensation is represented by $u_{seriesq+}^+$, $u_{seriesq-}^-$.

In contrast to unbalanced voltage sags commonly encountered in power systems, a symmetrical three-phase voltage swell allows the stator-side grid-side converter SGSC to regulate the terminal voltage of the stator, thereby supporting continuous power transmission from the DFIG to the utility grid [19]. Furthermore, through appropriate modulation of the RSC, the DFIG is capable of delivering reactive power assistance during the voltage swell event [20]. Throughout this condition, the PGSC maintains its primary function of stabilizing the DC-link voltage, which is essential to ensure reliable high-voltage ride-through capability of the DFIG system.

Since the grid voltage under a symmetrical overvoltage condition consists solely of fundamental balanced components, the control method designed for decomposing and handling unbalanced sequence components during asymmetric faults is not required [21]. At this stage, the control objective for the stator voltage can be expressed as:

$$\begin{cases} u_{sd} = u_{gd} = u_{sd}' \\ u_{sq} = 0 = u_{sq}' \end{cases} \tag{18}$$

In the above equation, u_{sd}' and u_{sq}' represent the stator terminal voltage vectors in the dq coordinate system during a grid voltage surge, while u_{sd} and u_{sq} represent the stator terminal voltage vectors during steady-state grid conditions. By combining Equation (1), The voltage expression at the stator side, when represented in the dq reference frame, can be reformulated as follows:

$$u_{sdq} = u_{seriesdq} - u_{gdq} - R_i i_{sdq} - j\omega_e L_i i_{sdq} - L_i \frac{di_{sdq}}{dt} \tag{19}$$

Based on the above, the stator voltage control equation in the dq coordinate system is established as follows:

$$\begin{cases} u_{sd} = \left[\frac{K_{ps}(\tau_{is}s+1)}{\tau_{is}s}\right]\left(u_{gd} - u_{sd}\right) \\ u_{sq} = \left[\frac{K_{ps}(\tau_{is}s+1)}{\tau_{is}s}\right]\left(0 - u_{sq}\right) \end{cases} \tag{20}$$

In Equation (20), K_{ps} and τ_{is} are the PI control gains for the SGSC in the $\alpha\beta$ rotating frame after the voltage surge.

3.2. MPCC Strategy Applied to RSC During Grid Failure

In the event that the generator continues to supply a significant amount of power during a grid fault, the positive-sequence current of the stator will rise due to the abrupt decrease in the generator's stator voltage [22]. The rotor cutting stator positive sequence magnetic field of the generator will also generate a large slip frequency current component in the rotor winding [23], which is also an important cause of overcurrent on the rotor side. Therefore, if the RSC control strategy can be improved during the fault process, the purpose of suppressing the stator overcurrent of the generator can be achieved, thereby improving the low-voltage transient operation of the DFIG system.

3.2.1. Building a Predictive Model

To implement the MPCC during low voltage ride-through, the rotor voltage and current are first discretely sampled, leading to the derivation of the following equation:

$$\frac{d}{dt}\begin{bmatrix} u \\ i \end{bmatrix} = \frac{1}{T_s}\begin{bmatrix} u(k) - u(k-1) \\ i(k) - i(k-1) \end{bmatrix} \tag{21}$$

A comparison of the MPCC strategy applied to RSC in this paper with the delayed positive and negative sequence separation control applied to SGSC reveals that the former does not require the separation of positive and negative sequences of voltage and current. Furthermore, the MPCC strategy does not entail a time delay under PI control [16]. The state-space model that accounts for a linear discrete-time system is as follows: In order to realize the MPCC during low-voltage ride-through, the voltage and current of the rotor must first be sampled discretely. This leads to the derivation of the subsequent mathematical expression:

$$\begin{aligned} x(k+1) &= Ax(k) + B_u u(k) + B_d d(k) \\ y_c(k+1) &= C_c x(k) \end{aligned} \tag{22}$$

In Equation (19), the matrix A is the state transition matrix, discretized based on the system's electrical parameters—primarily the stator and rotor inductance and resistance—along with the sampling interval T_s. The control input matrix B_u characterizes the influence of the control input $u(k)$, which in this case represents the modulated rotor voltage vector $u_{r\alpha\beta}$, on the system state evolution. External disturbances, such as variations in stator voltage u_s, grid faults, or flux oscillations, are modeled via the disturbance vector $d(k)$ and its associated input matrix B_d, which maps the disturbance to its effect on the system dynamics. The controlled output $y_c(k+1)$ corresponds to the predicted output, typically the rotor current to be regulated, and is derived through the output matrix C_c, which selects relevant elements from the state vector—often implemented as an identity or selection matrix depending on the control objective.

The current information, such as $i_{r\alpha}(k)$ and $i_{r\beta}(k)$, obtained from MATLAB/Simulink, is used to define the state variable vector $x(k)$, while the mathematical model $u(k)$ of the doubly fed machine is treated as the input control variable. Using this, the future current values $i_{r\alpha}(k+1)$ and $i_{r\beta}(k+1)$, which represent the next time step of the controlled output $y_c(k)$, are forecasted. Assuming all system states are measurable, Equation (19) is then restructured into an incremental model.

$$\begin{aligned} \Delta x(k+1) &= A\Delta x(k) + B_u \Delta u(k) + B_d \Delta d(k) \\ y_c(k) &= C_c \Delta x(k) + y_c(k-1) \end{aligned} \tag{23}$$

where, $\Delta x(k) = x(k) - x(k-1); \Delta u(k) = u(k) - u(k-1); \Delta d(k) = d(k) - d(k-1)$.

The current value of k is represented by $x(k)$, and the change in $x(k)$ from its previous value, $x(k-1)$, is calculated as $\Delta x(k) = x(k) - x(k-1)$. The quantity known as "$\Delta x(k)$" can be utilized as a point of departure for the purpose of predicting the future dynamics of the system, thereby enabling the estimation of the state that will be reached at time $k+1$. This process, in essence, constitutes a state increment. Equation (20) can be rewritten as:

$$\Delta x(k+1|k) = A\Delta x(k) + B_u \Delta u(k) + B_d \Delta d(k) \tag{24}$$

In the given equation, the notation "k+1 | k" indicates a one-step-ahead prediction from the current time k to the next time instant k+1, where the value at k+1 does not influence the system output at time k. Once the state-space model is constructed, the corresponding current prediction expression is formulated. Subsequently, various reference signals are assigned to the rotor-side converter, subject to predefined constraints.

3.2.2. Current Predictive Control Algorithm

When an asymmetric fault occurs in the power grid, the generator's stator voltage decreases, leading to the emergence of a negative-sequence component. As a result, the stator current contains both positive and negative sequence components. The stator flux consists of three components, in addition to the components corresponding to the positive and negative sequences, there is also a DC component of the stator flux formed by the inability to suddenly change due to faults. The total stator flux can be represented by the equation: $\varphi_s = \int (u_s - R_s i_s)dt$. Due to the relatively slow variation in the DC component of the magnetic flux, the free component may be obtained by means of filtering the magnetic flux through a low-pass filter. During the period of asymmetric voltage drop in the power grid, transient DC flux linkage of the stator will be induced. By using a low-pass filter to filter the magnetic flux, it was found that this component is small and gradually decays, so it can be ignored. In order to facilitate comprehension, the influence of the stator transient DC flux linkage is temporally disregarded in the analysis. That is to say, solely the positive and negative sequence flux linkages and currents of the stator are taken into account.

Equation (7) can be reformulated as follow:

$$R_r i_r + \left(L_r - \frac{L_m^2}{L_s}\right)\frac{di_r}{dt} = u_r - j\omega_r \varphi_r \tag{25}$$

The discretization of Equation (22) involves the implementation of the forward Euler method, which replaces the difference quotient with the discrete difference value. The result should be outputted as the current control increment at time $k+1$, as demonstrated in the subsequent equation:

$$\begin{aligned} i_{r\alpha}^*(k+1) &= T_s \frac{u_{r\alpha}(k) - R_r i_{r\alpha}(k) + \omega_r \varphi_{r\beta}(k)}{L_r - L_m^2/L_s} + i_{r\alpha}(k) \\ i_{r\beta}^*(k+1) &= T_s \frac{u_{r\beta}(k) - R_r i_{r\beta}(k) - \omega_r \varphi_{r\alpha}(k)}{L_r - L_m^2/L_s} + i_{r\beta}(k) \end{aligned} \tag{26}$$

3.2.3. The Establishment of the Value Function

A cost function needs to be defined in the predictive control scheme, so the motor model established above is used to predict the current. The error of the reference current and the predicted current must be considered in the value function, and then the value function is established:

$$g = \left| i_{r\alpha ref}(k+1) - i_{r\alpha}^*(k+1) \right| + \left| i_{r\beta ref}(k+1) - i_{r\beta}^*(k+1) \right| \tag{27}$$

As illustrated by Equation (24), $i_{r\alpha ref}(k+1)$ and $i_{r\beta ref}(k+1)$ correspond to the reference values in the $\alpha\beta$ coordinate system. These reference values are derived subsequent to the coordinate transformation of the original control strategy. In this study, we sought to determine the optimal switching sequence that would yield the desired control signal for the RSC. To this end, we compared the reference value with the predicted value of the current. We then identified the switch combination that corresponded to the current vector, which resulted in the minimization of the cost function. Finally, we selected this combination as the RSC trigger pulse signal. A two-level inverter is viewed as a nonlinear discrete system that can generate eight switching states and seven different voltage vectors. as shown in the Table 1 below:

Table 1. Switch state and voltage vector.

Sa	Sb	Sc	Voltage Vector V
0	0	0	$V_0 = 0$
1	0	0	$V_1 = \frac{2}{3}V_{dc}$
1	1	0	$V_2 = \frac{1}{3}V_{dc} + j\frac{\sqrt{3}}{3}V_{dc}$
0	1	0	$V_3 = -\frac{1}{3}V_{dc} + j\frac{\sqrt{3}}{3}V_{dc}$
0	1	1	$V_4 = -\frac{2}{3}V_{dc}$
0	0	1	$V_5 = -\frac{1}{3}V_{dc} - j\frac{\sqrt{3}}{3}V_{dc}$
1	0	1	$V_6 = \frac{1}{3}V_{dc} - j\frac{\sqrt{3}}{3}V_{dc}$
1	1	1	$V_7 = 0$

Ultimately, to realize current regulation, the switching vector that yields the minimum value of the cost function is determined and applied as the control output. The specific implementation structure of the MPCC strategy for the (RSC) is illustrated in the control block diagram presented in Figure 2, while the corresponding procedural steps are outlined in the flowchart depicted in Figure 3.

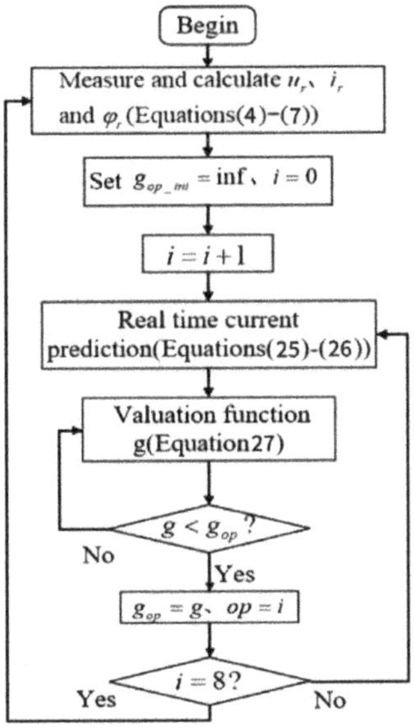

Figure 3. Model predicts the current control flow.

4. Simulation Verification

To validate the performance of the designed control scheme under conditions of voltage sags, a 2 MW doubly fed wind turbine is used to simulate the occurrence of symmetrical and asymmetrical faults in the power grid. As a comparison, the simulation results under the conventional control algorithm are also given in this paper. DFIG system parameters are as Table 2.

Table 2. Parameters of DFIG.

Parameter	Numerical Value
Stator and rotor mutual inductance (mH)	2.5
Rotor leakage inductance (mH)	0.087
Stator leakage inductance (mH)	0.087
DC bus voltage reference value (kV)	1.2
Rated Capacity (MW)	2
Rated voltage (kV)	0.69
Stator resistance (mΩ)	2.6
Rotor resistance (mΩ)	2.9
SGSC series transformer ratio	2:1
Capacity of the SGSC (MW)	2

Note: Data are collected from [15] according to the specific parameter.

At the moment of fault detection, the excitation converter detects a voltage dip at the grid connection point. The system switches the RSC from the original PI control to a model predictive current control strategy. The PGSC adopts a traditional PI control strategy, while the SGSC employs positive–negative sequence separation control during asymmetric faults and PI control during symmetric conditions. Figure 4 illustrates the simulated waveform when the system voltage drops to 60% for a three-phase symmetrical fault, Figure 5 illustrates the simulated waveform when the system voltage drops to 60% for a single-phase ground fault, Figure 6 displays the waveform for a 70% voltage drop during a two-phase short circuit. Panels 4 (a–e) depict rotor/stator currents, grid currents/voltages, torque, and DC bus voltage in the dq frame. Figures 5 and 6a–l cover key system responses, including phase-a currents, SGSC voltage/current components, grid-side current, torque, DC voltage, active power, and reactive power flow. In addition, the PI marked in the Figure represents the conventional PI control (RSC adopts the directional control based on the stator flux linkage, and SGSC adopts the T/4 positive and negative sequence voltage separation control). The MPCC represents the use of improved coordinated control (RSC uses MPCC, SGSC still uses positive and negative sequence separation control). Specifically, in Figure 4, due to a symmetrical fault, the voltage is only positive sequence. Therefore, SGSC adopts PI control instead of the T/4 delay positive and negative sequence component control.

After a three-phase fault occurs, the voltage components drop symmetrically and only have positive sequence components. When the fault occurs at 4 s, through the control of RSC, the rotor excitation voltage is properly controlled according to the degree of grid voltage drop, and the stator current of the generator shown in Figure 4b will also be reduced accordingly, so as to further limit the stator current. In the MPCC scheme without space vector control, the PWM control signal is sent directly to the RSC by means of minimization of the cost function. This enables the control of the rotor current to be directly tracked under faults, so that the rotor current fluctuation of Figure 4a is suppressed. The enhancement of the grid-side power waveforms depicted in Figure 4c–d facilitates the successful implementation of voltage dip recovery functionality for the DFIG system. At this point, as shown in Figure 4e,f, the power quality of the grid is significantly improved in contrast to traditional control methods, and the coordinated control of SGSC and RSC effectively mitigates fluctuations in electromagnetic torque and bus voltage. In Figure 4, due to a symmetrical fault, the voltage is only positive sequence. Therefore, SGSC adopts PI control instead of the T/4 delay positive and negative sequence component control. Thus, by comparing Figure 4 with Figures 5 and 6, it can be observed that when the fault is removed, the switching of the control strategy in Figure 4 will cause significant voltage and current fluctuations, which cannot be avoided.

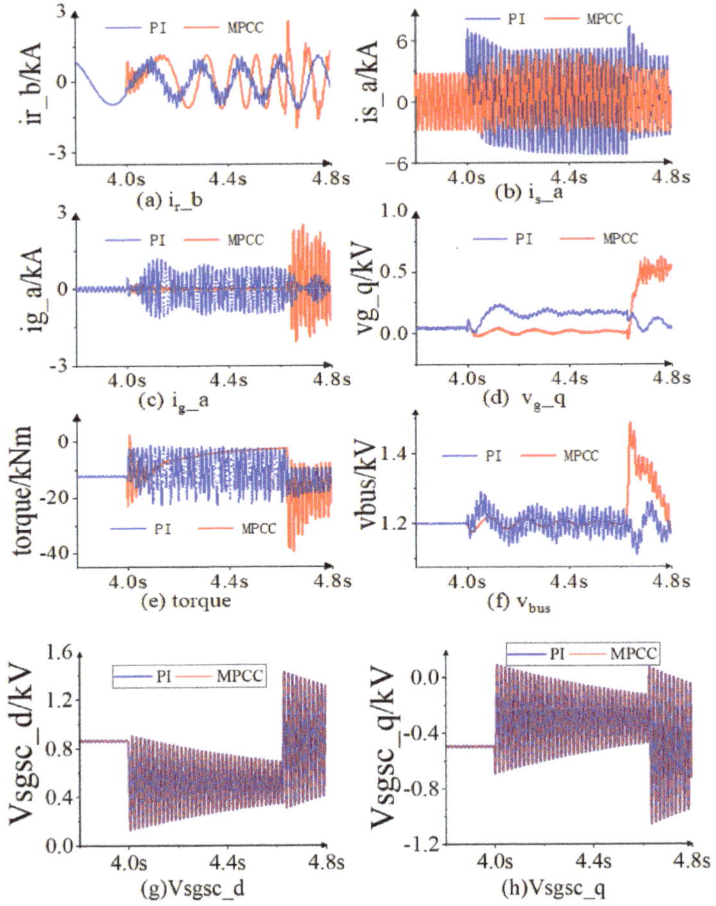

Figure 4. Comparison of LVRT performance between MPCC and PI for three-phase short-circuit fault.

Figures 5a and 6 show that the quality of the rotor current curve is stable if no fault occurs before 4 s. After the fault occurs, The rotor generates a starting current due to the negative-sequence component in the stator flux. Conventional PI control increases the rotor current to a peak of 2.1 kA at 0.076 s after a single-phase fault, and to a peak of 1.9 kA at 0.049 s after a two-phase fault. This means that under asymmetric faults, conventional PI control increases the rotor current to twice the steady-state value. However, under the improved collaborative control using MPCC, the maximum rotor current in Figures 5 and 6 is around 1.1 kA. After 0.383 s of two-phase fault, the rotor current gradually decays after reaching the peak value of 1.14 kA. In the later stage of fault, it is basically between −0.89 and 0.76 kA, and compared with the fluctuation range before fault −1.0 to 1.0 kA, the exceeding limit does not exceed 24%.

As the terminal voltage drops, the stator voltage will also drop and the flux linkage will change in a corresponding manner. At this point, there is also an inrush current in the stator current, as shown in (b) of Figures 5 and 6. With MPCC, the flux linkage is suppressed and the stator current amplitude is reduced. Compared to traditional PI control, MPCC is more effective in suppressing the stator and rotor currents when a fault is detected. This has the advantage that the stator can be controlled in such a way that reactive power can be sent via the rotor current after the voltage has dropped, thus providing reactive power support for the grid.

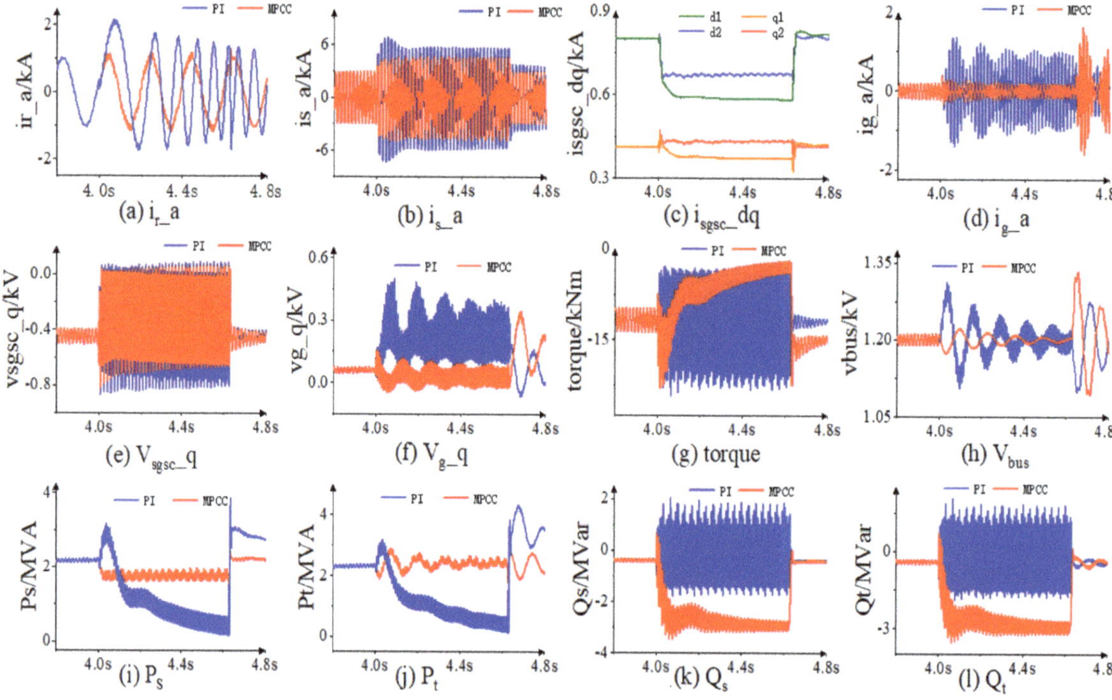

Figure 5. Comparison of LVRT performance between MPCC and PI for single-phase short-circuit fault.

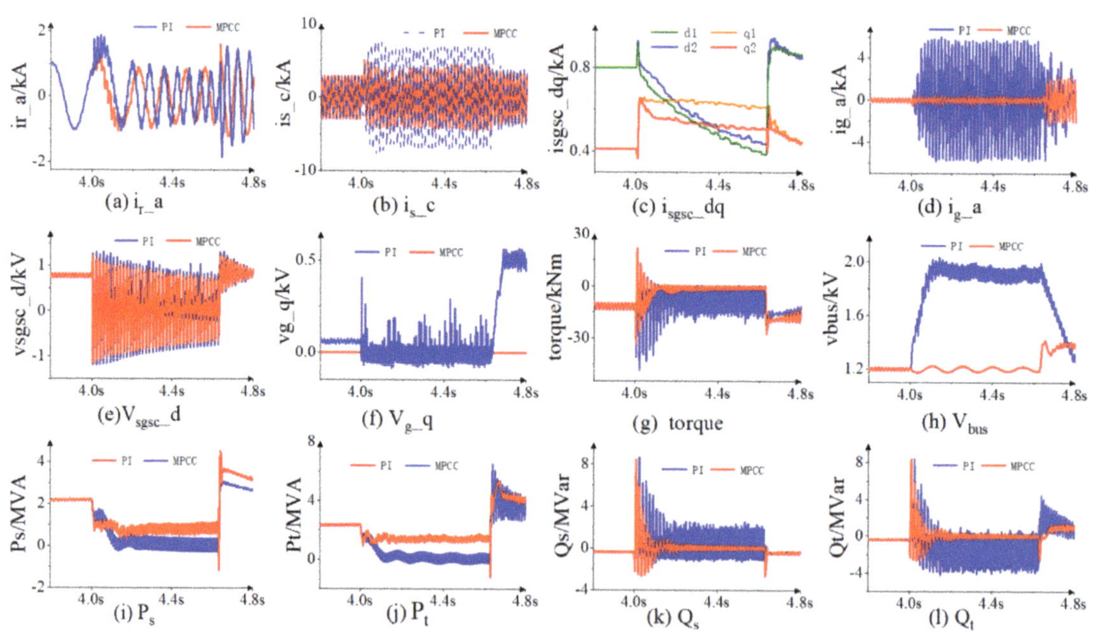

Figure 6. Comparison of LVRT performance between MPCC and PI for two-phase short-circuit fault.

The key to fault control is suppressing the flux linkage AC component that the defect causes, since the DC component and its impact are not taken into account in the study above. Figures 5 and 6's (g) illustrate how linkage cutting produces a sizable number of double-frequency AC components in conventional PI control electromagnetic torque. To some extent, MPCC limits the electromagnetic torque variation and keeps it at a relatively constant amount by exerting an inhibitory impact. Switching between the fault algorithm and the steady-state vector control technique creates some electromagnetic oscillation, especially when the mains voltage recovers.

Regardless of the control strategy employed by the three converters during a disturbance, maintaining the stability of the DC bus voltage remains a fundamental requirement for the proper functioning of the exciter converter and the enhancement of control effectiveness. Figures 5 and 6, specifically part (h), illustrate the variations in DC bus voltage under PI control and MPCC control, respectively. As demonstrated, in the absence of any fault, the DC bus voltage remains stable at approximately 1.2 kV prior to the 4-s mark. However, upon the occurrence of a fault, the energy begins to accumulate within the DC link, thereby inducing voltage fluctuations. It has been demonstrated that, among the various fault types, the voltage deviation under a two-phase short-circuit is more pronounced than that under a single-phase ground fault. The MPCC strategy has been demonstrated to effectively mitigate these oscillations and reduce harmonic distortion. This finding suggests that the MPCC approach, as implemented for the RSC in this study, exhibits superiority over the conventional PI method in preserving the stability of the DC bus voltage during disturbances, whether these disturbances involve a ground fault in one phase, a short circuit between two phases, or a balanced three-phase fault.

Throughout the fault duration, regardless of the specific control method implemented by the converter, maintaining a stable DC bus voltage—reflecting the effective regulation capability of the GSC—is a fundamental prerequisite for ensuring the reliable operation of the converter and the validity of control enhancement strategies. As indicated by the preceding analysis, the voltage deviation of the DC bus under MPCC regulation remains within a 100 V margin. The GSC output current and voltage shown in (d) and (f) of Figures 5 and 6 are also within a controllable range. This guarantees the effective regulation of the GSC, while markedly improving the quality of the voltage and current waveforms at the grid interface.

During a voltage sag, the DFIG compensates by injecting reactive power into the power network, and the level of reactive support provided rises correspondingly. The proposed scheme has less electromagnetic torque oscillation at the time of fault occurrence and removal, which has less mechanical impact on the wind turbine shaft system. As depicted in (k) and (l) of Figures 5 and 6, the generator only minimally takes in reactive power from the grid in the later phase. Under the traditional control, an increased demand for reactive power at the generator stator terminals can impede the restoration of voltage levels within the power grid. At 4.625 s, the fault is removed, and following a short transient response, the reactive power promptly stabilizes to its steady-state value prior to the disturbance. It can be seen from (i)–(l) of Figures 5 and 6 that through the effective control of the RSC, the output power fluctuation of the generator can be quickly suppressed in a relatively stable range when a fault occurs.

5. Conclusions

This paper presents a fault-tolerant strategy for DFIG-based wind energy systems. The proposed control mode enables DFIG to smoothly achieve low-voltage ride-through under various fault conditions. A MPCC method for RSC and a positive and negative sequence delay control scheme for SGSC are designed. The effectiveness of the control strategy presented is thoroughly assessed under normal, three-phase sag, single-phase grounding, and two-phase short-circuit voltage sag conditions for the DFIG. By utilizing the approach outlined in this study, the DFIG can withstand faults and maintain a stable connection to the grid. The improved control strategy proposed in this article is mainly applied during the fault period, but there are still strong impact issues caused by control strategy switching after the fault ends. Therefore, further research is needed on collaborative control from the beginning of the fault to the stage after the fault ends.

Quantitative analysis of the conclusions:

1. Three-phase symmetrical fault: In a three-phase short-circuit fault, when the grid voltage drops to 60%, the q-axis grid-side voltage under traditional PI control reaches its peak value of 0.25 kV after 0.125 s, while the improved coordinated control stabilizes it around 0.023 kV. Additionally, the improved coordinated control reduces the electromagnetic torque fluctuation range from −0.8 to 21.6 kNm to a near-stable value. The bus voltage fluctuation range is also reduced from 1.14–1.29 kV to 1.17–1.21 kV;
2. Single-phase grounding fault: For single-phase grounding faults, the comparison between traditional PI control and the improved coordinated control demonstrates clear performance differences. For example, in the case of a single-phase short-circuit fault, the stator current fluctuation range under traditional PI control is from −7.14 kA to 7.14 kA, while the improved coordinated control based on SGSC and RSC reduces this range to −4.3 kA to 4.3 kA, representing a reduction of approximately 40%. Furthermore, the improved coordinated control increases the minimum electromagnetic torque from −23.4 kNm to −22.1 kNm and quickly suppresses it to a relatively stable level;
3. Two-phase short-circuit fault: For a two-phase short-circuit fault, the improved coordinated control exhibits superior performance. For instance, in a two-phase short-circuit fault, the DC-link voltage under traditional PI control peaks at 2.03 kV after 0.143 s, while the improved coordinated control limits it to between 1.17 kV and 1.26 kV. Additionally, the improved coordinated control ensures that the reactive power stabilizes after 0.21 s, with the fluctuation range being only −0.4 MVar to 0.38 MVar. The quantitative analysis demonstrates that the improved coordinated control strategy significantly enhances system stability and response speed under various fault conditions, providing more reliable low-voltage ride-through capability for DFIG wind energy systems during grid faults.

Author Contributions: Conceptualization, X.Q. and W.C.; methodology, X.Q.; software, X.Q.; validation, C.D.; formal analysis, X.Q. and Q.W.; investigation, J.Z.; resources, J.Z.; data curation, W.C.; writing—original draft preparation, W.C.; writing—review and editing, C.D.; visualization, C.D.; supervision, C.D.; project administration, Q.W. All authors have read and agreed to the published version of the manuscript.

Funding: This research received no external funding.

Data Availability Statement: The original contributions presented in the study are included in the article, further inquiries can be directed to the corresponding author.

Conflicts of Interest: Xin Qi, Jun Zhang and Quan Wang were employed by the China Electric Power Research Institute Wuhan Branch. The remaining authors declare that the research was conducted in the absence of any commercial or financial relationships that could be construed as a potential conflict of interest.

References

1. Cheng, Q.M.; Wang, Y.J.; Cheng, Y.M.; Wei, L.; Shen, L. Coordinated control strategy of DFIG wind power system based on SGSC and PBC under unbalanced grid voltage. *Acta Energiae Solaris Sin.* **2021**, *42*, 351–358.

2. Cheng, Q.M.; Wei, L.; Cheng, Y.M.; Wei, L.; Wang, Y.J. Coordinated control of DFIG Based on SGSC and lyapunov controller in unbalanced power grid. *Electr. Mach. Control* **2020**, *24*, 27–37.
3. Suppioni, V.P.; Grilo, A.P.; Teixeira, J.C. Improving network voltage unbalance levels by controlling DFIG wind turbine using a dynamic voltage restorer. *Int. J. Electr. Power Energy Syst.* **2018**, *96*, 185–193. [CrossRef]
4. Suppioni, V.P.; Grilo, A.P.; Teixeira, J.C. Control methodology for compensation of grid voltage unbalance using a series-converter scheme for the DFIG. *Electr. Power Syst. Res.* **2016**, *133*, 198–208. [CrossRef]
5. Yao, J.; Li, H.; Chen, Z. Enhanced control of a DFIG-Based wind-power generation system with series grid-side converter under unbalanced grid voltage conditions. *IEEE Trans. Power Electron.* **2013**, *28*, 3167–3181. [CrossRef]
6. Kang, Y.L.; Zheng, T.T.; Miao, S.H.; Liu, Z.W.; Liu, Y.L. Coordinated control strategy of series and parallel networked side converters for DFIG system under unbalanced grid voltage. *Trans. Electr. Technol.* **2018**, *33*, 193–204.
7. Hiremath, R.; Moger, T. Grid-Connected DFIG Driven Wind System for Low Voltage Ride Through Enhancement using Neural Predictive Controller. *J. Inst. Eng. India Ser.* **2022**, *103*, 1577–1588. [CrossRef]
8. Petersson, A. Analysis, modeling and control of doubly-fed induction generators for wind turbines. Doctoral Thesis, Chalmers University of Technology, Gothenburg, Sweden, 2005.
9. Flannery, P.S.; Venkataramanan, G. A grid fault tolerant doubly fed induction generator wind turbine via series connected grid side converter. *arXiv* **2022**, arXiv:2201.08879.
10. Omar, S.; Helal, A.; Elarabawy, I. Stator voltage sensorless DFIG with low voltage ride-through capability using series and parallel grid side converters. In Proceedings of the 2016 7th International Renewable Energy Congress (IREC), Hammamet, Tunisia, 22–24 March 2016.
11. Jahanbakhsh, B.; Xu, D. Improved low voltage ride through capability of doubly fed induction generator using series grid side converter. In Proceedings of the 7th International Power Electronics and Motion Control Conference, Harbin, China, 2–5 June 2012.
12. Morari, M.; Lee, J.H. Model predictive control: Past, present and future. *Comput. Chem. Eng.* **1999**, *23*, 667–683. [CrossRef]
13. Kouro, S.; Cortes, P.; Vargas, R. Model Predictive Control—A Simple and Powerful Method to Control Power Converters. *IEEE Trans. Ind. Electr.* **2000**, *56*, 1826–1838. [CrossRef]
14. Ding, C.; Zhang, H.; Chen, Y. Research on Control Strategy of Solid State Transformer Based on Improved MPC Method. *IEEE Access* **2023**, *11*, 9431–9440. [CrossRef]
15. Ding, C.; Chen, Y.; Nie, T. LVRT Control Strategy for Asymmetric Faults of DFIG Based on Improved MPCC Method. *IEEE Access* **2021**, *9*, 165207–165218. [CrossRef]
16. Hu, S.J.; Meng, Y.F.; Li, F.L. AC voltage sensorless Control strategy of grid-connected inverter under unbalanced grid voltage. *Trans. China Electrotech. Soc.* **2017**, *32*, 146–152.
17. Yao, J.; Liao, Y.; Li, H. Coordinated control of a doubly fed induction generator wind turbine with series grid-side converter under unbalanced grid voltage conditions. *Electr. Mach. Control* **2010**, *14*, 8.
18. Liu, R.Z.; Chen, Z.; Tang, W.B. Control strategy of an LCL grid-connected Inverter with the influence of a phase-locked loop under a weak power grid. *Power Syst. Prot. Control* **2022**, *50*, 10.
19. Makrini, A.E.; Karkri, Y.E.; Boukhriss, Y.; Markhi, H.E.; Moussaoui, H.E. LVRT Control Strategy of DFIG based Wind Turbines Combining Passive and Active Protection. *Int. J. Renew. Energy Res.* **2017**, *7*, 1258–1269.
20. Li, B.; Zheng, D.; Li, B. Analysis of low voltage ride-through capability and optimal control strategy of doubly-fed wind farms under symmetrical fault. *Prot. Control Mod. Power Syst.* **2023**, *8*, 1–15. [CrossRef]
21. Li, B.T.; Zheng, D.C.; Li, B.; Ji, L.; Hong, Q.T.; Meng, Q.L. Research on low voltage ride-through strategies for doubly-fed wind farms during asymmetric faults. *Int. J. Electr. Power Energy Syst.* **2024**, *160*, 110138.
22. Cheng, X.K.; Sun, X.D.; Chai, J.Y.; Liu, H.; Song, P. Virtual synchronous control strategy for doubly-fed induction generator under asymmetrical gris faults. *Autom. Electr. Power Syst.* **2018**, *42*, 120–126.
23. Lin, R.X.; Ma, X.Y.; Chang, X.Q.; Xu, L.; Ding, L.J.; Yang, H.G. Analysis of DFIG rotor current considering chopper protection under asymmetrical grid voltage sags. *Proc. CSU-EPSA* **2019**, *31*, 106–117.

energies

MDPI

Article

A Fast SOC Balancing Method for MMC-BESS Based on Nonlinear Model-Predictive Control

Xiaofan Ji [1], Fengxiang Xie [2], Yuantang Qi [2], Yongdong Ji [2], Decun Niu [3,*] and Qizhong Yan [3]

[1] National Institute of Clean-and-Low-Carbon Energy of CHN Energy, Beijing 102209, China; 20024057@ceic.com

[2] CHN Energy Gonghe New Energy Development Co., Ltd., Hainan Tibetan Autonomous Prefecture 813000, China; 12068234@ceic.com (F.X.); 20109584@ceic.com (Y.Q.); 20084510@ceic.com (Y.J.)

[3] School of Automation and Electronic Engineering, Qingdao University of Science and Technology, Qingdao 266061, China; 2023041005@mails.qust.edu.cn

* Correspondence: niudecun@qust.edu.cn; Tel.: +86-15069128650

Abstract: In modular multilevel converter battery energy storage systems (MMC-BESS), state-of-charge (SOC) balancing is essential for ensuring safe and reliable operation. Existing methods based on linear controllers or conventional model-predictive control (MPC) often suffer from slow balancing speed, difficult parameter tuning, and high computational burden. To address these challenges, this paper proposes a fast SOC balancing strategy based on nonlinear MPC. A nonlinear state-space model is first developed and then linearized to enable discrete single-step prediction of arm- and phase-level SOC values. A two-stage control scheme is introduced to coordinate inter-arm and inter-phase SOC balancing, significantly reducing the number of state variables involved in the MPC formulation. The proposed method eliminates the need for circulating current reference calculation and control parameter tuning. Simulation results demonstrate that the proposed method takes approximately 17.5 s and 39 s for inter-arm and inter-phase SOC balancing, respectively, while traditional three-level SOC balancing takes approximately 42 s and 88 s.

Keywords: MMC-BESS; SOC-balancing control; nonlinear model-predictive control; state-space model; cost function

1. Introduction

The intermittency and volatility of renewable energy pose significant challenges to power system stability at scale [1,2]. Battery energy storage systems effectively mitigate power fluctuations from renewable sources, making the integration of BESSs with renewable generation essential. Among various converter topologies, the modular multilevel converter-based battery energy storage system (MMC-BESS) stands out as a promising solution for large-capacity storage due to its unique advantages [3,4]. The MMC-BESS distributes batteries across submodules, which prevents the 'bucket effect' and eliminates parallel circulating currents commonly found in traditional battery arrays with extensive series-parallel connections. However, balancing the State of Charge (SOC) of submodule batteries is critical for safe and reliable MMC-BESS operation [5,6].

Research on SOC-balancing control for MMC-BESS has advanced considerably. Reference [7] introduced a traditional three-level control scheme, but its open-loop structure and reliance on proportional components resulted in slower SOC balancing. References [8,9] implemented independent PI controllers for each submodule storage system, enhancing SOC-balancing performance while simultaneously increasing system complexity and complicating parameter tuning. References [10,11] developed a three-level SOC-balancing

https://doi.org/10.3390/en18102502

strategy by combining PI controller, PR controller, and modulation wave reconstruction techniques, though the carrier phase shift modulation method employed exhibits limited applicability in high-voltage, high-power scenarios. Furthermore, conventional PI/PR controllers are highly dependent on parameter tuning, directly affecting system dynamic response and stability, while demonstrating poor adaptability to system nonlinearities and time-varying parameters [12].

Model-predictive control (MPC) enables real-time management of complex systems through dynamic prediction and cost function optimization, effectively eliminating control-parameter-tuning challenges [13,14]. Recent years have witnessed notable progress in MPC strategies for MMC-BESS systems. Reference [15] introduces a modulation-based MPC approach that directly computes duty cycles between two predefined voltage levels, eliminating weighting factors while preserving constant switching frequency. References [16,17] develop a switching state selection MPC method, where combinatorial complexity remains challenging (C_{2N}^{N} valid states) despite limiting active submodules to N per phase. Several enhanced MPC variants address specific challenges: Reference [18] achieves superior current quality by expanding voltage levels from N + 1 to 2N + 1 with minimal computational overhead. For SOC balancing, Reference [19] employs a dual-mode strategy using fixed circulating currents for large imbalances and adaptive coefficients for minor corrections, albeit with increased switching losses. Reference [20] implements a precision dual-loop architecture combining SOC prediction with MPC current tracking, though requiring exact parameter matching.

Advanced control architectures include Reference [21]'s computationally demanding but effective dual-layer solution for low-frequency operation and [22]'s hierarchical approach separating global MPC balancing from local virtual resistance control, offering speed–accuracy compromises. Reference [23] further refines balancing through a three-level algorithm with dynamic coefficients, enhancing stability at the cost of control complexity. Finally, Reference [24] presents an adaptive-weight MPC that dynamically adjusts cost functions based on submodule voltages for optimized MMC performance. Reference [25] proposed a hybrid approach using MPC for output current and circulating current control at the submodule level while still relying on PI control for battery-side SOC balancing, failing to achieve direct MPC over SOC equilibrium. References [26,27] employed traditional double-loop PI control for output currents while incorporating MPC for SOC balancing, indirectly achieving SOC balancing through circulating-current-cost functions but lacking direct control over SOC. References [25,28] introduced an MPC methodology for MMC-embedded storage systems that successfully balanced storage submodule SOC and suppressed current ripple, though this approach is limited to storage submodules with supplementary DC-DC converters and cannot be extended to conventional half bridge energy storage submodules. Also, the cost function is developed based on the current of the energy storage unit. Reference [29] presented an enhanced inner- and outer-loop MPC strategy for output-current control, double-loop PI control for DC/DC converters within storage submodules, and improved inner–outer-loop MPC for SOC balancing. While this method demonstrated effective SOC balancing, it still required PR controllers to calculate circulating-current reference values, maintaining dependency on control parameter tuning.

The existing SOC-balancing control methods for MMC-BESS can be summarized in Table 1. It is evident that the current MPC method for SOC balancing has two significant limitations. First, the predictive model and cost function are based on current rather than SOC, which indirectly reflects SOC characteristics. Second, MPC must be combined with PI/PR control because the current reference required for the cost function needs to be calculated. It can be seen from the above that existing MPC research focused on SOC balancing in MMC-BESS primarily addresses storage submodules equipped with DC-DC converters, while comparatively limited attention has been given to MPC for SOC balancing in traditional half-bridge storage submodules. Most approaches indirectly achieve SOC balancing by minimizing circulating current reference value cost functions, necessitating parameter tuning for proportional, PI, or PR control components. Additionally, these control methodologies share an inherent limitation where balancing speed diminishes as SOC deviation decreases.

Table 1. Comparison of the existing SOC-balancing control methods for MMC-BESS.

Control Method	Technical Features	Main Limitations.
Traditional three-level control [7]	Open-loop proportional control and sequencing strategy	Slow speed and sensitive parameters
PI/PR composite control [8–12]	Multi-stage closed-loop architecture	High complexity and poor dynamic response
Indirect MPC [25–27]	The SOC is indirectly balanced through the circulation cost function	The PI/PR controller is needed
Direct MPC [25,28]	Based on the SOC prediction model	Only applicable to the submodule with DC-DC converter
Hybrid MPC [15,29]	MPC with conventional controller cooperation	Still dependent on parameter tuning

To address these challenges, this paper proposes an advanced SOC-rapid-balancing technology for MMC-BESSs based on nonlinear MPC. Key innovations include the following: first, establishing a nonlinear state-space model for average SOC values between upper and lower arms and developing a single-step prediction model with deviation cost function for arm and phase average SOC values through linearization and discretization; second, optimizing control stage division and state variable configuration within the overall MPC scheme to substantially reduce the computational burden. The proposed control strategy offers several advantages: complete SOC control without calculating the circulating current; elimination of traditional control-parameter-tuning requirements; and maintenance of consistent SOC-balancing speed regardless of the SOC deviation magnitude.

2. MMC-BESS Topology and Its Traditional Three-Level SOC-Balancing Control

2.1. Topological Structure

The topology of the MMC-BESS and its Energy Submodule (ESM) is shown in Figure 1. Each phase consists of the upper and lower arms. Each arm includes an inductor L_{arm}, an equivalent resistor R_{arm}, and N ESMs. The parameters are as follows:

Grid-side parameters: u_{sk} (k = a, b, c) represents the three-phase grid voltage; L_s and R_s are the equivalent inductance and resistance between the MMC and the grid.

MMC parameters: u_k (k = a, b, c) represents the output AC voltage of the MMC; i_{vk} (k = a, b, c) represents the output AC current of the MMC.

Arm electrical parameters: $u_{\mathrm{p}k}$ and $u_{\mathrm{n}k}$ (k = a, b, c), respectively, represent the output voltage of the upper and lower arms of phase-k; $i_{\mathrm{p}k}$ and $i_{\mathrm{n}k}$, respectively, represent the current of the upper and lower arms of phase k.

ESM parameters: The half-bridge topology is adopted, consisting of two IGBTs and one battery pack, where U_{cell} is the battery pack voltage; u_{bat} and i_{bat} are the output voltage and current of the ESM, respectively.

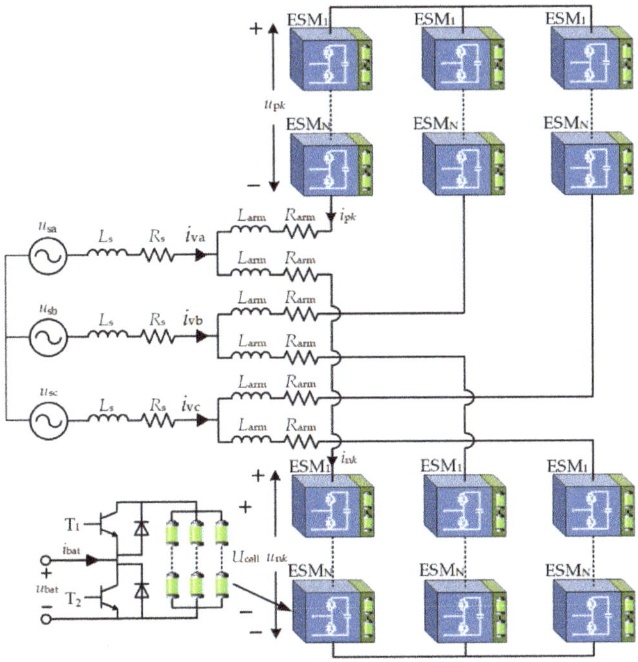

Figure 1. Schematic representation of the MMC-BESS topology.

2.2. Model-Predictive Control of Output Current

The MPC algorithm predicts the single-step output current of MMC-BESS, as shown in Equation (1):

$$i_{vk}(t+T_s) = \frac{1}{R_{eq} + L_{eq}/T_s}\left[u_k(t+T_s) - u_{sk}(t+T_s) + \frac{L_{eq}}{T_s} i_{vk}(t)\right] \tag{1}$$

where $i_{vk}(t + T_s)$ and $i_{vk}(t)$ are the predicted value and measured value of the output current of the phase-k, respectively; $u_k(t + T_s)$ is the predicted value of the AC-side voltage of the kth phase of MMC; $u_{sk}(t + T_s)$ is the predicted value of the grid voltage; L_{eq} and R_{eq} are the equivalent inductance and resistance of the AC side, respectively, with $L_{eq} = L_s + L_{arm/2}$, $R_{eq} = R_s + R_{arm/2}$; T_s is the simulation time step.

The cost function of output current is as follows:

$$J_{ivk} = |i_{vk} * (t+T_s) - i_{vk}(t+T_s)| \tag{2}$$

where $i_{vk} * (t + T_s)$ is the reference value of the output current.

2.3. Traditional SOC-Balancing Control

Traditional SOC-balancing control employs a three-level hierarchical architecture that encompasses inter-phase SOC balancing, inter-arm SOC balancing, and submodule SOC balancing within each arm. An open-loop sorting strategy is typically implemented to minimize controller complexity. The process follows a systematic approach, outlined as follows:

(1) Calculate the required number of submodules (n*) that need to be activated;
(2) Sort all submodules within the arm according to their estimated SOC values;

(3) Apply a current-direction-based selection algorithm: when the arm current causes the battery to discharge, the n submodules with the highest SOC values are activated; when it causes charging, those with the lowest SOC values are prioritized.

The SOC value estimation of the submodule utilizes the Coulomb counting method [30], with the specific calculation methodology detailed in Equation (3):

$$SOC(t) = SOC(t_0) - \frac{1}{3600Q_{\max}} \int_{t_0}^{t} i_{\text{bat}}(t)dt \tag{3}$$

where $Q_{\max}$ represents the battery capacity.

The block diagram of inter-arm SOC-balancing control is presented in Figure 2. Here, $\overline{\text{SOC}}_{pk}$ and $\overline{\text{SOC}}_{nk}$ represent the average SOC values of the upper and lower arms of phase k, respectively; K_3 is the gain coefficient; and $u^*_{\text{SOC}_{\text{arm}k}}$ can be obtained, which is the reference voltage for achieving SOC balancing between arms.

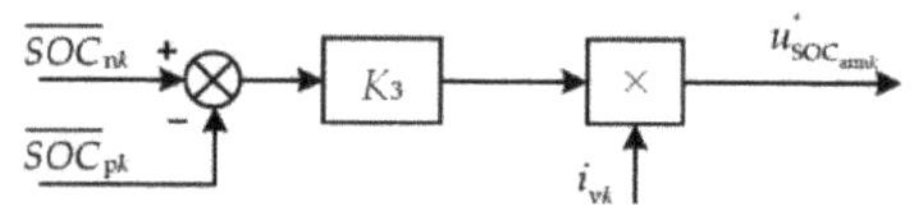

Figure 2. Control block diagram of inter-arm SOC balancing.

The block diagram of inter-phase SOC-balancing control is presented in Figure 3. Here, $\overline{\text{SOC}}_k$ represents the average SOC value of phase-k, $\overline{\text{SOC}}$ stands for average SOC value of all submodules in the three phases, K_1 and K_2 are gain coefficients, and LPF refers to a low-pass filter. Finally, $u^*_{\text{SOC}_k}$ can be obtained, which is the reference voltage value for achieving the inter-phase SOC balancing.

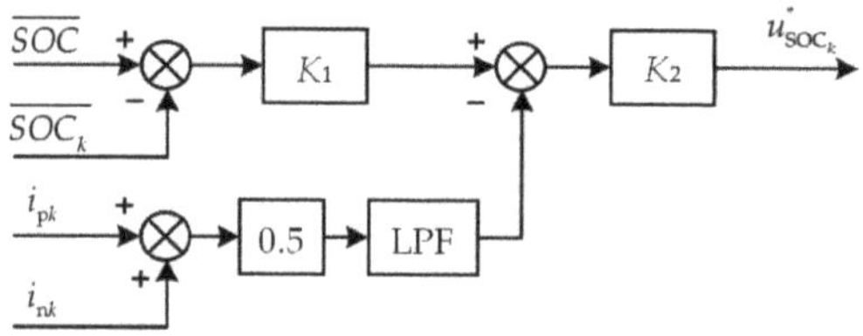

Figure 3. Control block diagram of inter-phase SOC balancing.

3. Nonlinear State-Space Model and Its Linearization

The traditional three-level SOC-balancing control for MMC-BESS has inherent deficiencies, such as slow balancing speed and difficult parameter tuning. To overcome these technical challenges, this paper proposes an SOC-balancing control strategy based on nonlinear model-predictive control, aimed at achieving fast and precise SOC balancing. The following section first establishes a nonlinear state space model for the MMC-BESS system.

This strategy significantly reduces computational burden by decreasing the number of MPC state variables, eliminates the need for calculating arm circulating current references, and achieves fast SOC balancing by only requiring the setting of maximum submodule numbers.

3.1. Nonlinear State-Space Model of MMC-BESS

As shown in Equations (4) and (5), construct the differential dynamic equations of the upper and lower arms.

$$L_{\text{arm}}\frac{\text{d}i_{pk}}{\text{d}t} + R_{\text{arm}}i_{pk} + u_{pk} = \frac{1}{2}U_{\text{DC}} - u_k \tag{4}$$

$$L_{\mathrm{arm}}\frac{\mathrm{d}i_{\mathrm{n}k}}{\mathrm{d}t}+R_{\mathrm{arm}}i_{\mathrm{n}k}+u_{\mathrm{n}k}=u_k-\frac{1}{2}U_{\mathrm{DC}} \tag{5}$$

where U_{DC} represents the DC-side voltage of MMC-BESS.

Subtracting Equation (5) from Equation (4) yields the dynamic equation governing the MMC output current. This fundamental relationship enables the development of output-current MPC, as formulated in Equations (1) and (2), which will not be discussed in further detail at this juncture.

When Equations (4) and (5) are summed, the differential dynamic equation can be obtained to characterize the arm circulating current:

$$2L_{\mathrm{arm}}\frac{\mathrm{d}i_{\mathrm{cir}k}}{\mathrm{d}t}+R_{\mathrm{arm}}i_{\mathrm{cir}k}+u_{\mathrm{p}k}+u_{\mathrm{n}k}=U_{\mathrm{DC}} \tag{6}$$

where $i_{\mathrm{cir}k}$ represents the circulating current of phase k, calculated as $i_{\mathrm{cir}k} = 0.5 \times (i_{\mathrm{p}k} + i_{\mathrm{n}k})$. The values of $u_{\mathrm{p}k}$ and $u_{\mathrm{n}k}$ depend on the number of submodules activated in the upper and lower arms, as shown below:

$$u_{\mathrm{p}k}=(N-n_{1k}^{*}+n_{2k}^{*})\overline{U_{\mathrm{cp}k}} \tag{7}$$

$$u_{\mathrm{n}k}=(n_{1k}^{*}+n_{2k}^{*})\overline{U_{\mathrm{cn}k}} \tag{8}$$

where N denotes the total number of submodules within each arm; n_{1k}^{*} and n_{2k}^{*} represent the reference quantity of submodule activation required by output current MPC and inter-arm SOC-balancing control, respectively; $\overline{U_{\mathrm{cp}k}}$ and $\overline{U_{\mathrm{cn}k}}$ correspond to the average voltage values across all submodules in the upper and lower arms of phase k, respectively, and must satisfy the conditions $\overline{U_{\mathrm{cp}k}} = 3 + 1.2\overline{\mathrm{SOC}}_{\mathrm{p}k}$, $\overline{U_{\mathrm{cn}k}} = 3 + 1.2\overline{\mathrm{SOC}}_{\mathrm{n}k}$ [5]. It should be noted that the relationship between SOC and the voltage of the battery is nonlinear due to many factors, such as battery material, temperature, etc. In engineering applications, it is necessary to first measure the SOC characteristics of the batteries and fit their nonlinear expressions. Here, the SOC characteristic is simplified to linear expression for analyzing easily. It is important to note that while $n_{1k}^{*} \geq 0$, n_{2k}^{*} can be negative values, indicating the flexibility to increase or decrease the number of activated submodules relative to the baseline established by n_{1k}^{*}.

Substituting Equations (7) and (8) into Equation (6), the differential dynamic equation of the phase-k arm circulating current can be obtained.

$$\begin{aligned}2L_{\mathrm{arm}}\tfrac{\mathrm{d}i_{\mathrm{cir}k}}{\mathrm{d}t} &= U_{\mathrm{DC}}-2R_{\mathrm{arm}}i_{\mathrm{cir}k}-1.2\left(N-n_{1k}^{*}+n_{2k}^{*}\right)\overline{SOC}_{\mathrm{p}k}\\ &+1.2\left(n_{1k}^{*}+n_{2k}^{*}\right)\overline{SOC}_{\mathrm{n}k}-3\left(N+2n_{2k}^{*}\right)\end{aligned} \tag{9}$$

Taking the upper arm of phase-k as an example, assume that the number of submodules put into operation at time t is $\left(N-n_{1k}^{*}+n_{2k}^{*}\right)$, then it can be obtained that

$$Q_{\max}\frac{\mathrm{d}\sum\limits_{i=1}^{N}SOC_{\mathrm{p}ki}}{\mathrm{d}t}=(N-n_{1k}^{*}+n_{2k}^{*})i_{\mathrm{p}k} \tag{10}$$

Due to

$$\frac{1}{N}\sum_{i=1}^{N}SOC_{\mathrm{p}ki}=\overline{SOC}_{\mathrm{p}k} \tag{11}$$

Substitute Equation (11) into Equation (10), and considering that $i_{\mathrm{p}k} = i_{\mathrm{cir}k} - 0.5i_{\mathrm{v}k}$ at the same time, this yields

$$Q_{\max}\frac{\mathrm{d}\overline{SOC}_{\mathrm{p}k}}{\mathrm{d}t}=\frac{N-n_{1k}^{*}+n_{2k}^{*}}{N}(i_{\mathrm{cir}k}-0.5i_{\mathrm{v}k}) \tag{12}$$

Equation (12) indicates that the average SOC dynamics are primarily governed by three factors: the number of activated submodules, the arm-circulating current, and the output current. Considering that the output current can accurately track its reference value (i.e., $i_{vk} = i^*_{vk}$), to simplify the design of the nonlinear MPC controller, Equation (12) can be rewritten as follows:

$$Q_{\max}\frac{\mathrm{d}\overline{SOC}_{\mathrm{p}k}}{\mathrm{d}t} = \frac{N - n^*_{1k} + n^*_{2k}}{N}(i_{\mathrm{cir}k} - 0.5i^*_{\mathrm{v}k}) \tag{13}$$

Equation (13) demonstrates that, after the aforementioned simplification, the dynamic characteristics of the arm's average SOC have been completely decoupled from the output current. Following the same derivation process, the dynamic equation for the lower arm's average SOC can be expressed as in Equation (14):

$$Q_{\max}\frac{\mathrm{d}\overline{SOC}_{\mathrm{n}k}}{\mathrm{d}t} = \frac{n^*_{1k} + n^*_{2k}}{N}(i_{\mathrm{cir}k} + 0.5i^*_{\mathrm{v}k}) \tag{14}$$

In summary, Equations (9), (13) and (14) constitute the state space model for the average SOC values of the upper and lower arms. Next, we will proceed with the linearization of this state space model.

3.2. Linearization Processing of State-Space Model

Based on the analysis of Equations (9), (13) and (14), it is evident that in the MMC-BESS, there exists a complex coupling relationship between the control variable n^*_{1k} and n^*_{2k} for output current and SOC balancing, and the state variables $i_{\mathrm{cir}k}$, $\overline{\mathrm{SOC}}_{\mathrm{p}k}$, and $\overline{\mathrm{SOC}}_{\mathrm{n}k}$ This results in significant nonlinear characteristics in the state-space model of the average SOC values for the upper and lower arms. Therefore, it is not possible to perform single-step prediction directly based on the discretized differential dynamic equations.

To effectively predict the state of the MMC-BESS, it is necessary to linearize the nonlinear state-space model of the system at each control cycle [31]. The specific linearization process first requires reconstructing Equations (9), (13) and (14) into a standard nonlinear state space model form:

$$\dot{\boldsymbol{x}} = f(\boldsymbol{x}, \boldsymbol{u}) \tag{15}$$

where $\boldsymbol{x} = [i_{\mathrm{cir}k}, \overline{\mathrm{SOC}}_{\mathrm{p}k}, \overline{\mathrm{SOC}}_{\mathrm{p}k}]\mathrm{T}$, $\boldsymbol{u} = [n^*_{2k}]$.

It is worth noting that, due to the previous decoupling of the arm SOC dynamics from the output current dynamics, there is no need to treat n^*_{1k} as an input variable in this model. In practice, it can be viewed as a time-varying system parameter that always can be treated as a constant.

By applying Taylor expansion to Equation (15) at the operating point ($\boldsymbol{x}_0$, $\boldsymbol{u}_0$) at time t and neglecting the higher order terms beyond second order, the following can be obtained:

$$\Delta\dot{x} = \left.\frac{\partial f}{\partial \boldsymbol{x}}\right|_{\boldsymbol{x}_0}\Delta \boldsymbol{x} + \left.\frac{\partial f}{\partial \boldsymbol{u}}\right|_{\boldsymbol{u}_0}\Delta \boldsymbol{u} = \boldsymbol{A}\Delta \boldsymbol{x} + \boldsymbol{B}\Delta \boldsymbol{u} \tag{16}$$

where $\Delta\boldsymbol{x} = \boldsymbol{x} - \boldsymbol{x}_0$, and $\Delta\boldsymbol{u} = \boldsymbol{u} - \boldsymbol{u}_0$. Let $\boldsymbol{A}$ be the system matrix and $\boldsymbol{B}$ be the input matrix. By substituting Equations (9), (13) and (14) into Equation (16), the following can be obtained:

$$\boldsymbol{A} = \begin{bmatrix} \frac{R_{\mathrm{arm}}}{L_{\mathrm{arm}}} & \frac{0.6\left(n^*_{1k} - n^*_{2k} - N\right)}{L_{\mathrm{arm}}} & \frac{0.6\left(n^*_{1k} + n^*_{2k}\right)}{L_{\mathrm{arm}}} \\ \frac{N - n^*_{1k} + n^*_{2k}}{NQ_{\max}} & 0 & 0 \\ \frac{n^*_{1k} + n^*_{2k}}{NQ_{\max}} & 0 & 0 \end{bmatrix} \tag{17}$$

$$
\boldsymbol{B} = \begin{bmatrix} \frac{-3-0.6\overline{SOC}_{pk}+0.6\overline{SOC}_{nk}}{L_{arm}} \\ \frac{i_{cirk}-0.5i^*_{vk}}{NQ_{max}} \\ \frac{i_{cirk}+0.5i^*_{vk}}{NQ_{max}} \end{bmatrix} \tag{18}
$$

Equations (16) and (18) represent the linearized state space model of the arm's average SOC at time t_0. Similarly to the output-current MPC, we apply the forward Euler method [32] to approximate the discretization of this model. The discretized expression is shown in Equation (19):

$$
\Delta \boldsymbol{x}(t+T_s) = (T_s\boldsymbol{A}+\boldsymbol{I})\Delta \boldsymbol{x}(t) + T_s\boldsymbol{B}\Delta \boldsymbol{u}(t+T_s) \tag{19}
$$

It should be noted that at time t, when the system is linearized at the operating point and has not been disturbed by changes in input variables, $\Delta x(t) = 0$. Therefore, Equation (19) can be simplified to the following:

$$
\Delta \boldsymbol{x}(t+T_s) = T_s\boldsymbol{B}\Delta \boldsymbol{u}(t+T_s) \tag{20}
$$

Due to

$$
\boldsymbol{x}(t+T_s) = \boldsymbol{x}(t) + \Delta \boldsymbol{x}(t+T_s) \tag{21}
$$

Based on Equations (20) and (21), single-step prediction of the target variables that need to be controlled can be performed, which subsequently facilitates the design of the cost function.

4. SOC-Balancing Control Strategy Based on Model-Predictive Control

4.1. MPC for Inter-Arm SOC Balancing

Based on Equations (20) and (21), the SOC-balancing controller between the upper and lower arms is first constructed. Single-step predictions of average SOC values for upper and lower arms of phase k are performed separately. The prediction equations are shown in Equations (22) and (23):

$$
\overline{SOC}_{pk}(t+T_s) = \frac{T_s\left(i_{cirk}(t) - 0.5i^*_{vk}(t)\right)}{NQ_{max}}\left(n^*_{2k}(t+T_s) - n^*_{2k}(t)\right) + \overline{SOC}_{pk}(t) \tag{22}
$$

$$
\overline{SOC}_{nk}(t+T_s) = \frac{T_s\left(i_{cirk}(t) + 0.5i^*_{vk}(t)\right)}{NQ_{max}}\left(n^*_{2k}(t+T_s) - n^*_{2k}(t)\right) + \overline{SOC}_{nk}(t) \tag{23}
$$

The cost function of the inter-arm SOC balancing can be designed as follows:

$$
J_{SOC\_arm} = \left|\overline{SOC}_{pk}(t+T_s) - \overline{SOC}_{nk}(t+T_s)\right| \tag{24}
$$

Considering the computational efficiency requirements for practical controller applications, this paper retains the sorting-control method described in Section 2.3 for SOC-balancing control among submodules within each arm. This design strategy significantly reduces the number of switching state combinations that need to be considered, thereby effectively simplifying the complexity of the model-predictive controller design and making it more suitable for practical simulation verification and hardware implementation.

4.2. MPC for Inter-Phase SOC Balancing

For inter-phase SOC-balancing control, by adding Equations (22) and (23), the single-step prediction value of the average SOC for a phase can be obtained as follows:

$$\begin{aligned}\overline{SOC}_k(t+T_s) &= 0.5\left(\overline{SOC}_{pk}(t+T_s)+\overline{SOC}_{nk}(t+T_s)\right)\\ &= \frac{T_s i_{cirk}(t)}{NQ_{max}}\left(n_{2k}^*(t+T_s)-n_{2k}^*(t)\right)+\overline{SOC}_k(t)\end{aligned} \tag{25}$$

The average value of the three phase SOC can be expressed as follows:

$$\overline{SOC}(t+T_s)=\frac{\overline{SOC}_a(t+T_s)+\overline{SOC}_b(t+T_s)+\overline{SOC}_c(t+T_s)}{3} \tag{26}$$

For the inter-phase SOC balance control, its cost function can be designed as follows:

$$\begin{aligned}J_{SOC\_pha} = &\left|\overline{SOC}(t+T_s)-\overline{SOC}_a(t+T_s)\right|+\\ &\left|\overline{SOC}(t+T_s)-\overline{SOC}_b(t+T_s)\right|+\\ &\left|\overline{SOC}(t+T_s)-\overline{SOC}_c(t+T_s)\right|\end{aligned} \tag{27}$$

4.3. MPC Scheme Design Considering Inter-Arm and Inter-Phase SOC Balancing

By integrating the strategies proposed in Sections 2.2, 4.1 and 4.2, an overall MPC strategy for MMC-BESS that considers both inter-arm and inter-phase SOC balancing can be constructed. Based on the different number of control states, this paper proposes two design schemes.

(1) Design Scheme One: Unified Control Approach

This scheme adopts a unified cost function to simultaneously achieve both inter-arm and inter-phase SOC-balancing control. Specifically, n_{2k}^* is obtained by optimizing the cost function and then added to the n_{1k}^* to determine the number of submodules to be inserted in the upper and lower arms of each phase. Combining Equations (24) and (27), the unified SOC-balancing cost function can be expressed as follows:

$$J_{SOC}=\lambda_1 J_{SOC\_arm}+\lambda_2 J_{SOC\_pha} \tag{28}$$

where λ_1 and λ_2 are the weighting coefficients for inter-arm and inter-phase SOC balancing, respectively.

This scheme enables simultaneous inter-arm and inter-phase SOC balancing between arms and phases, with the flexibility to control the balancing speed at different levels by adjusting the weighting coefficients. However, this design scheme also presents significant computational burden issues. The specific analysis is as follows:

Assume the maximum number of submodules for output current control and SOC balancing are N_1 and N_2, respectively, with $N_1 + N_2 = N$, then n_{2k}^* must satisfy the following:

$$-N_2 \le n_{2k}^* \le N_2,\ n_{2k}^* \in \mathbf{N}^+ \tag{29}$$

From Equation (29), it can be seen that n_{2k}^* includes ($2N_2$ + 1) state variables. When calculating the cost function in Equation (27), nested loops through n_{2k}^* of all three phases are required, resulting in $(2N_2 + 1)^3$ states to be processed at each control step. The prediction of average SOC values for upper and lower arms and phase units must be calculated at each step. When a faster SOC balancing speed is required, a larger value of N_2 causes the number of states to grow exponentially as a cubic function, imposing a significant computational burden.

(2) Design Scheme Two: Staged Control Approach

To reduce the computational complexity of the MPC, this paper proposes a staged control design scheme. This approach implements inter-arm and inter-phase SOC balancing separately, with the following control strategy: The system prioritizes SOC-balancing control between the upper and lower arms of each phase. When $\left|\overline{SOC_{pk}} - \overline{SOC_{nk}}\right|$ is detected to be less than a preset error threshold δ_{arm}, the controller automatically switches to inter-phase SOC-balancing mode, while temporarily suspending inter-arm SOC balancing. If $\left|\overline{SOC_{pk}} - \overline{SOC_{nk}}\right|$ subsequently exceeds the preset error threshold δ_{arm}, inter-arm SOC-balancing control is reactivated.

In this scheme, the maximum number of submodules for upper- and lower-arm SOC balancing and inter-phase SOC balancing is set as N_{21} and N_{22}, respectively, with $N_{21} + N_{22} = N_2$. According to the analysis in Equation (24), the number of states to consider for inter-arm SOC-balancing control is $2N_{21} + 1$, and the three phases can be computed in parallel without nested loops; inter-phase SOC-balancing control considers $(2N_{22} + 1)^3$ states. Compared to the first scheme, this design significantly reduces the number of switching states.

Further analysis shows that the number of states for inter-phase SOC-balancing control can be further optimized. This is because inter-phase SOC balancing is essentially achieved by adjusting the DC component of the circulating current in each phase arm. When additional submodules are simultaneously inserted in both the upper and lower arms of a phase, the DC-side voltage of that phase momentarily increases relative to the other two phases, causing the incremental circulating current to accelerate the discharge process of the ESM, thereby reducing the average SOC value of that phase.

Based on this analysis, in inter-phase SOC-balancing control, it is only necessary to insert additional submodules in the two phases with higher average SOC values, while the phase with the lowest SOC average does not require extra control, to achieve convergence of the three phase SOC averages. This optimization means that the two phases with higher SOC only need to consider the number of additional submodules to be inserted, without considering removal operations, while the phase with the lowest SOC requires no additional control. Therefore, the number of switching states for inter-phase SOC-balancing control can be further reduced to $(N_{22} + 1)^2$.

Through in depth analysis of the control performance of the two design schemes, research has found that the second design scheme offers significant computational efficiency advantages, reducing the computational complexity from cubic exponential to square exponential, effectively alleviating the computational burden on the controller. Although this scheme has certain limitations in terms of the utilization efficiency of submodule regulation capability and requires additional optimization of the N_{21} and N_{22} parameters, considering the key requirements of the system's actual operational performance, especially the important impact of computational burden on the control system, this paper ultimately chooses to adopt this staged control strategy as the overall MPC scheme for SOC balancing in MMC-BESS. The specific implementation process of this control scheme is shown in Figure 4.

As shown in Figure 4, the stage control scheme consists of two parts: MPC for output current and MPC for SOC balancing. These two components operate in parallel without interfering with each other. The output current MPC can be divided into three steps:

Step 1: Initialize $J_{\_ivk}$ as infinity.

Step 2: Iteratively compute $i_{\text{vk}}(t + \text{T}_\text{s})$ and $J_{\_ivk}$ within the interval $[0, N_1]$ based on Equations (1) and (2).

Step 3: Obtain the control input n^*_{1k} that minimizes $\text{J}_{\_ivk}$.

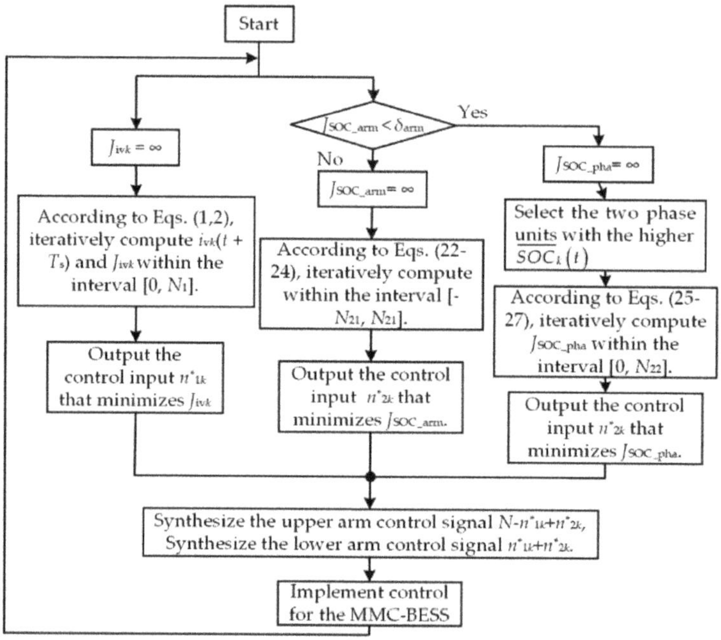

Figure 4. Staged control scheme for SOC balancing.

Similarly, the SOC balancing MPC can also be divided into three steps:

Step 1: Initially, perform a conditional judgment of J_{SOC_arm}. If J_{SOC_arm} is not less than δ_{arm}, the inter-arm SOC is not considered balanced, and J_{SOC_arm} is continuously updated through a simple for-loop over $[-N_{21}, N_{21}]$ according to Equations (22)–(24) until the control input n^*_{2k} that minimizes J_{SOC_arm} is found and $J_{SOC\_arm} < \delta_{arm}$.

Step 2: If $J_{SOC\_arm} < \delta_{arm}$, the inter-arm SOC is balanced. Then, the two phases with higher average SOC_k are selected to insert additional sub-modules. J_{SOC_pha} is continuously updated through double for-loops over [0, N_{22}] according to Equations (25)–(27) to find the control input n^*_{2k} that minimizes J_{SOC_pha}.

Step 3: The upper and lower arm control signals ($N - n^*_{1k} + n^*_{2k}$) and ($n^*_{1k} + n^*_{2k}$) are obtained, respectively. It should be noted that the control modes for inter-arm and inter-phase SOC balancing dynamically switch during the operation of the MMC-BESS to ensure stable operation.

5. Simulation Verification

5.1. Simulation Parameters

To verify the effectiveness of the proposed overall MPC strategy and demonstrate its advantages over traditional SOC-balancing control methods, this section builds a complete MMC-BESS system simulation model based on the MATLAB(R2022b)/SIMULINK simulation platform. The topology of the simulation system is shown in Figure 1, and the main simulation parameters are shown in Table 2.

Table 2. Simulation parameters of MMC-BESS system.

Type	Items	Values
Circuit parameters	Grid voltage	35 kV
	Rated output power	50 MW
	Equivalent inductance (L_s)	2.2 mH
	Equivalent resistance (R_s)	10 mΩ
	No. of ESMs within each arm (N)	80
	Arm inductance (L_{arm})	0.6 mH
	Arm equivalent resistance (R_{arm})	200 mΩ
	Rated capacity of battery pack (Q_{max})	1000 Ah
	Rated battery pack voltage (U_{cell})	800 V
Control parameters	Control step size	100 μs
	No. of ESMs for output current (N_1)	75
	No. of ESMs for SOC balancing (N_2)	5
	No. of ESMs for inter-arm SOC balancing (N_{21})	3
	No. of ESMs for inter-phase SOC balancing (N_{22})	2

The parameter selection in Table 2 was carefully designed based on both theoretical requirements and practical engineering considerations:

Circuit Parameters: The 35 kV/50 MW configuration represents a popular medium-voltage application for renewable energy integration in China. The arm inductance (0.6 mH) and resistance (200 mΩ) were calculated to limit the circulating-current ripple below 10% of the rated current. The 800 V battery voltage was selected to maintain a safe 2:1 margin below the 1700 V IGBT rating, ensuring robust switching operation. The 80 submodules per arm provide sufficient voltage resolution for 35 kV grid integration while allowing redundancy for fault tolerance.

Control Optimization: The 100 μs control step balances computational feasibility with control precision. The submodule allocation (N_1 = 75, N_2 = 5) maximizes output current control (N_1 covers > 90% normal operation) while reserving adequate modules (N_2) for balancing. As demonstrated in Section 5.4, the N_{21}:N_{22} = 3:2 ratio achieves optimal balancing efficiency.

5.2. Simulation Results of the Proposed SOC-Balancing Control

The initial simulation conditions are set as follows: the average SOC values of the upper- and lower-arms of phase-a are $\overline{SOC}_{pa}(0)$ = 100% and $\overline{SOC}_{na}(0)$ = 99.5%, respectively; the average SOC values of the upper and lower arms of phase b are $\overline{SOC}_{pb}(0)$ = 99.5% and $\overline{SOC}_{nb}(0)$ = 99%, respectively; and the average SOC values of the upper and lower arms of phase-c are $\overline{SOC}_{pc}(0)$ = 99% and $\overline{SOC}_{nc}(0)$ = 98.5%, respectively. The overall MPC strategy is activated at 0 s, and the simulation results are shown in Figures 5 and 6. Figure 5 shows the dynamic response characteristics of the AC-side current and voltage of the MMC-BESS system, and Figure 6 presents the dynamic change processes of the upper- and lower-arm average SOC values, inter-phase average SOC values, and circulating currents of each bridge arm.

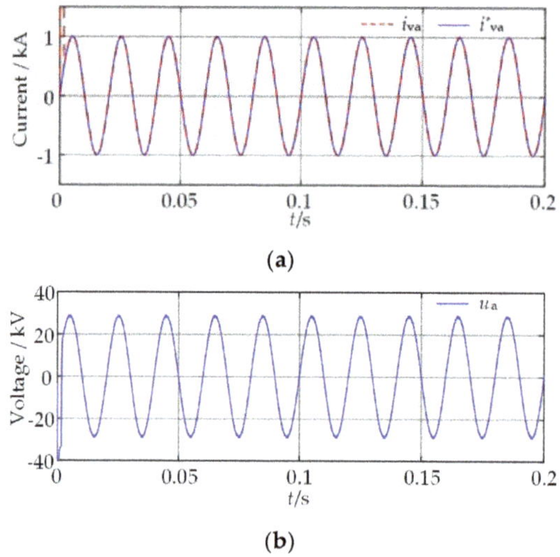

Figure 5. Simulation waveforms of (**a**) phase-a output current and (**b**) phase-a output voltage under proposed control strategy.

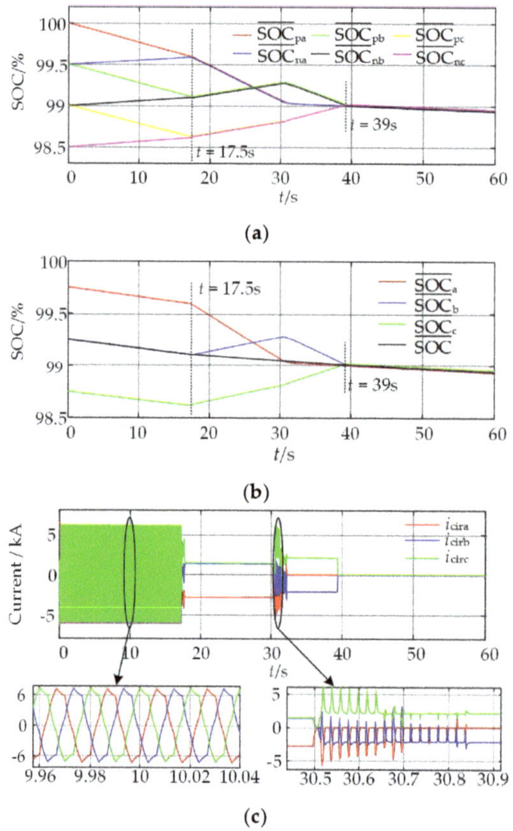

Figure 6. Simulation waveforms of (**a**) arm SOC average value, (**b**) phase SOC average value, and (**c**) arm-circulating-current variation during two-stage control.

As shown in Figure 5, the proposed control strategy can ensure the stable operation of MMC-BESS. Specifically, MMC-BESS can output stable fundamental frequency output voltage and output current. It is worth noting that during the inter-arm SOC-balancing control process, the basic electrical characteristics of MMC-BESS are not significantly disturbed, maintaining good dynamic performance.

The overall MPC strategy proposed in this paper demonstrates good dynamic characteristics during the SOC-balancing process, as shown in Figure 6. The control process can be divided into two main stages: inter-arm SOC balancing and inter-phase SOC balancing.

In the first stage (0–17.5 s), the system prioritizes SOC balancing between the upper and lower arms of each phase. Figure 6a,b show that the arms with higher average SOC values exhibit a decreasing trend, while those with lower values gradually increase. When the deviation of average SOC values between the upper and lower arms of each phase drops below the preset threshold δ_{arm} (0.001%), it marks the completion of inter-arm SOC balancing.

The second stage (17.5–39.0 s) mainly achieves inter-phase SOC balancing. Since phase-a and phase-b have higher initial average SOC values, the system prioritizes inserting additional submodules in these two phases. First, phase-a, which has the highest average SOC value, inserts additional submodules, causing its average SOC value to decrease while driving the average SOC values of other phases to increase. At 30.0 s, the average SOC value of phase-a approaches the total average SOC value of the three phases. Subsequently, phase-b becomes the new phase with the highest SOC and begins to insert additional submodules. Since phase-a maintains the insertion of additional submodules, phase-b submodules do not discharge to phase-a, keeping phase a SOC stable while phase-b SOC decreases and phase-c SOC continues to increase. Finally, at 39.0 s, the system achieves complete inter-phase SOC balancing.

The dynamic characteristics of the circulating currents (Figure 6c) clearly demonstrate the four stages of the entire control process. During the inter-arm SOC-balancing stage, the circulating currents are dominated by output-current components, which are used to eliminate SOC deviations between arms. After 17.5 s, the circulating currents transform to primarily DC components, manifested as negative values for phase a and positive values for phases b and c, indicating that phase a is charging phases b and c. Around 30.0 s, when phase b inserts additional submodules, since the SOC deviation temporarily exceeds the threshold δ_{arm}, the system briefly resumes inter-arm balancing control, and output current components reappear in the circulating currents. After the inter-arm SOC is rebalanced, the system continues with inter-phase balancing control, and finally achieves comprehensive balancing at 39 s, with the three phase-circulating currents reduced to zero.

5.3. Comparison with Traditional SOC-Balancing Control Method

To comprehensively verify the advantages of the MPC scheme proposed in this paper over traditional three-level SOC-balancing control, this section conducts a comparative analysis in the same MMC-BESS simulation system. Using the traditional control system described in Section 2.3, maintaining the initial average SOC values of each arm consistent with Section 5.1, and setting control parameters $K_1 = 100$, $K_2 = 5$, $K_3 = 4$. After enabling the three-level SOC-balancing control at 0 s, the dynamic response characteristics of the system are shown in Figure 7.

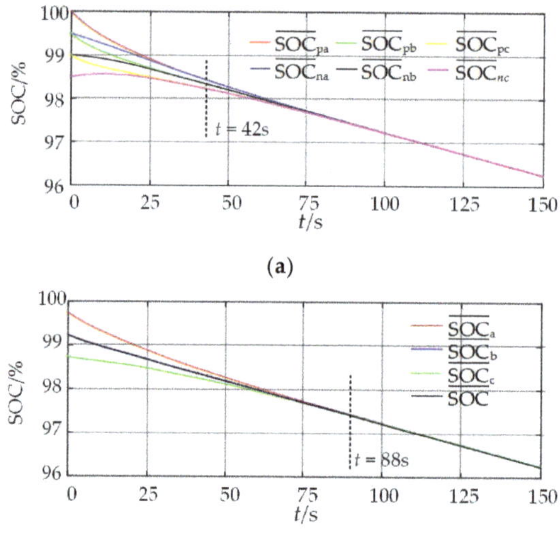

(a)

(b)

Figure 7. Simulation waveforms of (**a**) arm SOC average value and (**b**) phase SOC average value under traditional three-level control scheme.

The results indicate that the traditional scheme suffers from considerably longer balancing times: SOC balancing between the upper and lower arms of each phase takes approximately 42.0 s, while the proposed scheme requires only 17.5 s; SOC balancing between phases takes approximately 88 s, while the proposed scheme requires only 39.0 s. This comparison clearly highlights the superiority of the MPC strategy proposed in this paper in improving SOC-balancing rates.

Detailed analysis reveals the following limitations of traditional three-level SOC-balancing control: First, the selection of control parameters faces a dilemma. Although the balancing rate can be changed by adjusting K_1, K_2, and K_3, excessive adjustment will cause overshoot in the reference values of arm voltage, affecting system operational stability. In contrast, the MPC strategy proposed in this paper does not require parameter tuning, only needing to select the maximum number of submodules for each control stage under the premise of satisfying total quantity constraints, fundamentally avoiding the overshoot problem. Second, due to the use of proportional control, the balancing speed of the traditional scheme is positively correlated with SOC deviation, resulting in slower speed as the deviation decreases. Even introducing an integral component cannot completely solve this problem. However, the MPC strategy proposed in this paper, as shown in Figure 6, achieves linear SOC reduction in each control stage, maintaining a stable balancing speed. These analysis results fully demonstrate the comprehensive advantages of the control strategy proposed in this paper in terms of balancing speed, parameter design, and dynamic characteristics.

5.4. Comparison of SOC Balancing Rates with Different Submodule Number Proportions

Although the MPC system proposed in this paper cannot adjust the SOC balancing speed through traditional proportional or integral components, it can achieve balancing rate adjustment by optimizing the number of MPC submodules for inter-arm and inter-phase SOC balancing (N_{21} and N_{22}).

To thoroughly investigate the impact of submodule number proportions on balancing performance, this section conducted a detailed analysis of four different proportion schemes while keeping N_1 and N_2 constant: (N_{21} = 1, N_{22} = 4), (N_{21} = 2, N_{22} = 3), (N_{21} = 3, N_{22} = 2), and (N_{21} = 4, N_{22} = 1). The simulation system used the same initial SOC conditions as in Section 4.2, focusing on examining the variation characteristics of the inter-arm SOC cost function J_{SOC_arm} (using phase a as an example) and the inter-phase cost function J_{SOC_pha}. The results are shown in Figure 8.

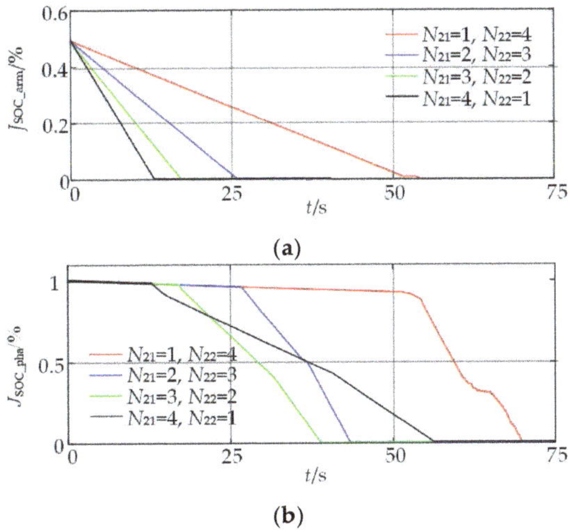

Figure 8. Dynamic characteristics of (**a**) inter-arm cost function and (**b**) inter-phase cost function under different submodule quantity ratios.

From Figure 8, three key findings can be found as follows: First, increasing N_{21} and decreasing N_{22} result in shorter inter-arm SOC-balancing time and improved inter-phase SOC-balancing speed. Table 3 lists detailed balancing time data for each proportion scheme, showing that the maximum number of submodules for each control section is positively correlated with the balancing rate. However, this enhancement effect exhibits diminishing returns, meaning that as the number of submodules increases, the magnitude of reduction in balancing time gradually decreases. Also, it can be seen from Figure 8b that the declining rate of J_{SOC_pha} is not always the same which has a turning point. The reason for this phenomenon is two phase units reach SOC balancing after a certain time during the inter-phase SOC-balancing control mode.

Table 3. Comparison of SOC-balancing time.

Condition	Inter-Arm	Inter-Phase	Overall
(N_{21} = 1, N_{22} = 4)	54.5 s	15.5 s	70.0 s
(N_{21} = 2, N_{22} = 3)	26.0 s	17.5 s	43.5 s
(N_{21} = 3, N_{22} = 2)	17.5 s	21.5 s	39.0 s
(N_{21} = 4, N_{22} = 1)	13.0 s	43.0 s	56.0 s

Second, in terms of overall SOC balancing time, simply increasing N_{21} or N_{22} cannot effectively shorten the system's overall balancing time. Research has found that when the values of N_{21} and N_{22} are close, the system can achieve relatively optimal overall balancing performance. This finding provides important guidance for parameter selection in engineering practice.

Finally, it is worth noting that although increasing N_2 can also shorten SOC balancing time, this will correspondingly reduce the value of N_1, thereby affecting the maximum output power of MMC-BESS. Therefore, the value of N_2 should be determined based on the maximum output power requirements of MMC-BESS.

5.5. Comparison of Computation Burden Between Unified and Staged-Control Approach

In order to verify the advantages of staged control over unified control in terms of computational burden, Simulink Profiler is used to analyze the computation time of the MPC part in the simulation model under the two control schemes. The operating system is Windows 10, the CPU is i5-13600KF, and the simulation duration is set to 15 s. The CPU time results are presented in Table 4 below. It can be observed that the CPU time increases slightly as N_{22} increases gradually, indicating that the computational burden of staged control is marginally influenced by N_{22}. In contrast, the CPU time for unified control is nearly 45 s, which is approximately 10 times higher than that of staged control, imposing a significantly greater computational burden on the CPU. Therefore, staged control substantially reduces the computational burden compared to unified control.

Table 4. Comparison of CPU time between unified and staged-control scheme.

Control Scheme	CPU Time
Unified Control	45.682 s
Staged Control (N_{21} = 1, N_{22} = 4)	4.882 s
Staged Control (N_{21} = 2, N_{22} = 3)	4.837 s
Staged Control (N_{21} = 3, N_{22} = 2)	4.808 s
Staged Control (N_{21} = 4, N_{22} = 1)	4.796 s

6. Conclusions

This paper proposes a fast SOC-balancing strategy for MMC-BESS based on nonlinear model-predictive control. By constructing and linearizing a nonlinear state-space model, the proposed method enables direct prediction of inter-arm and inter-phase SOC estimations with minimal computational overhead. The staged-control strategy eliminates the need for circulating current reference calculation and parameter tuning while maintaining consistent balancing speed. Simulation results indicate that the proposed method achieves inter-arm SOC balancing in approximately 17.5 s and inter-phase SOC balancing in approximately 39.0 s. In comparison, traditional three-level SOC balancing requires approximately 42.0 s and 88 s, respectively, representing reductions of 58% and 56% in balancing times for inter-arm and inter-phase SOC, respectively. Additionally, the analysis of various submodule allocation strategies offers practical insights into parameter optimization. However, the Coulomb counting method employed for SOC estimation in this study is prone to accumulating errors over time, which may lead to inaccurate SOC predictions and degrade the performance of NMPC. Future research will focus on integrating more precise SOC estimation models to enhance the balancing accuracy.

Author Contributions: Methodology, X.J.; Software, Y.Q.; Validation, D.N.; Formal analysis, Y.J. and D.N.; Investigation, F.X.; Data curation, Y.Q.; Writing—original draft, D.N.; Writing—review & editing, Q.Y. All authors have read and agreed to the published version of the manuscript.

Funding: This research was funded by China Energy Investment Corporation Co., Ltd. grant number GJNY-22-109 and the APC was funded by GJNY-22-109.

Data Availability Statement: The original contributions presented in this study are included in the article. Further inquiries can be directed to the corresponding author.

Conflicts of Interest: Authors Fengxiang Xie, Yuantang Qi and Yongdong Ji were employed by CHN Energy Gonghe New Energy Development Co., Ltd. The remaining authors declare that the research was conducted in the absence of any commercial or financial relationships that could be construed as a potential conflict of interest.

References

1. Zhang, X.; Feng, G. MMC-based PV grid-connected system with SMES-battery hybrid energy storage system. *IEEE Trans. Appl. Supercond.* **2024**, *34*, 1–4. [CrossRef]
2. Babu, T.S.; Vasudevan, K.R.; Ramachandaramurthy, V.K.; Sani, S.B.; Chetmud, S.; Laljim, R.M. A comprehensive review of hybrid energy storage systems: Converter topologies, control strategies and future prospects. *IEEE Access* **2020**, *8*, 148702–148721. [CrossRef]
3. Kasari, P.R.; Bhattacharjee, S. An HESS based on a modified MMCC-TSBC converter integrating energy storages. *CSEE J. Power Energy Syst.* **2023**, *9*, 1698–1709.
4. Wang, H.; Yang, G.; Xiao, F.; Fan, X. Wide-range operation optimization strategy of bidirectional energy storage converter for 10 kV medium voltage DC integrated power system. *IEEE Trans. Transp. Electrif.* **2024**, *10*, 4604–4615. [CrossRef]
5. Ma, Z.; Yu, F.; Zhao, X.; Gao, F. The multidimensional battery management strategy for MMC battery energy storage system. In Proceedings of the 2022 IEEE 7th Southern Power Electronics Conference (SPEC), Nadi, Fiji, 5–8 December 2022; IEEE: Piscataway, NJ, USA, 2022; pp. 1–5.
6. He, R.; Tian, K.; Ling, Z. Offline equalization control of modular multilevel converter-based battery energy storage system. *IEEE Trans. Ind. Electron.* **2023**, *71*, 9003–9012. [CrossRef]
7. Quraan, M.; Yeo, T.; Tricoli, P. Design and control of modular multilevel converters for battery electric vehicles. *IEEE Trans. Power Electron.* **2016**, *31*, 507–517. [CrossRef]
8. Gao, F.; Zhang, L.; Zhou, Q.; Chen, M.; Xu, T.; Hu, S. State-of-charge balancing control strategy of battery energy storage system based on modular multilevel converter. In Proceedings of the 2014 IEEE Energy Conversion Congress and Exposition (ECCE), Pittsburgh, PA, USA, 14–18 September 2014; IEEE: Piscataway, NJ, USA, 2014; pp. 2567–2574.
9. Maharjan, L.; Inoue, S.; Akagi, H.; Asakura, J. State-of-charge (SOC)-balancing control of a battery energy storage system based on a cascade PWM converter. *IEEE Trans. Power Electron.* **2009**, *24*, 1628–1636. [CrossRef]
10. Luo, H.; Xu, C.; Dai, K.; Cheng, C.; Huang, Y.; Pan, F. Balance control of SOC for MMC-BESS with power fluctuation suppression, PCC voltage regulation, and harmonic mitigation in grid-connected wind farm. *IEEE Access* **2022**, *10*, 117732–117744. [CrossRef]
11. Cheng, C.; Xu, C.; Dai, K.; Huang, Y.; Pan, F. Three-level equalization control strategy for state of charge of MMC-BESS. *Power Syst. Prot. Control.* **2021**, *49*, 100–108.
12. Premchand, M.; Gudey, S.K. Analysis of PI, PR and SMC controllers for bidirectional converter used in electric vehicle. In *International Conference on Automation, Signal Processing, Instrumentation and Control*; Springer: Singapore, 2020; pp. 449–466.
13. Xu, S.; Pei, T.; Wang, S. Model predictive control strategy of MMC based on energy decoupling. *Power Electron.* **2024**, *58*, 115–118.
14. Cheng, Z.; Tian, F.; Yang, H.; Yang, T. Model predictive optimal control strategy for battery energy storage system of modular multilevel composite converter. *Acta Energiae Solaris Sinica* **2023**, *44*, 59–66.
15. Wang, J.; Liu, X.; Xiao, Q.; Zhou, D.; Qiu, H.; Tang, Y. Modulated model predictive control for modular multilevel converters with easy implementation and enhanced steady-state performance. *IEEE Trans. Power Electron.* **2020**, *35*, 9107–9118. [CrossRef]
16. Böcker, J.; Freudenberg, B.; The, A.; Dieckerhoff, S. Experimental comparison of model predictive control and cascaded control of the modular multilevel converter. *IEEE Trans. Power Electron.* **2014**, *30*, 422–430. [CrossRef]
17. Qin, J.; Saeedifard, M. Predictive control of a modular multilevel converter for a back-to-back HVDC system. *IEEE Trans. Power Deliv.* **2012**, *27*, 1538–1547.
18. Liu, Z.; Li, Q.; Fu, J.; Zhou, H.; Chen, J.; Xia, Z. An extended-level model predictive control method with reduced current THD for modular multilevel converter battery energy storage system. *IEEE J. Emerg. Sel. Top. Power Electron.* **2025**. [CrossRef]
19. Liu, Z.; Li, Q.; Zhou, H.; Fu, J.; Ren, D.; Zhang, K.; Qin, H. An improved SOC balancing strategy based on reduced computational burden voltage level model predictive control for modular multilevel converter-battery energy storage system. *Electr. Eng.* **2025**. [CrossRef]
20. Xu, J.; Wang, F.; Wang, J.; Li, G. SOC balancing control based on predictive power model amongst supercapacitor packs in MMC with embedded energy storage system. *IEEE Trans. Power Deliv.* **2023**, *38*, 2641–2649. [CrossRef]
21. Zhao, G.; Shen, K.; Luo, G.; Zhao, D. A novel control strategy for modular multilevel converter with integrated supercapacitor energy storage system. In Proceedings of the IECON 2019-45th Annual Conference of the IEEE Industrial Electronics Society, Lisbon, Portugal, 14–17 October 2019; IEEE: Piscataway, NJ, USA, 2019; Volume 1, pp. 6037–6042.
22. Sankar, S.G.; Dreke, V.D.R.; Lazar, M. Hierarchical Model Predictive Control for Modular Multilevel Converters: A Linear Parameter Varying Approach. *IFAC-PapersOnLine* **2024**, *58*, 126–131. [CrossRef]
23. Cao, S.; Wang, Y.; Yu, X.; Yu, Y.; Yu, Z. SOC adaptive balance control strategy based on MMC-BESS. *Power Supply* **2024**, 1–11.
24. Tang, C.; Thiringer, T. A model predictive control method with adaptive weighting factors for enhancing performance of modular multilevel converters. *IEEE J. Emerg. Sel. Top. Power Electron.* **2024**, *12*, 3887–3899. [CrossRef]

25. Akbari, E.; Shadlu, M.S. Integrated battery energy storage with modular multilevel converter for grid-connected photovoltaic system: Design and control. In Proceedings of the 2023 13th Smart Grid Conference (SGC), Tehran, Iran, 5–6 December 2023; pp. 1–6.
26. Li, N.; Xie, Y.; Sha, Z.; Gao, F.; Wang, W. Hybrid predictive control method for battery integrated modular multilevel converter. In Proceedings of the 2016 IEEE 8th International Power Electronics and Motion Control Conference (IPEMC-ECCE Asia), Hefei, China, 22–26 May 2016; IEEE: Piscataway, NJ, USA, 2016; pp. 3777–3782.
27. Li, N.; Gao, F. Hybrid model predictive control method for battery-energy-storage-based modular multilevel converter. *Trans. China Electrotech. Soc.* **2017**, *32*, 165–174.
28. Wang, J.; Xu, J. Optimized control strategy for state-of-charge balancing of distributed energy-storage-based MMC batteries. *Electr. Power Autom. Equip.* **2023**, *43*, 44–50.
29. Ma, W.; Sun, W.; Wang, Y.; Zhang, W.; Li, H.; Zhu, Y. MMC-based distributed energy storage system and its fast SOC balancing control strategy. *Power Syst. Prot. Control.* **2024**, *52*, 1–11.
30. Chen, J.; Xiao, Y.; Zhou, G.; Li, X.; Luo, Y.; Zhang, W.; Wang, P.; Zhou, Y. A Hybrid Estimation Method for State-of-Charge (SOC) of Lithium-Ion Batteries. Chinese Patent CN201910460801.9; CN110133510A, 16 August 2019.
31. Grüne, L.; Pannek, J.; Grüne, L.; Pannek, J. *Nonlinear Model Predictive Control*; Springer International Publishing: Berlin/Heidelberg, Germany, 2017.
32. Vatani, M.; Bahrani, B.; Saeedifard, M.; Hovd, M. Indirect finite control set model predictive control of modular multilevel converters. *IEEE Trans. Smart Grid* **2015**, *6*, 1520–1529. [CrossRef]

MDPI AG
Grosspeteranlage 5
4052 Basel
Switzerland
Tel.: +41 61 683 77 34

Energies Editorial Office
E-mail: energies@mdpi.com
www.mdpi.com/journal/energies

www.ingramcontent.com/pod-product-compliance
Lightning Source LLC
Chambersburg PA
CBHW061447210926
51850CB00008B/1640
9783725873883